Optoelectronics in Machine Vision-Based Theories and Applications

Moises Rivas-Lopez
Universidad Autónoma de Baja California, Mexico

Oleg Sergiyenko
Universidad Autónoma de Baja California, Mexico

Wendy Flores-Fuentes
Universidad Autónoma de Baja California, Mexico

Julio Cesar Rodríguez-Quiñonez
Universidad Autónoma de Baja California, Mexico

A volume in the Advances in Computational
Intelligence and Robotics (ACIR) Book Series

Published in the United States of America by
  IGI Global
  Engineering Science Reference (an imprint of IGI Global)
  701 E. Chocolate Avenue
  Hershey PA, USA 17033
  Tel: 717-533-8845
  Fax: 717-533-8661
  E-mail: cust@igi-global.com
  Web site: http://www.igi-global.com

Library of Congress Cataloging-in-Publication Data

Names: Rivas-Lopez, Moises, 1960- editor. | Sergiyenko, Oleg, 1969- editor. |
  Flores-Fuentes, Wendy, 1978- editor. | Rodriguez-Quinonez, Julio C.,
  1985- editor.
Title: Optoelectronics in machine vision-based theories and applications /
  Moises Rivas-Lopez, Oleg Sergiyenko, Wendy Flores-Fuentes, and Julio Cesar
  Rodriguez-Quinonez, editors.
Description: Hershey, PA : Engineering Science Reference (an imprint of IGI
  Global), [2019] | Includes bibliographical references and index.
Identifiers: LCCN 2017055227| ISBN 9781522557517 (hardcover) | ISBN
  9781522557524 (ebook)
Subjects: LCSH: Computer vision--Industrial applications. | Optoelectronic
  devices. | Industrial electronics.
Classification: LCC TA1634 .O687 2019 | DDC 621.39/93--dc23 LC record available at https://lccn.loc.gov/2017055227

This book is published in the IGI Global book series Advances in Computational Intelligence and Robotics (ACIR) (ISSN: 2327-0411; eISSN: 2327-042X)

British Cataloguing in Publication Data
A Cataloguing in Publication record for this book is available from the British Library.

All work contributed to this book is new, previously-unpublished material. The views expressed in this book are those of the authors, but not necessarily of the publisher.

For electronic access to this publication, please contact: eresources@igi-global.com.

Advances in Computational Intelligence and Robotics (ACIR) Book Series

Ivan Giannoccaro
University of Salento, Italy

ISSN:2327-0411
EISSN:2327-042X

MISSION

While intelligence is traditionally a term applied to humans and human cognition, technology has progressed in such a way to allow for the development of intelligent systems able to simulate many human traits. With this new era of simulated and artificial intelligence, much research is needed in order to continue to advance the field and also to evaluate the ethical and societal concerns of the existence of artificial life and machine learning.

The **Advances in Computational Intelligence and Robotics (ACIR) Book Series** encourages scholarly discourse on all topics pertaining to evolutionary computing, artificial life, computational intelligence, machine learning, and robotics. ACIR presents the latest research being conducted on diverse topics in intelligence technologies with the goal of advancing knowledge and applications in this rapidly evolving field.

COVERAGE

- Cyborgs
- Algorithmic Learning
- Machine Learning
- Computer Vision
- Brain Simulation
- Synthetic Emotions
- Pattern Recognition
- Computational Intelligence
- Automated Reasoning
- Fuzzy systems

IGI Global is currently accepting manuscripts for publication within this series. To submit a proposal for a volume in this series, please contact our Acquisition Editors at Acquisitions@igi-global.com or visit: http://www.igi-global.com/publish/.

Titles in this Series

For a list of additional titles in this series, please visit: www.igi-global.com/book-series

Critical Developments and Applications of Swarm Intelligence
Yuhui Shi (Southern University of Science and Technology, China)
Engineering Science Reference • copyright 2018 • 478pp • H/C (ISBN: 9781522551348) • US $235.00 (our price)

Handbook of Research on Biomimetics and Biomedical Robotics
Maki Habib (The American University in Cairo, Egypt)
Engineering Science Reference • copyright 2018 • 532pp • H/C (ISBN: 9781522529934) • US $325.00 (our price)

Androids, Cyborgs, and Robots in Contemporary Culture and Society
Steven John Thompson (University of Maryland University College, USA)
Engineering Science Reference • copyright 2018 • 286pp • H/C (ISBN: 9781522529736) • US $205.00 (our price)

Developments and Trends in Intelligent Technologies and Smart Systems
Vijayan Sugumaran (Oakland University, USA)
Engineering Science Reference • copyright 2018 • 351pp • H/C (ISBN: 9781522536864) • US $215.00 (our price)

Handbook of Research on Modeling, Analysis, and Application of Nature-Inspired Metaheuristic Algorithms
Sujata Dash (North Orissa University, India) B.K. Tripathy (VIT University, India) and Atta ur Rahman (University of Dammam, Saudi Arabia)
Engineering Science Reference • copyright 2018 • 538pp • H/C (ISBN: 9781522528579) • US $265.00 (our price)

Concept Parsing Algorithms (CPA) for Textual Analysis and Discovery Emerging Research and Opportunities
Uri Shafrir (University of Toronto, Canada) and Masha Etkind (Ryerson University, Canada)
Information Science Reference • copyright 2018 • 139pp • H/C (ISBN: 9781522521761) • US $130.00 (our price)

Handbook of Research on Applied Cybernetics and Systems Science
Snehanshu Saha (PESIT South Campus, India) Abhyuday Mandal (University of Georgia, USA) Anand Narasim-hamurthy (BITS Hyderabad, India) Sarasvathi V (PESIT- Bangalore South Campus, India) and Shivappa Sangam (UGC, India)
Information Science Reference • copyright 2017 • 463pp • H/C (ISBN: 9781522524984) • US $245.00 (our price)

Handbook of Research on Machine Learning Innovations and Trends
Aboul Ella Hassanien (Cairo University, Egypt) and Tarek Gaber (Suez Canal University, Egypt)
Information Science Reference • copyright 2017 • 1093pp • H/C (ISBN: 9781522522294) • US $465.00 (our price)

701 East Chocolate Avenue, Hershey, PA 17033, USA
Tel: 717-533-8845 x100 • Fax: 717-533-8661
E-Mail: cust@igi-global.com • www.igi-global.com

Table of Contents

Preface xvii

Acknowledgment xxv

Section 1
Optical Detectors for Selected Applications

Chapter 1
Zeolite-Based Optical Detectors 1

Fabian N. Murrieta-Rico, Centro de Investigación Científica y de Educación Superior de
Ensenada (CICESE), Mexico
Vitalii Petranovskii, Universidad Nacional Autónoma de México (UNAM), Mexico
Rosario I. Yocupicio-Gaxiola, Universidad Nacional Autónoma de México (UNAM), Mexico
Vera Tyrsa, Universidad Autónoma de Baja California (UABC), Mexico

Chapter 2
Analog Circuit of Light Detector for CMOS Image Sensor 17

Arjuna Marzuki, Universiti Sains Malaysia, Malaysia
Mohd Tafir Mustaffa, Universiti Sains Malaysia, Malaysia
Norlaili Mohd Noh, Universiti Sains Malaysia, Malaysia
Basir Saibon, Universiti Kuala Lumpur, Malaysia

Chapter 3
Optoelectronic Differential Cloudy Triangulation Method for Measuring Geometry of Hot
Moving Objects 49

Sergey Vladimirovich Dvoynishnikov, Novosibirsk State University, Russia
Vladimir Genrievich Meledin, Novosibirsk State University, Russia

Chapter 4
Surface Measurement Techniques in Machine Vision: Operation, Applications, and Trends 79

Oscar Real Real, Universidad Autónoma de Baja California, Mexico
Moises J. Castro-Toscano, Universidad Autónoma de Baja California, Mexico
Julio Cesar Rodríguez-Quiñonez, Universidad Autónoma de Baja California, Mexico
Oleg Serginyenko, Universidad Autónoma de Baja California, Mexico
Daniel Hernández-Balbuena, Universidad Autónoma de Baja California, Mexico
Moises Rivas-Lopez, Universidad Autónoma de Baja California, Mexico
Wendy Flores-Fuentes, Universidad Autónoma de Baja California, Mexico
Lars Lindner, Universidad Autónoma de Baja California, Mexico

Section 2
Machine Vision for Modeling the Human Approach

Chapter 5
Leveraging Models of Human Reasoning to Identify EEG Electrodes in Images With Neural
Networks 106
    Alark Joshi, University of San Francisco, USA
    Phan Luu, Philips Neuro, USA
    Don M. Tucker, Philips Neuro, USA
    Steven Duane Shofner, Philips Neuro, USA

Chapter 6
Object-Oriented Logic Programming of Intelligent Visual Surveillance for Human Anomalous
Behavior Detection 134
    Alexei Alexandrovich Morozov, Kotel'nikov Institute of Radio Engineering and Electronics
        of RAS, Russia
    Olga Sergeevna Sushkova, Kotel'nikov Institute of Radio Engineering and Electronics of
        RAS, Russia
    Alexander Fedorovich Polupanov, Kotel'nikov Institute of Radio Engineering and
        Electronics of RAS, Russia

Chapter 7
Moving Object Classification Under Illumination Changes Using Binary Descriptors 188
    S. Vasavi, V. R. Siddhartha Engineering College, India
    Ayesha Farha Shaik, V. R. Siddhartha Engineering College, India
    Phani chaitanya Krishna Sunkara, V. R. Siddhartha Engineering College, India

Chapter 8
Machine Vision Application on Science and Industry: Machine Vision Trends 233
    Bassem S. M. Zohdy, Institute of Statistical Studies and Research, Egypt
    Mahmood A. Mahmood, Institute of Statistical Studies and Research, Egypt
    Nagy Ramadan Darwish, Institute of Statistical Studies and Research, Egypt
    Hesham A. Hefny, Institute of Statistical Studies and Research, Egypt

Section 3
Machine Vision for Structural Health Monitoring

Chapter 9
Application of Computer Vision Technology to Structural Health Monitoring of Engineering
Structures 256
    X. W. Ye, Zhejiang University, China
    T. Jin, Zhejiang University, China
    P. Y. Chen, Zhejiang University, China

Chapter 10
Machine Vision-Based Application to Structural Analysis in Seismic Testing by Shaking Table..... 269
Ivan Roselli, ENEA, Italy
Vincenzo Fioriti, ENEA, Italy
Marialuisa Mongelli, ENEA, Italy
Alessandro Colucci, ENEA, Italy
Gerardo De Canio, ENEA, Italy

Chapter 11
Methods to Reduce the Optical Noise in a Real-World Environment of an Optical Scanning
System for Structural Health Monitoring 301
Jesus E. Miranda-Vega, Universidad Autónoma de Baja California, Mexico
Moises Rivas-Lopez, Universidad Autónoma de Baja California, Mexico
Wendy Flores-Fuentes, Universidad Autónoma de Baja California, Mexico
Oleg Sergiyenko, Universidad Autónoma de Baja California, Mexico
Julio Cesar Rodríguez-Quiñonez, Universidad Autónoma de Baja California, Mexico
Lars Lindner, Universidad Autónoma de Baja California, Mexico

Section 4
Technical Vision for Avoiding Obstacles and Autonomous Navigation

Chapter 12
Mobile Robot Path Planning Using Continuous Laser Scanning 338
Mykhailo Ivanov, Universidad Autónoma de Baja California, Mexico
Lars Lindner, Universidad Autónoma de Baja California, Mexico
Oleg Sergiyenko, Universidad Autónoma de Baja California, Mexico
Julio Cesar Rodríguez-Quiñonez, Universidad Autónoma de Baja California, Mexico
Wendy Flores-Fuentes, Universidad Autónoma de Baja California, Mexico
Moises Rivas-Lopez, Universidad Autonoma de Baja California, Mexico

Chapter 13
Methods and Algorithms for Technical Vision in Radar Introscopy 373
Oleg Sytnik, A. Usikov Institute for Radio Physics and Electronics of the National Academy
    of Sciences of Ukraine, Ukraine
Vladimir Kartashov, Kharkiv's National University of Radio and Electronics, Ukraine

Compilation of References 392

About the Contributors 422

Index 431

Detailed Table of Contents

Preface... xvii

Acknowledgment xxv

Section 1
Optical Detectors for Selected Applications

The detection of physical magnitudes is a common and required task in several areas of technology for measurement and control purposes. Optical detectors are currently applied in a wide range of disciplines. In this section, Chapter 1 introduces the optical detectors as an alternative method for detection and characterization of chemicals and materials. Chapter 2 describes the designing of a CMOS sensor to achieve good sensitivity while achieving low noise and low power simultaneously for an optical detector in agriculture application. Chapter 3 is focused on laser cloudy triangulation method for measurement of thickness of hot dynamic objects. Chapter 4 provides useful information about the different types of non-contact surface measurement techniques, theory, basic equations, system implementation, actual research topics, engineering applications, and future trends.

Chapter 1
Zeolite-Based Optical Detectors 1

    Fabian N. Murrieta-Rico, Centro de Investigación Científica y de Educación Superior de
       Ensenada (CICESE), Mexico
    Vitalii Petranovskii, Universidad Nacional Autónoma de México (UNAM), Mexico
    Rosario I. Yocupicio-Gaxiola, Universidad Nacional Autónoma de México (UNAM), Mexico
    Vera Tyrsa, Universidad Autónoma de Baja California (UABC), Mexico

The detection of chemical species is a common and required task in several areas of technology. Currently, measurements in dedicated labs are the predominant tools for detection and characterization of chemicals and materials. Although these techniques are available in specialized equipment, their use is often bounded by cost of application or the operator's expertise. Also, in many applications rather than an analysis of all the detectable chemical species, it is only of interest to determine the presence of a particular chemical compound, and if it is present, to quantify its concentration. For these reasons, alternative methods for detecting specific chemical species are required. One case of such methods are the optical chemical sensors, particularly the ones based on the materials known as zeolites. In a broad sense, these sensors are constituted by an optical detector that is modified with zeolites. This combination allows the detection of specific chemical compounds if the zeolitic materials is properly modified to have an optical response for the analyte.

Chapter 2

Analog Circuit of Light Detector for CMOS Image Sensor 17

Arjuna Marzuki, Universiti Sains Malaysia, Malaysia
Mohd Tafir Mustaffa, Universiti Sains Malaysia, Malaysia
Norlaili Mohd Noh, Universiti Sains Malaysia, Malaysia
Basir Saibon, Universiti Kuala Lumpur, Malaysia

Plant phenotyping studies represent a challenge in agriculture application. The studies normally employ CMOS optical and image sensor. One of the most difficult challenges in designing the CMOS sensor is the need to achieve good sensitivity while achieving low noise and low power simultaneously for the sensor. At low power, the CMOS amplifier in the sensor is normally having a lower gain, and it becomes even worse when the frequency of the interest is in the vicinity of flicker noise region. Using conventional topology such as folded cascode will result in the CMOS amplifier having high gain, but with the drawback of high power. Hence, there is a need for a new approach that improves the sensitivity of the CMOS sensor while achieving low power. The objective of this chapter is to update CMOS sensors and to introduce a modified light integrating circuit which is suitable for CMOS image sensor.

Chapter 3

Optoelectronic Differential Cloudy Triangulation Method for Measuring Geometry of Hot
Moving Objects 49

Sergey Vladimirovich Dvoynishnikov, Novosibirsk State University, Russia
Vladimir Genrievich Meledin, Novosibirsk State University, Russia

The new differential cloudy triangulation method for quick and precise measurement of thickness of hot dynamic objects is proposed. The method is based on laser cloudy triangulation, used in the synchronous differential mode. The experimental model of the laser system for thickness measurements, which implements the proposed method, has been developed. Laboratory and industrial tests of the laser system have been performed. They confirm efficiency and possibilities of differential cloudy triangulation in metallurgy. The differential cloudy triangulation method may be used for geometry measurement of hot dynamic objects in a wide range of strong optical refractions in media.

Chapter 4

Surface Measurement Techniques in Machine Vision: Operation, Applications, and Trends 79

Oscar Real Real, Universidad Autónoma de Baja California, Mexico
Moises J. Castro-Toscano, Universidad Autónoma de Baja California, Mexico
Julio Cesar Rodríguez-Quiñonez, Universidad Autónoma de Baja California, Mexico
Oleg Serginyenko, Universidad Autónoma de Baja California, Mexico
Daniel Hernández-Balbuena, Universidad Autónoma de Baja California, Mexico
Moises Rivas-Lopez, Universidad Autónoma de Baja California, Mexico
Wendy Flores-Fuentes, Universidad Autónoma de Baja California, Mexico
Lars Lindner, Universidad Autónoma de Baja California, Mexico

Surface measurement systems (SMS) allow accurate measurements of surface geometry for three-dimensional computational models creation. There are cases where contact avoidance is needed; these techniques are known as non-contact surface measurement techniques. To perform non-contact surface measurements there are different operating modes and technologies, such as lasers, digital cameras, and integration of both. Each SMS is classified by its operation mode to get the data, so it can be divided into three basic groups: point-based techniques, line-based techniques, and area-based techniques. This chapter provides useful information about the different types of non-contact surface measurement techniques, theory, basic equations, system implementation, actual research topics, engineering applications, and future trends. This chapter is particularly valuable for students, teachers, and researchers that want to implement a vision system and need an introduction to all available options in order to use the most convenient for their purpose.

Section 2
Machine Vision for Modeling the Human Approach

With the beginning of the deep learning era, the rate of advancement in machine vision for modeling the human approach has been astonishingly increased. Classical computer vision techniques have been enhanced and some of them replaced by machine learning techniques. In this section, Chapter 5 makes use of these techniques to perform a localization and identification task over a set of EEG Electrodes images. Chapter 6 describes a 3D vision system based on logic programming for analysis, recognition, and understanding of human anomalous behavior detection. Chapter 7 describes the design of a system based on classical computer vision techniques enhanced with artificial intelligence algorithms for moving object classification under illumination changes. Chapter 8 resumes the machine vision trends and its application on science and industry.

Chapter 5

Leveraging Models of Human Reasoning to Identify EEG Electrodes in Images With Neural Networks 106

Alark Joshi, University of San Francisco, USA
Phan Luu, Philips Neuro, USA
Don M. Tucker, Philips Neuro, USA
Steven Duane Shofner, Philips Neuro, USA

Humans have very little trouble recognizing discrete objects within a scene, but performing the same tasks using classical computer vision techniques can be counterintuitive. Humans, equipped with a visual cortex, perform much of this work below the level of consciousness, and by the time a human is conscious of a visual stimulus, the signal has already been processed by lower order brain regions and segmented into semantic regions. Convolutional neural networks are modeled loosely on the structure of the human visual cortex and when trained with data produced by human actors are capable of emulating its performance. By black-boxing the low-level image analysis tasks in this way, the authors model solutions to problems in terms of the workflows of expert human operators, leveraging both the work performed pre-consciously and the higher-order algorithmic solutions employed to solve problems.

Chapter 6

Object-Oriented Logic Programming of Intelligent Visual Surveillance for Human Anomalous Behavior Detection 134

Alexei Alexandrovich Morozov, Kotel'nikov Institute of Radio Engineering and Electronics of RAS, Russia
Olga Sergeevna Sushkova, Kotel'nikov Institute of Radio Engineering and Electronics of RAS, Russia
Alexander Fedorovich Polupanov, Kotel'nikov Institute of Radio Engineering and Electronics of RAS, Russia

The idea of the logic programming-based approach to the intelligent visual surveillance is in usage of logical rules for description and analysis of people behavior. New prospects in logic programming of the intelligent visual surveillance are connected with the usage of 3D machine vision methods and adaptation of the multi-agent approach to the intelligent visual surveillance. The main advantage of usage of 3D vision instead of the conventional 2D vision is that the first one can provide essentially more complete information about the video scene. The availability of exact information about the coordinates of the parts of the body and scene geometry provided by means of 3D vision is a key to the automation of behavior analysis, recognition, and understanding. This chapter supplies the first systematic and complete description of the method of object-oriented logic programming of the intelligent visual surveillance, special software implementing this method, and new trends in the research area linked with the usage of novel 3D data acquisition equipment.

Chapter 7

Moving Object Classification Under Illumination Changes Using Binary Descriptors 188
S. Vasavi, V. R. Siddhartha Engineering College, India
Ayesha Farha Shaik, V. R. Siddhartha Engineering College, India
Phani chaitanya Krishna Sunkara, V. R. Siddhartha Engineering College, India

Object recognition and classification has become important in a surveillance video situated at prominent areas such as airports, banks, military installations, etc. Outdoor environments are more challenging for moving object classification because of incomplete appearance details of moving objects due to illumination changes and large distance between the camera and moving objects. As such, there is a need to monitor and classify the moving objects by considering the challenges of video in the real time. Training the classifiers using feature-based approaches is easier and faster than pixel-based approaches in object classification. Extraction of a set of features from the object of interest is most important for classification. Viewpoint and sources of light illumination plays major role in the appearance of an object. Abrupt transitions are identified using Chi-square and corners are detected using Harris corner detection. Silhouettes are captured using background subtraction and feature extraction is done using ORB. k-NN classifier is used for classification.

Chapter 8

Machine Vision Application on Science and Industry: Machine Vision Trends 233
Bassem S. M. Zohdy, Institute of Statistical Studies and Research, Egypt
Mahmood A. Mahmood, Institute of Statistical Studies and Research, Egypt
Nagy Ramadan Darwish, Institute of Statistical Studies and Research, Egypt
Hesham A. Hefny, Institute of Statistical Studies and Research, Egypt

Machine vision studies opens a great opportunity for different domains as manufacturing, agriculture, aquaculture, medical research, also research studies and applications for better understanding of processes and operations. As scientists' efforts had been directed towards deep understanding of the particular material systems or particular classes of types of specific fruits, or diagnosis of patients through medical images classification and analysis, also real time detection and inspection of malfunction piece, or process, as various domains witnessed advancement through using machine vision techniques and methods.

Section 3
Machine Vision for Structural Health Monitoring

The most recent and successful experiences in the application of machine vision-based techniques for structural health monitoring are described in this section. Chapter 9 introduces an unmanned bridge inspection and evaluation plane system, a railway foreign body intrusion recognition system, and a concrete crack tracking and evaluation system. Chapter 10 is devoted to the analysis of seismic testing using an integration of 3D motion capture methodologies with motion magnification analysis. Chapter 11 describes different methods and devices that can be used in optical scanning systems for noise reduction.

Chapter 9

Application of Computer Vision Technology to Structural Health Monitoring of Engineering
Structures 256

X. W. Ye, Zhejiang University, China
T. Jin, Zhejiang University, China
P. Y. Chen, Zhejiang University, China

The computer vision technology has gained great advances and applied in a variety of industry fields. It has some unique advantages over the traditional technologies such as high speed, high accuracy, low noise, anti-electromagnetic interference, etc. In the last decade, the technology of computer vision has been widely employed in the field of structure health monitoring (SHM). Many specific hardware and algorithms have been developed to meet different kinds of monitoring demands. This chapter presents three application scenarios of computer vision technology for health monitoring of engineering structures, including bridge inspection and evaluation with unmanned aerial vehicle (UAV), recognition and surveillance of foreign object intrusion for railway system, and identification and tracking of concrete cracking. The principles and procedures of three application scenarios are addressed following with the experimental study, and the possibilities and ideas for the application of computer vision technology to other monitoring items are also discussed.

Chapter 10

Machine Vision-Based Application to Structural Analysis in Seismic Testing by Shaking Table 269

Ivan Roselli, ENEA, Italy
Vincenzo Fioriti, ENEA, Italy
Marialuisa Mongelli, ENEA, Italy
Alessandro Colucci, ENEA, Italy
Gerardo De Canio, ENEA, Italy

In the present chapter the most recent and successful experiences in the application of machine vision-based techniques to structural analysis with main focus on seismic testing by shaking table are described and discussed. In particular, the potentialities provided by 3D motion capture methodologies and, more recently, by motion magnification analysis (MMA) emerged as interesting integrations, if not alternatives, to more conventional and consolidated measurement systems in this field. Some examples of laboratory applications are illustrated with the aim of providing evidence and details on the practical potentialities and limits of these methodologies for vibration motion acquisition, as well as on data processing and analysis.

Chapter 11

Methods to Reduce the Optical Noise in a Real-World Environment of an Optical Scanning
System for Structural Health Monitoring 301

Jesus E. Miranda-Vega, Universidad Autónoma de Baja California, Mexico
Moises Rivas-Lopez, Universidad Autónoma de Baja California, Mexico
Wendy Flores-Fuentes, Universidad Autónoma de Baja California, Mexico
Oleg Sergiyenko, Universidad Autónoma de Baja California, Mexico
Julio Cesar Rodríguez-Quiñonez, Universidad Autónoma de Baja California, Mexico
Lars Lindner, Universidad Autónoma de Baja California, Mexico

This chapter describes different methods and devices that can be used in optical scanning systems (OSS), especially applied to structural health and monitoring (SHM) in order to reduce the interference and losing of resolution in the measurements of the displacements and coordinates calculated by the OSS of a specific structure to be monitored. The principal parts of the OSS are a photo-detector, non-rotating emitter source of light, a DC electrical motor, lens, and mirror. All the measurements and experiments have been realized in a controlled environmental; the optical noise was simulated with a similar intensity than the intensity of the reference signal of the emitter source. Applying analogue filters has disadvantages because part of signal with important information for the performance of the system is removed, but particularly the components will often be too costly. However, there are digital filters and techniques of computational statistics that can solve these problems.

Section 4
Technical Vision for Avoiding Obstacles and Autonomous Navigation

One of the main goals of technical vision systems is to detect and identify targets covered by obstacles. Section 4 also describes the application of technical vision systems to autonomous mobile robots (AMR), which represent mobile machines and which can move independently in their surroundings, carrying out specific tasks. Chapter 12 presents mobile robot path planning using continuous laser scanning, while Chapter 13 describes a technical vision based on radar technology.

Chapter 12

Mobile Robot Path Planning Using Continuous Laser Scanning 338

Mykhailo Ivanov, Universidad Autónoma de Baja California, Mexico
Lars Lindner, Universidad Autónoma de Baja California, Mexico
Oleg Sergiyenko, Universidad Autónoma de Baja California, Mexico
Julio Cesar Rodríguez-Quiñonez, Universidad Autónoma de Baja California, Mexico
Wendy Flores-Fuentes, Universidad Autónoma de Baja California, Mexico
Moises Rivas-Lopez, Universidad Autonoma de Baja California, Mexico

The main object of this book chapter is an introduction and presentation of mobile robot path planning using continuous laser scanning, which has significant advantages compared with discrete laser scanning. A general introduction to laser scanning systems is given, whereby a novel technical vision system (TVS) using the dynamic triangulation measurement method for 3D coordinate determination is found suitable for accomplishing this task of mobile robot path planning. Furthermore, methods and algorithms for mobile robot road maps and path planning are presented and compared.

Chapter 13
Methods and Algorithms for Technical Vision in Radar Introscopy 373
    Oleg Sytnik, A. Usikov Institute for Radio Physics and Electronics of the National Academy
        of Sciences of Ukraine, Ukraine
    Vladimir Kartashov, Kharkiv's National University of Radio and Electronics, Ukraine

Optimization of technical characteristics of radio vision systems is considered in the radars with ultra-wideband sounding signals. Highly noisy conditions, in which such systems operate, determine the requirements that should be met by the signals being studied. The presence of the multiplicative noise makes it difficult to design optimal algorithms of echo-signal processing. Consideration is being given to the problem of discriminating objects hidden under upper layers of the ground at depths comparable to the probing pulse duration. Based upon the cepstrum and textural analysis, a subsurface radar signal processing technique has been suggested. It is shown that, however the shape of the probing signal spectrum might be, the responses from point targets in the cepstrum images of subsurface ground layers make up the texture whose distinctive features enable objects to be detected and identified.

Compilation of References 392

About the Contributors 422

Index 431

Preface

INTRODUCTION

The physical phenomena´s related to the combination of optics and electrics is the fundament of technological development of optical sensors and optical communication, allowing sensing and propagation of optoelectronics signals. Benefiting multiples machine vision-based applications with the advantages of optoelectronics signals acquisition and processing techniques developed in recent years. A transformation between the electrical and optical signal is performed using electronic devices, and the manipulation of them is achieved through sophisticated signal processing algorithms and hardware. The signals are conditioned and handled according to application needs, given origin to a wide range of machine vision-based theories and applications.

AN OVERVIEW OF OPTOELECTRONICS AND MACHINE VISION

Optoelectronics have given origin to machine vision and its multidisciplinary applications. Although it is true, that the history of optoelectronics started in the nineteenth century, it was not until the twenty-first century, that the machine vision-based on optoelectronics emerged at big scale.

In the beginning, the optical phenomena´s were simply related to thermal phenomena´s. The first optical radiation measurement was done using its thermal effects over metals. However, it has been discovered that the optical radiation could be measured by an indirect method, based on other phenomena called thermoelectricity. When two metals are brought together in two points with different temperatures, an electrical potential appears between the two metals allowing direct measurement of temperature and the indirect measurement of radiation, obtaining a transformation from optical-thermal signal to an electric signal. Prior, it was discovered that piezoelectric crystals respond to temperature changes producing a measurable electrical voltage. Also the photoelectric effect was discovered, arising from the interaction between photons and material. From the photoelectric effect surged the use of semiconductors, due to the conductance increases in semiconductors by the formation of free charge carrier pairs in form of electrons and holes, using the energy of an absorbed photon. Been this the principle of moderns' optical detectors, which represents the main source of devices for machine vision-based theories and applications.

The first machine vision applications concepts were developed after the introduction of the optical flow concept. Optical mathematical models were then developed, and researchers start to work on classical algorithms.

It was the great development of electronic devices in the early 21th century when surged enhanced and powerful computers and machine vision systems based on photoresistors, photodiodes, phototransistors and CCD sensors. Followed by optical character recognition (OCR), face recognition, and 3D coordinates extraction from 2D scenes, which initialized a massive growth of machine vision motivation to innovations in academic and industry research areas.

The majority of theories and applications of machine vision are based on the two big branches: Image Processing, and Signal Processing, due to they represent the two different main outputs obtained from optical detectors.

For image processing, the first algorithms were developed for image enhancement purpose to improve human perception and interpretability of image information´s or to provide more useful input for other automated image processing techniques. So, giving origin to the following algorithms: gray scale transformation, piecewise linear transformation, bit plane slicing, histogram equalization, histogram specification, enhancement by arithmetic operations, smoothing filter, sharpening filter, image blur correction, etc. Consequently, it was of main interest to acquire a quantitative description of geometric structures and shapes inside images. A well-known general approach was provided by mathematical morphology, where the images being analyzed are considered as sets of points, applying the set theory to morphological operations. This approach is based upon logical relations between pixels and can extract geometric features by choosing a suitable structuring shape as a probe. Useful for tasks like, object surface defections and assembly defects inspection, due to mathematical morphology can extract image shape features, such as edges, fillets, holes, corners, wedges, cracks, enabling fast object recognition and defect inspection. Giving origin to morphological algorithms, as: binary morphology, opening and closing, hit-or-miss transform (HMT), grayscale morphology, basic morphological algorithms, and several variations of morphological filters, including alternating sequential filters (ASFs), recursive morphological filters, soft morphological filters, order-statistic soft morphological (OSSM) filters, recursive soft morphological filters, recursive order statistic soft morphological filters, regulated morphological filters, and fuzzy morphological filters. To extract objects from the background, various segmentation techniques have surged: thresholding, boundary-based, region-based, and hybrid techniques. Algorithms that take advantage of the complementary nature of such information, component labeling, locating object contours by the snake model, edge detection, linking edges by adaptive mathematical morphology, automatic seeded region growing (SRG), and top–down region dividing (TDRD), Markov model, Bayesian methods, genetic algorithm, artificial neural network, and multicue partial differential equation. Distance transformation has wide applications in image analysis. This basically allows performing measurements inside an image, and even measurements of object trajectories, when the time between images is known. Consequently stimulating advances in robotics and artificial intelligence. Distance transformation is performed by mathematical morphology, an approximation of Euclidean distance, and distance transformations by acquiring and deriving approaches. One of the challenges has been the big data handling, and storage from images databases, emerging codes for efficient storage during image compression, as example: Run-length coding, binary tree and quadtree representations, contour representation schemes (chain code, crack code, and mid-crack code), medial axis transformation, and maxima tracking on distance transformation. Regarding feature extractions algorithms, it should be mentioned: Fourier descriptor (FD) and moment invariants, shape number and hierarchical features, Hough transform, principal component analysis (PCA), and linear discriminate analysis (LDA). Regarding to pattern recognition, we can recognize algorithms such as: the unsupervised clustering algorithm, Bayes classifier, support vector machine (SVM), neural networks (NN), the adaptive resonance theory network (ART), fuzzy sets in image analysis, and hybrids to mention some.

For image processing the big issues are defined by the description and characterization of images and by target recognition, which is one of the most important tasks for applications. Among the most famous applications are presented the following: human face and body processing and analysis, document image processing and classification, image watermarking, and 2D and 3D monitoring of industrial and structural health.

Regarding signal processing from optical detectors, the most common signals are obtained from optoelectronic scanning, a measurement method using at least one rotating optical detector, searching a light source. Optoelectronic scanners are used for multiple applications, most of them are based on the triangulation principle. Mainly formed by three subsystems, that is an emitter, a receiver and an electronic processor, the latter is responsible of signal processing and its interpretation. Concerning the emitter, it can be an incoherent light source, such as LEDs or a coherent light source like LASER. When we talk about the receiver, we are particularly referring to optical detectors, photodiodes, phototransistors, photodetectors (photoconductive, photovoltaic, and photoemissive), photoresistors, infrared sensors, thermopiles, position-sensitive detectors (PSD), including charge coupled detector arrays (CCD) and Metal-Oxide-Semiconductor (MOS) capacitors, even though CCD and MOS are more commonly used in applications like image sensors. The optoelectronic processor is linked with the mechanism allowing scanning and formation of the optoelectronic signal during an instant in the search of a light source. The signal processing depends on the nature of the signal due to the characteristics of the mechanism of the measurement system and the light source used. The signal processing also depends of the nature of the application and the environment where it develops. Optoelectronic scanners are mainly used for measuring of 3D coordinates, which implements the measurement of fundamental physical magnitudes, used by a variety of other applications. Optoelectronic scanners for measurement can be classified regarding using a passive or an active measurement principle. When using the passive principle, the sensor is static and with the help of a rotating mirror, the sensor receives the light emitted from the object under monitoring. When using the active principle, an optical signal is emitted and when it finds an obstacle, it returns to the sensor with different characteristics. The most typical scanners systems are: Scanners with position triangulation sensors using CCD or PSD (although the output of these scanners is an image), Scanners with rotating mirrors and remote sensing, like as: polygonal scanners, pyramidal and prismatic facets, holographic scanners, galvanometer and resonant scanners, and the 45° cylindrical mirror scanner.

OPTOELECTRONICS IN MODERN MACHINE VISION-BASED SYSTEMS

Vision-based methods to provide different measurements has developed in a wide range of technologies for multiples fields of applications such as robot navigation, medical scanning, and structural monitoring, to mention some. Computer vision has guided the machine vision to the tendency of duplicating the abilities of human vision by electronically perceiving and understanding an image with a high a dimensional data, increasing the data storage requirement, and the time processing due to the complexity of algorithms to extract important patterns and trends. Optoelectronics in Machine Vision-based systems plays an important role in 2D & 3D measuring and monitoring applications. In this sense, vision-based theories and applications are important to modern science and industrial practical implementation in the world today. Modern and sophisticated life could not be possible without the development of opto-electronics in machine vision-based systems, as the following.

- Image and signal sensors for computing and machine vision, as Semiconductor, Photodetectors, Photodiodes for Visible and Infrared light, Phototransistors, Ultraviolet Photodetectors, Nano Optoelectronic Sensors, CCD Image Sensors, CMOS Image Sensors.
- Scanning technologies for indoor and outdoor stereo vision, as Laser scanners, Time-of-flight scanners, Structured Light scanners, Modulated Light scanners, Stereo Vision-based scanners, Photogrammetry, Alternative vision systems.
- 2D & 3D Information reconstruction and applications, as Point Cloud, Polygon Mesh, Surface Models, and Volume Rendering Reconstruction.
- Machine vision application on science and industry.
- Instrumentation and control for vehicles, robot navigation, and structural health monitoring.
- Vision-based control schemes, as Vision Sensing and Data Processing for Human Assistive Systems, Optimization of Camera Position for Vehicles and Robots Vision, Intelligent Vision Guide for Automatic Navigation, Accuracy Improvement of Vision System, Object Recognition and Manipulation by Motion and Vision Data, Active Stereoscopic View to Build an Occupancy Grid for Autonomous Navigation, Mobile Vehicles and Robots Navigation System Configuration Space, Navigation Systems Design and Implementation, Autonomous Navigation Systems Capabilities and Constraints.

DEDICATED TO READERS

This book covers both theories and application of optoelectronics in machine vision-based systems, since it is intended to be used as text and reference work on advanced topics in optoelectronics. It is dedicated to academics, researchers, advanced-level students and technology developers, who will find this text book useful for furthering their research exposure to pertinent topics in Optoelectronics in Machine Vision-Based Theories and Applications, and assisting in their own future research in this field.

THE IMPORTANCE OF THE OPTOELECTRONICS IN MACHINE VISION-BASED THEORIES AND APPLICATIONS

It must be agreed, that the development of image processing without the development of appropriate optoelectronics would not have been possible. The combination of these two disciplines has produced a big number of applications, which have broken down into different areas of study. Covering all areas in one book would be impossible; however the most significant have been selected. Present book contains thirteen chapters, which have been classified into four sections: 1) Optical Detectors for Selected Applications, 2) Machine Vision for the Modeling the Human Approach, 3) Machine Vision for the Structural Health Monitoring, and 4), Technical Vision for Obstacles Avoid and Autonomous Navigation, as briefly described in the following:

Section 1 is dedicated to optical detectors and selected applications. The detection of physical magnitudes is a common and required task in several areas of technology for measurement and control purposes. Currently, the optical detectors are applied in a wide range of disciplines, as shown in the selected four chapters included in this section.

Chapter 1 gives an introduction to the detection of chemical components as a common and required task in several areas of technology. It describes how measurements are currently being carried out in specialized laboratories for the detection and characterization of chemical components and materials. Although these techniques are available in specialized equipment, their use is often limited by the total cost of application or the operator's expertise. For many applications, the main concern is only the presence of a particular chemical compound and, if any, its concentration. For these reasons, alternative methods for detecting specific chemical components are required. One case of such methods is presented by optical chemical sensors, particularly, the ones based on the materials known as zeolites. In a wide sense, these sensors are constituted by an optical detector that is modified with zeolites. This combination allows the detection of specific chemical compounds if the zeolitic materials are properly modified to have an optical response for the analyte.

Chapter 2 describes the designing of a CMOS sensor to achieve good sensitivity, with low noise and low power levels, simultaneously for an optical detector in agriculture application. It Chapter 2 explains, how plant phenotyping studies represent a challenge in modern agriculture application. The studies normally employ CMOS optical and image sensor. One of the most difficult challenges in designing CMOS sensor is the need to achieve good sensitivity, while having both low noise and low power for the sensor. At low power, the CMOS amplifier in the sensor is normally has less gain and is even worse, when the frequency of interest is near the flicker noise region. Using conventional topology such as folded cascode will result in the CMOS amplifier having higher gain, but with the drawback of higher power consumption. Hence, there is a need for a new approach that improves the sensitivity of the CMOS sensor while achieving low power. The objective of this chapter is to update type of the CMOS sensors, to introduce modified light integrating circuit which is suitable for CMOS image sensor.

Chapter 3 is focused on a laser cloudy triangulation method for measurement of the thickness of hot dynamic objects. It presents a new differential cloudy triangulation method for fast and precise measurement of the thickness of hot dynamic objects proposed. The method is based on laser cloudy triangulation, used in synchronous differential mode. The experimental model of the laser system for thickness measurements, which implements the proposed method, has been developed. Laboratory and industrial tests of the laser system have been performed. They confirm efficiency and possibilities of differential cloudy triangulation in metallurgy. The differential cloudy triangulation method can be used for geometry measurement of hot dynamic objects in a wide range of strong optical refractions in media.

Chapter 4 provides useful information about different types of non-contact surface measurement techniques, theory, basic equations, system implementation, actual research topics, engineering applications and future trends. It summarizes how Surface Measurement Systems (SMS) allows performing accurate measurements of surface geometry used for creation of 3D computational models. There are cases, where non-contact surface measurements techniques are required. To perform these non-contact surface measurements different operating modes and technologies are used, such as lasers, digital cameras, and integration of both. Each SMS is classified by its operation mode to obtain the measurement data, so it can be divided into three basic groups: Point-based techniques, Line based techniques, and Area-based techniques. This chapter provides useful information about the different types of non-contact surface measurement techniques, theory, basic equations, system implementation, actual research topics, engineering applications and future trends. It is particularly valuable for students, teachers, and researchers that want to implement a vision system and need an introduction to all available options and use the most convenient for their purpose.

Section 2 describes how with the beginning of the deep learning era the rate of advancement in machine vision for modeling the human approach has been astonishing increased. Classical computer vision techniques have been enhanced and some of them replaced by machine learning techniques as detailed in the below four chapters that compound this section.

Chapter 5 makes use of Neural Networks techniques to perform a localization and identification task over a set of EEG Electrodes images. It also describes how humans can easily recognize discrete objects within a scene, while using classic computer vision techniques can lead to counterintuitive results. Humans, using the visual cortex, perform much of this work on the sub-consciousness level and by the time a human is conscious of a visual stimulus, the signal has already been processed by lower order brain regions and segmented into semantic regions. Convolutional neural networks are modeled based on the structure of the human visual cortex and when trained with data produced by human actors are capable of emulating its performance. By black-boxing the low-level image analysis tasks in this way, we can model our solutions to problems in terms of the workflows of expert human operators, leveraging both the work performed pre-consciously and the higher-order algorithmic solutions employed to solve problems.

Chapter 6 describes a 3D vision system based on logic programming for analysis, recognition, and understanding of human anomalous behavior detection. The idea behind the logic programming-based approach to intelligent visual surveillance is to use logical rules to describe and analyze people's behavior. New prospects in the logic programming of the intelligent visual surveillance are connected with the usage of 3D machine vision methods and adaptation of the multi-agent approach to the intelligent visual surveillance. The main advantage of the usage of 3D vision instead of the conventional 2D vision is that the first one can provide essentially more complete information about the video scene. The availability of accurate information about body part coordinates and scene geometry provided by 3D vision is the key to automating behavioral analysis, detection, and understanding. This chapter supplies the first systematic and complete description of the method of object-oriented logic programming of the intelligent visual surveillance, special software implementing this method, and new trends in the research area linked with the usage of novel 3D data acquisition equipment.

Chapter 7 describes the design of a system based on classical computer vision techniques enhanced with artificial intelligence algorithms for moving object classification under illumination changes. Object recognition and classification has become important in a surveillance video situated at prominent areas such as airports, banks, military installations etc. Outdoor environments are more challenging for moving object classification because of incomplete appearance details of moving objects due to illumination changes and the large distance between the camera and moving objects. As such, there is a need to monitor and classify the moving objects by considering the challenges of video in real time. Training the classifiers using feature-based is easier and faster than pixel-based approaches in object classification. Extraction of a set of features from the object of interest is most important for classification. Viewpoint and sources of light illumination play a major role in the appearance of an object. Abrupt transitions are identified using Chi-square and corners are detected using Harris corner detection. Silhouettes are captured using background subtraction and feature extraction is done using ORB. The k-NN classifier is used for classification.

Chapter 8 resumes machine vision trends and its application in science and industry. Machine vision studies open a great opportunity for different domains as manufacturing, agriculture, aquaculture, medical research, also research studies and applications for a better understanding of processes and operations.

As scientists efforts had been and still directed towards deep understanding of the particular material systems or particular classes of types of specific fruits, or diagnosis of patients through medical images classification, and analysis, also real-time detection and inspection of malfunction piece, or process, as various domains witnessed advancement through using machine vision techniques and methods.

Section 3 shows the most recent and successful experiences in the application of machine vision-based techniques to structural health monitoring, three chapters are part of this section, as described in the following:

Chapter 9 introduces an unmanned bridge inspection and evaluation plane system, a railway foreign body intrusion recognition system, and a concrete crack tracking and evaluation system. The concept of computer vision systems emerged around the 1950s and matured around the 1990s. It is featured by its advantages include high speed, high accuracy, low noise, excellent anti-interference ability, convenience, long-term monitoring duration, etc. Due to the advantages above, this technology is applied gradually to the field of structural health monitoring. Plenty of techniques are developed to meet all kinds of demands in structural health monitoring industry for a variety of structures. This chapter is going to introduce three developing applications of computer vision technology in structural health monitoring of engineering structures. Which are: unmanned bridge inspection and evaluation plane system, railway foreign body intrusion recognition system, concrete crack tracking and evaluation system. Brief principles and procedures of the above developing applications are introduced, followed by experiments.

Chapter 10 is devoted to the analysis seismic testing using an integration of 3D motion capture methodologies with motion magnification analysis.

Chapter 11 describes different methods and devices that can be used in optical scanning systems for noise reduction.

Section 4 presents one of the main goals of technical vision systems, which is defined by detection and identifying targets covered by an obstacle. Section 4 also presents, Autonomous Mobile Robots (AMR), which represents mobile machines, which can move independently in their surroundings and can carry out specific tasks. This section is formed by two chapters.

Chapter 12 presents mobile robot path planning using continuous laser scanning, which has significant advantages compared with discrete laser scanning. A general introduction to laser scanning systems is given, whereby a novel Technical Vision System (TVS) using the Dynamic Triangulation measurement method for 3D coordinate determination is found suitable for accomplishing this task of mobile robot path planning. Furthermore, methods and algorithms for mobile robot roadmaps and path planning are presented and compared.

Chapter 13 describes a technical vision application based on radar technology. Optimization of technical characteristics of radiovision systems is considered in the radars using ultra wideband sounding signals. Highly noisy conditions, in which such systems operate, determine the requirements that should be met by the signals being studied. The presence of the multiplicative noise makes it difficult to design optimal algorithms for echo-signal processing. Consideration is being given to the problem of discriminating objects hidden under upper layers of the ground at depths comparable to the probing pulse duration. Based upon the cepstrum and textural analysis a subsurface radar signal processing technique has been suggested. It is shown that, whatever the shape of the probing signal spectrum might be, the responses from point targets in the cepstrum images of subsurface ground layers make up the texture whose distinctive features enable objects to be detected and identified.

IMPACT AND CONTRIBUTION OF THE BOOK

Optoelectronics in Machine Vision-Based Theories and Applications allows engineers and students, as well as advanced specialists in the area, to take advice in contributions focused on optoelectronic sensors, 2D and 3D machine vision technologies, robot navigation, control schemes, motion controllers, intelligent algorithms, and vision systems. The book aims to cover particular topics such as Optical Detectors, and applications of Machine Vision for Modeling the Human Approach, Machine Vision for the Structural Health Monitoring, Technical Vision for Obstacles Avoid, and Autonomous Navigation

Acknowledgment

We want to thank all the authors that have contributed with a chapter to this book, who have given their best effort in writing his most recent research findings, and to all the reviewers who have done a great job reading and suggesting improvements for each chapter. Acknowledgments go also for the editorial board and the officials at IGI Global Publications for their invaluable efforts, great support and valuable advice for this project towards successful publication of this book. We also want to thank our institution Universidad Autónoma de Baja California, to provide us with a location and time where to develop this project.

Moises Rivas-Lopez
Universidad Autónoma de Baja California, Mexico

Oleg Sergiyenko
Universidad Autónoma de Baja California, Mexico

Wendy Flores-Fuentes
Universidad Autónoma de Baja California, Mexico

Julio Cesar Rodríguez-Quiñonez
Universidad Autónoma de Baja California, Mexico

Section 1
Optical Detectors for Selected Applications

The detection of physical magnitudes is a common and required task in several areas of technology for measurement and control purposes. Optical detectors are currently applied in a wide range of disciplines. In this section, Chapter 1 introduces the optical detectors as an alternative method for detection and characterization of chemicals and materials. Chapter 2 describes the designing of a CMOS sensor to achieve good sensitivity while achieving low noise and low power simultaneously for an optical detector in agriculture application. Chapter 3 is focused on laser cloudy triangulation method for measurement of thickness of hot dynamic objects. Chapter 4 provides useful information about the different types of non-contact surface measurement techniques, theory, basic equations, system implementation, actual research topics, engineering applications, and future trends.

Chapter 1
Zeolite–Based Optical Detectors

Fabian N. Murrieta-Rico
Centro de Investigación Científica y de Educación Superior de Ensenada (CICESE), Mexico

Vitalii Petranovskii
Universidad Nacional Autónoma de México (UNAM), Mexico

Rosario I. Yocupicio-Gaxiola
Universidad Nacional Autónoma de México (UNAM), Mexico

Vera Tyrsa
Universidad Autónoma de Baja California (UABC), Mexico

ABSTRACT

The detection of chemical species is a common and required task in several areas of technology. Currently, measurements in dedicated labs are the predominant tools for detection and characterization of chemicals and materials. Although these techniques are available in specialized equipment, their use is often bounded by cost of application or the operator's expertise. Also, in many applications rather than an analysis of all the detectable chemical species, it is only of interest to determine the presence of a particular chemical compound, and if it is present, to quantify its concentration. For these reasons, alternative methods for detecting specific chemical species are required. One case of such methods are the optical chemical sensors, particularly the ones based on the materials known as zeolites. In a broad sense, these sensors are constituted by an optical detector that is modified with zeolites. This combination allows the detection of specific chemical compounds if the zeolitic materials is properly modified to have an optical response for the analyte.

INTRODUCTION

The detection and quantification of chemical species is a task required in quite diverse applications, e.g. industrial processes (Lauwerys & Hoet, 2001), diagnosis of human diseases (Shirasu & Touhara, 2011), or homeland security (Pramanik et al., 2011). In some cases, the identification of chemical compounds can be accomplished after the use of techniques and equipment, which are available in dedicated laboratories. Examples of such instruments are spectroscopy-based techniques (Pavia et al., 2008), mass

DOI: 10.4018/978-1-5225-5751-7.ch001

spectrometry (Kind & Fiehn, 2007), size exclusion chromatography (Liu et al., 2006; Mori & Barth, 2013), high-performance liquid chromatography (HPLC), and refractometry (McDonagh et al., 2008). The lab instruments have a high reliability, but in most cases they are expensive; also their use requires highly specialized personnel; in addition, such equipment for analysis, due to their size and operation conditions, needs dedicated facilities.

In some cases, for the purposes of detecting and quantifying of chemicals, the use of "traditional" laboratory equipment is not possible, therefore new approaches are required. One option is to create detection and measurement systems for this task. The core part of such systems is the sensor that interacts with the desired analyte. Particularly, optical chemical sensors (OCS) are used to detect and quantify chemical species, without interruption of the measuring process. This allows the use of this kind of sensors in diverse applications. Below is a brief description of the advances related to OCS.

The presence of explosives in a particular environment could be achieved by the detection and quantification of chemical compounds, which are into their composition. These compounds include nitro-organics, peroxides, and nitroamines (Germain & Knapp, 2009). Typically, for this task, chemical compounds such as TNT and DNT are determined using chemical-physical means, such as the fluorescent quenching and calorimetry.

The detection of chemical and biological species using sensors based on surface plasmon resonance (SPR) has gained attention, due to the different kind of analytes that can be detected; an extensive review is available (Homola, 2008). Most of SPR sensors use unlocalized surface plasmons. In this case, the plasmon propagates along planar structures. For biosensing applications, the sensors rely on angular and prism coupling, or in other cases, the "wavelength spectroscopy of surface plasmons" (Homola, 2008).

Materials known as metal-organic frameworks (MOFs) are solids, which are hybrid crystalline materials (similar to zeolites that are completely inorganic) that contain mixed organic/inorganic composition (Lee et al., 2009). The MOFs have a wide range of applications. Particularly, for detection and sensing, the optical emission properties of the MOFs are of interest (Hu et al., 2014). The photoluminescence of the MOFs is given by the process of charge transfer. The sensors developed with MOFs allow the detection of volatile organic compounds (VOCs), ionic species, and small molecules.

The application for optical sensing of "single mode subwavelength-diameter" silica nanowires was proposed in Lou et al. (2005). Its functioning is due to the properties of nanowire waveguides for evanescent-waveguiding. This works offers a theoretical analysis where there is an index change in a nanowire, which could be caused by the addition of chemical species when the nanowire is in an aqueous solution. The index change is calculated using Maxwell's equation. The sensitivity is estimated after the calculation of phase shift of the guided mode, which is generated by index change.

An optical sensor for detecting ammonia is reported in Lee et al. (2003). UV-Vis-NIR spectroscopy was used to analyze the optical properties of the polymer sensing layers. This sensor was made of polyaniline layer that was deposited onto an inner glass pillar located in a gas flow. The experiments showed that the transmittance ratio is changed during the adsorption of NH_3 molecules on the surface of conducting polymer.

Sensors to detect hydrogen are presented in Yang et al. (2010). For these sensors, the end surface of a single-mode optical fiber, and a side-etched multimode fiber were modified using palladium and tungsten trioxide" (Pd/WO_3) thin films. The sensors show the modulation of "optical power intensity" when they were exposed to different concentrations of hydrogen. Other palladium optical sensors are reported in Zhao et al. (2004); Villatoro et al. (2005); Tittl et al., (2011).

A sensor for detecting methane consisted of cryptophane molecules confined to a "transparent polymeric cladding (polysiloxane)" (Benounis et al., 2005). The cladding was deposited on a fiber of plastic clad silica (PCS). Among the results of the experiments, it is shown that the refractive index of cladding was increased as a consequence of the methane absorption into cryptophane.

A sensor for calorimetric detection of VOCs was proposed in Endo et al. (2007). The detector consisted of a poly (dimethylsiloxane) elastomer and a three-dimensional colloid crystal with a glass substrate. In experiments, the interaction of the sensor with organic solvents (benzene, toluene, xylene, methanol, 2-propanol, acetone) and ultrapure water was reported. When one of these compounds reached the elastomer matrix, swelling of the matrix was observed at the contact spot. Since the latter occurs on the surface of the crystal, it affects the lattice parameter. As a consequence, "the wavelength of the Bragg diffracted light was increased" (Endo et al., 2007).

Other sensors for VOCs detection were proposed in Penza et al. (2005). These studies were focused on changing the mass and changing the refractive index represented by these sensors when gas molecules were adsorbed on modified quartz crystal microbalances (QCM) and optical fibers modified with single-walled carbon nanotubes (SWCNTs). The sensors were exposed to ethanol, methanol, isopropanol, ethylacetate, and toluene. The authors report that the sensors shown high sensitivity, good reversibility, and repeatability. In addition, the detection of the VOCs was improved by integrating both types of sensors.

A mathematical analysis of the process associated with *in vivo* detection of glucose is presented in Barone et al. (2005). This analysis considered the use of optical sensors based on nanotubes and nanoparticles used as fluorescent probes. The results of the research show that the optical sensor is more stable than an electrochemical sensor of equal area.

Small concentrations of transition metal ions Ni(II), Cd(II), and Cr(III) can be detected using a chromogenic supramolecular sensor that contains phenanthroline (Resendiz et al., 2004). This device was characterized using elementary analysis and UV-Vis absorption spectroscopy. Experimental results show that the sensor detects micromolar concentrations of the mentioned analytes.

Analysis of membranes containing luminophores was described in Huber et al. (2001). The membrane exhibited an optical response to chloride. This was due to the luminophores encapsulated into the membrane. Fluorescence was observed, depending on the interaction with the analyte. An important characteristic of this material is that the detection range can be adjusted after the variation of the luminophores ratio or the combination of optical fibers used in the experiment. The modification of the optical sensor by means of "heterogeneous integration" of an InGaAs thin film with a "digital microfluidic system" was reported by Luan et al. (2008). The sensor detected a chemiluminescent signal during interaction with a solution of pyrogallol and hydrogen peroxide.

The report of the immunosensor that operates on the principle of total internal reflection (TIRF) was reported in Tschmelak et al. (2005). This sensor was tested to measure estrone but its application can be extended to other organic compounds. In these experiments, the analyte is pumped through a flow cell, covered with an absorbing layer. This array is exposed to a laser beam, and the optical response is detected by polymer fibers and processed.

The use of polyaniline in the construction of an optical sensor was presented in Jin et al. (2000). These devices are capable to detect changes in pH in the range from 2 to 12, with this variation their color changes. These sensors showed a color change when they were exposed to a pH change. In addition, it was proven that after 12 hours of polymerization, stability in water and air was achieved.

Quantum dots (QDs) are particles with size ranging between 1 to 10 nm. Because of its optical properties, particularly size depending light absorption, the QDs are materials suitable for use as "electrochemical sensors" (Murphy 2002). Particularly, due their quantum yields and biocompatibility, the use of quantum dots is actively being researched for applications in biological systems (Ailivsatos, 2004). The modification of quantum dots with solvated polystyrene was presented in (Meng et al. 2011). This material was used to modify waveguiding nanofibers; this experiment allowed to study the properties of the fibers after the modification with the functionalized QDs; the modified fibers were used as humidity sensors. In Goldman et al. (2005), the use of QDs for detecting 2,4,6-trinitrotoluene (TNT) is proposed. In the experiments, TNT was detected in seawater. The QDs were used to make "nanoscale sensing assemblies" that worked under the principle of "fluorescence resonance energy transfer".

In general, for the OCS, a sensing element (i.e. the optical fiber) is modified with a material that somehow interacts with the target analyte; one of such materials may be zeolites. The OCS draw attention because of their operation properties, which include high sensitivity, fast response, no analyte consumption, and no electrical connection (Ye et al., 2015). Zeolites are materials used to interact with the analyte, or in other words, they are used as a sensitive layer. The use of zeolites in the modification of the capabilities of optical detectors has been studied, but to date, there has not been a clear summary and comparison of the applications of zeolites in optical detectors. In the current work, we present a brief review of the literature on optical detectors based on zeolites.

Zeolites

Zeolites are crystalline hydrated aluminosilicates that can be found in nature, or which can be synthesized (Smith, 1984). The structure of zeolites consists of a crystalline network, with pores and cavities (some examples are illustrated in Figure 1), consequently, the zeolites have a high surface area.

Until now, more than 200 zeolite structures have been reported (Baerlocher and McCusker, 2017), and each one with pores well defined in particular size and shape. The dimension of such pores is in the range of 0.2 to 1.5 nm. The latter permits to understand how the shape of the zeolites pores allows only the molecules with dimensions smaller than the size of pores and particular shapes enter inside the zeolitic structure. Some properties of zeolites include ionic exchange (Rodríguez-Iznaga et al., 2018), and selective adsorption (Murrieta-Rico et al., 2015). The last is why the zeolites are known also as molecular sieves (Smith, 1984).

Classical applications of zeolites include their use as gas sorbents, coating membranes, and supports for molecular sieving and catalytic reactions. Recently, the research about zeolites proved that such materials possess optical properties that depend on the adsorbed substances. Some of these optical effects can be quantified after the measurement of fluorescence, birefringence, Raman spectra, UV-Vis-IR absorption spectra, or refractive index (Tang et al., 2011).

After all the known properties of the zeolites, it is clear that these materials are useful for the detection of chemicals if the zeolite optical properties are monitored during adsorption of the analyte species. But the development of sensors modified with zeolites is bounded by the attachment the zeolite to the transducer part. Also, it is required a deeper understanding of the optical behavior during analyte loading.

Figure 1. Channels and pores representation of a unit cell of the zeolites a) FAU, b) LTA, c) MOR, d) HEU, pictures drawn using the software provided by Baerlocher and McCusker (2017)

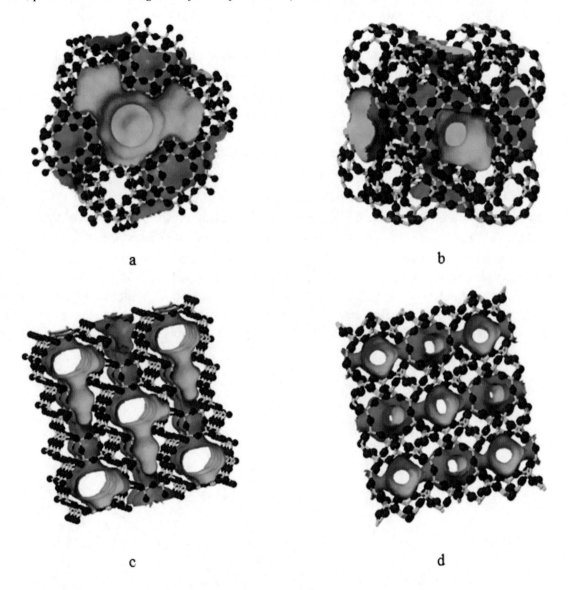

a

b

c

d

OPTICAL DETECTORS MODIFIED WITH ZEOLITES FOR THE RECOGNITION OF CHEMICAL SPECIES

In the literature, various cases of optical detectors modified with zeolites are described. This section is devoted to the brief discussion of some cases of such sensors, which are grouped by the analyte to detect. Also, the sensing principle is specified in this section.

Ammonia

Silicalite-1 (MFI-type zeolite) thin films were grown on a long period fiber grating (LPFG) (Zhang et al., 2006). For this sensor, the chemicals in the environment were adsorbed by the zeolite. This caused a change in the optical refractive index (n_z) of the LPFG. As a result, such a sensor is able to detect chemicals present in the environment. According to Zhang et al. (2006), the term LPFG refers to "an index module created by inscribing periodic index perturbations into the core of an optical fiber" after the use of laser irradiation. While the light travels through the LPFG, it is known that some wavelengths are guided from lossless to high-loss cladding modes. These wavelengths are known as resonant wavelengths (λ_R). The coupling results in a set of bands in the transmission spectrum. In the case, where the LPFG has a thin film as coating, the shift of λ_R is observed as consequence of changes in the refractive index of the overcoat (n_{ov}). Under the exposed principle, OCS can be constructed if the LPFG is coated with thin films of photonic materials, which are chemically sensitive. One case of this kind of sensors, are the LPFG coated with zeolites (Z-LPFG). In this kind of sensors, the shift of λ_R in the normal cladding mode is measured. This allows to quantify the concentration of adsorbed gases.

The experiments reported in Tang et al. (2011) shown the synthesis of a ZSM-5 film on an LPFG. The purpose of such research was to evaluate the influence of temperature and zeolite surface acidification, upon detection of NH_3, NO_2, CH_4. Figure 2 illustrates the results of these experiments. The authors, after the measurement of the shift of the resonant wavelength during the process of molecular adsorption in

Figure 2. Optical fiber coating with ZSM-5 for detection of NH_3, NO_2, and CH_4

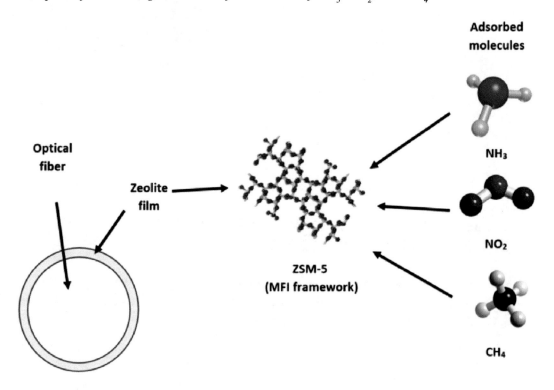

the zeolite, concluded that the Z-LPFG are suitable for gas sensing due to their high sensitivity. It was observed that the selectivity during sensing of the Z-LPFG depends on the selectivity of gas adsorption of the zeolite. Also, the modification of the zeolite surface in combination with the controlled temperature provided an efficient method to improve the selectivity for gas sensing and the response rate.

Acetone

The optical fiber sensors are a kind of photonic devices, where the fiber-based Mach-Zehnder interferometers (MZI) (Zetie et al., 2000) are foregrounded due to their physical dimensions, sensitivity to different sources of stimulus, and a fabrication process that is relatively simple (Li et al., 2011). It has been observed the excitation of high-order cladding modes in fiber MZIs. This is because such modes are directly exposed to the environment and led by the cladding-ambient interface. The last generates a shift in the refractive index that is attributed to the alteration in the environmental refractive index. This drastically changes the propagation of the cladding modes.

The fiber MZIs were used to measure the Refractive Index (RI) change, temperature, and angular displacement (Li et al., 2011). Also, the RI was improved after the use of a femtosecond laser. Moreover, the MZI was coated with zeolite and tested as a sensor of acetone vapor in N_2. A thin film of MFI-type zeolite was growth on the MZI, this granted to the interferometer the capacity to detect acetone vapors. The cylindrical surface of the MZI was modified with the zeolite film. This process was achieved by the crystallization of precursor solution free of aluminum, and the tetrapropylammonium ion was used as the structure-direct agent.

The resulting sensor was used to measure gas concentration (Li et al., 2011). The mechanism of such process was based on the fact that changes of the interference pattern position are induced by the molecular adsorption. The sensitivity of the sensor was affected by the type of the zeolite, the film thickness, the kind of gas to detect, and the characteristics of the interferometer. For the reported experiment, the length of the interferometer was 36 mm. After coating the fiber with zeolite, it was observed a change in the transmission spectrum. The experiment has shown that after an increase in the acetone concentration, a blue shift in the transmission spectra was observed. The corresponding sensitivities of the first and second attenuation peaks, after linear fitting, were found to be −3.83 pm/ppm, and −4.67 pm/ppm.

Methanol

A fiber optic intrinsic Fabry-Perot (Yoshino et al., 1982) interferometric chemical sensor was proposed for methanol detection (Liu et al., 2006). Such sensor was made by synthetizing and polishing a thin layer of polycrystalline MFI zeolite on the cleaved endface of a single mode fiber. This detector was used to monitor the optical thickness changes of the zeolite thin film, which were caused by the adsorption of organic molecules into the zeolite channels. White light interferometry was used to measure the optical thickness of the zeolite thin film. The sensor was used to detect methanol, isopropanol, and toluene dissolved in deionized water.

During the tests, it was observed that the sensor shown a high repeatability after repeated usage. The sensor had different responses in time and amplitude when it was exposed to different organic analytes. The detection limit was estimated in the range of 2 ppm for toluene, 5 ppm for 2-propanol, and 1000 ppm for methanol.

Isopropanol

In Xiao et al. (2005), it was reported the development of an optical fiber sensor modified with MFI-type zeolite, and also it was demonstrated that the sensor was sensible to chemical vapors, in this case, isopropanol. The sensor was made by growing a polycrystalline silicalite (MFI-type zeolite) thin film on the cleaved end face of a standard single-mode optical fiber. The output from the sensor was compared in two scenarios, one when the sensor was exposed only to N_2, and secondly when the sensor was exposed to a mixture of isopropanol and N_2. The sensing mechanism is based on the measurement of the change of the zeolite crystals optical reflectivity. For such property, it was observed that the change was reversible, and it was in function of the amount of chemical vapor adsorbed by the zeolitic structure.

The concept of a chemical sensor based on an optical microsphere coated with zeolitic material was proposed in Lin et al. (2010). The suggested sensing device detects the molecules of a particular analyte when they are adsorbed in the zeolitic material. During such adsorption, the zeolite refractive index increases. A reversible effect is observed during desorption, where the release of the analyte generates a decrement in the refractive index (Striebel et al., 1997). Since the wavelengths of the resonant bands have a high sensitivity to the changes in the refractive index, the detection of target chemicals can be done after the monitoring the resonant wavelength shift in the transmission spectrum. Also, another important characteristic reported in this work is that the operating temperature must be well controlled, the reason of this is that adsorbing selectivity of the zeolite is affected by temperature (Zhang et al., 2008). The result of this work (Lin et al., 2010) was a numerical model for a chemical sensor based on a zeolite-coated microsphere. Such results are a tool that permits to select and analyze this kind of sensor prior to its construction. Also, such model allows considering parameters for the design and optimization of this sensor.

Ethanol

A device for highly sensitive detection of chemical vapors was proposed in Lan et al. (2012). It was composed of fiber structure in the configuration of singlemode-multimode-singlemode (SMS). The fiber was coated with a thin film of zeolite MFI-type. This sensor was evaluated by using a fiber ring laser. The SMS structure is tailored in a section of two multimode fibers that have been into a fusion splicing process. The SMS structure works as an optical bandpass because of the self-imaging effect, and the multimodal interference. Such structure can be used as a detector of chemical species. The light propagates inside the SMS structure in case if it is uncoated, and interacts with the environment in which the structure is located. This effect allows monitoring the refractive index due to environmental changes. The SMS structure has a solid structure that can endure conditions of high pressure and high temperature, which exists during synthesis of zeolitic material, and the coating of the SMS. Also, the SMS generates a narrow optical band spectrum, therefore, there is a small Q factor that causes a poor resolution during the measurement of the spectral shift. In the work of Lan et al. (2012), a fiber ring laser configuration was proposed to improve the detection accuracy of the zeolite-coated SMS sensor. The zeolite layer behaves as a coating distributed in a thin film that changes its refractive index during the processes of adsorption, and desorption of molecules from the surroundings. When the adsorbate molecules go into the cavities of the zeolitic material, there is a change in the refractive index of the fundamental mode of the multimode fiber. Such change is explained by the Goos-Hänchen effect (Goos & Lindberg-Hänchen, 1949; Wild & Giles, 1982).

Oxygen

The detection/quantification of oxygen has several important applications. Some of them include the possibility to determine the suitability of water (where there is oxygen demand by biological organisms) analysis of gases emitted by the human blood, pressure measurement in highly sensitive parts, analysis of organisms *in vivo* conditions, combustion monitoring (Payra & Dutta, 2003), and the level of water pollution (Kannel et al., 2007). Like most of the optical sensing methods, the optical oxygen sensors have the properties of a quick response, no need to consume analytes, and unnecessary electrical wiring.

The oxygen concentration in liquids can be measured through fluorescence-based optical techniques. As stated in Remillard et al. (1999), such approaches involve the use of indicators based on organic fluorescent molecules that are contained inside the host, usually polymeric. A property of these systems is that in most cases, they only work in temperatures below 200-300 °C. This restriction is due to the instability of organic materials. A case where it is of interest to quantify the presence of oxygen is in the automobile exhaust. In this scenario, the oxygen sensors build with most organic materials are not of use, because they cannot handle high temperatures.

In Remillard et al. (1997), it was reported "an oxide-based ceramic system that showed fluorescence quenching" when it was exposed to oxygen. This behavior was observed in temperatures ranging from 400 to 575 °C. The material used in these experiments was a copper-exchanged zeolite (Cu-ZSM-5). During the quenching process, it was observed a reduction of Cu^{2+} in a presence of oxygen. When the Cu^{1+} ions were excited with blue or UV light, a strong fluorescence at 470 and 550 nm was observed, while the Cu^{2+} ions showed no fluorescent response. The results shown in Remillard et al. (1997) illustrate that the reaction $Cu^{1+} \leftrightarrow Cu^{2+}$ is quick and reversible. Also, it was noted that the fluorescence intensity depends proportionally on the concentration of Cu^{1+}, such concentration was controlled or determined by levels of oxygen during the gas flow. As a result of these experiments, it was shown how highly porous zeolites are can be used as sensitive materials for applications, where it is of interest to detect chemical species in gas phase; in this case the fluorescent response of a zeolite modified with Cu was used as sensing mechanism, this is possible because the zeolite provided a suitable way for the interaction of the fluorescent indicator (Cu), and the gas flow.

The work of the group of Remillard (Remillard et al., 1999) was extended by using the previously explored material (Cu-ZSM-5) to functionalize an optical fiber. In such experiments, for excitation of the detector, a 488-nm line of an Ar-ion laser was used. In addition, sol-gel-processed silica was used as an adhesive for the Cu-ZSM-5. In order to get the required thickness, the end side of the optical fiber segment was sprayed with a silica sol thick slurry that contained the zeolite, which was used as indicator phase. This method generated a ZSM-5 fluorescence indicator on the end of an all-silica optical fiber. It was able to work as a gas sensor at high temperature (475 °C), which allows the application of this sensor for exhaust-gas monitoring. With sensors fabricated under the method described, tests in dilute mixtures of several reductants and oxygen showed, how the tailored sensor was suitable to detect and quantify different analytes; namely, weak reductants, the oxygen content of the gas or the equivalence ratio of the gas for strong reductants (Remillard et al., 1999).

Since the pyrene has a high fluorescence quantum yield, a good photostability, and capacity to interact with the oxygen, it is useful as oxygen sensing probe for construction of OCS. For practical applications, it must be distributed in gas permeable organic polymers, the last allows the interaction of the pyrene with oxygen molecules (Ye et al., 2015). There are several examples of the use of pyrene as sensing element (Basu et al., 2003; Gregory et al., 2008; He et al., 2011; Shyamal et al., 2017; Sun et al., 2017; Turhan et al., 2018), but there are some drawbacks. The pyrene is a carcinogen, and environmental pollutant. Also, it aggregates or evaporates at high temperature or low pressures. In addition, the pyrene luminescence can be diminished not only by the oxygen; several molecules have this effect, for example, such as nitrogen-based compounds. In Ye et al. (2015), it was proposed the encapsulation of pyrene in a metal-organic zeolite SOD. Such process generated a paintable material with tunable photoluminescence properties and oxygen sensitivities. In addition, the authors report that the luminescent properties show long-term stability, even at 160 °C. This demonstrated high photostability, the absence of aggregation and leaking of pyrene.

An oxygen sensor based on zeolite Y with ruthenium was reported in Meier et al. (1995). After dealumination, the zeolite Y was loaded with ruthenium trichloride trihydrate. Subsequently, two different materials were synthesized: Ru-bipy (a complex of ruthenium bipyridyl) encapsulated in the cavities of zeolite Y, and another one, where the Ru-bipy was adsorbed on the surface of the zeolite. Both materials were mixed with silicone and spread onto a polyester support. Two kinds of sensing layers were obtained, one with the Ru-bipy encapsulated, and other with the Ru-bipy on the zeolitic surface. These membranes were immobilized inside a "flow-through cell", posteriorly a wall was generated. The response curves and luminescence spectra were quantified using a spectrofluorometer. A light source of 480 nm was used to excite the luminescence of the films, and the emission spectra were also measured at 610 nm. In these experiments, gas mixtures were used. They were prepared using pure oxygen and nitrogen. For both membranes, it was observed that their luminosity was quenched by oxygen, but it was significantly different under nitrogen exposition. The observed response was reversible, and no hysteresis was reported. The time of response of the sensor was short, around 6 to 10, but since the measured time depends mainly on the duration of complete gas exchange inside the sample cell, and the homogenization of the mixture of two gases. According to the experiments, such response time could be shortened.

The use of highly siliceous zeolites (Si/Al>100) to detect dissolved oxygen was reported in Payra and Dutta, (2003). This was done after the encapsulation of the dye molecules $Ru(bpy)_3^{2+}$ into zeolite-Y. Films of PDMS with zeolites were fabricated, and the quenching, or emitting dismissing of $Ru(bpy)_3^{2+}$ was analyzed. The results of this work shown that the O_2 dissolved in water generated the emission quenching, due to the photoexcitation of $Ru(bpy)_3^{2+}$. In addition, it is expected to have a long-term stability of the zeolite modified with $Ru(bpy)_3^{2+}$ because there will be no dye loss.

CONCLUSION

Optical detectors based on zeolites are devices that are able to detect and quantify target analytes, in other words, they are sensors capable of analyzing the content of specific chemical compounds in the medium. This task is achieved by modifying the optical transducer using a zeolitic material. This leads to a change in such optical properties as resonance wavelength, interference response, and optical thickness. When the modified sensor is exposed to the target analyte, its properties change, and a response occurs in the optical property. For such sensors, their sensitivity and selectivity are determined primarily by the type of zeolite used for modification. Therefore, by choosing the type of zeolite and its cationic composition, it is possible to achieve an increased sensitivity to the target analyte that the sensor is able to detect. In addition, the combination of a certain type of optical transducer with a certain type of zeolite is a factor that determines the desired parameters of the optical detector. Optical fibers are by far the most widely used for the creation of optical sensors modified with zeolites. Like any other new measurement method, the sensors mentioned in this review are only laboratory instruments. For a broad industrial application, in addition to research in the field of materials science, the more detailed metrological analysis is required.

ACKNOWLEDGMENT

This work was partially supported by DGAPA-UNAM IN107817 Grant. Fabian N. Murrieta-Rico acknowledges a scholarship from CONACYT for Ph. D. studies. The authors thank the extended support of CICESE that permit to finish this document.

REFERENCES

Alivisatos, P. (2004). The use of nanocrystals in biological detection. *Nature Biotechnology, 22*(1), 47–52. doi:10.1038/nbt927 PMID:14704706

Baerlocher, Ch., & McCusker, L. B. (2017). *Database of Zeolite Structures*. Retrieved from http://www.iza-structure.org/databases/

Barone, P. W., Parker, R. S., & Strano, M. S. (2005). In vivo fluorescence detection of glucose using a single-walled carbon nanotube optical sensor: Design, fluorophore properties, advantages, and disadvantages. *Analytical Chemistry, 77*(23), 7556–7562. doi:10.1021/ac0511997 PMID:16316162

Basu, B. J., Anandan, C., & Rajam, K. S. (2003). Study of the mechanism of degradation of pyrene-based pressure sensitive paints. *Sensors and Actuators. B, Chemical, 94*(3), 257–266. doi:10.1016/S0925-4005(03)00450-7

Benounis, M., Jaffrezic-Renault, N., Dutasta, J. P., Cherif, K., & Abdelghani, A. (2005). Study of a new evanescent wave optical fibre sensor for methane detection based on cryptophane molecules. *Sensors and Actuators. B, Chemical, 107*(1), 32–39. doi:10.1016/j.snb.2004.10.063

Endo, T., Yanagida, Y., & Hatsuzawa, T. (2007). Colorimetric detection of volatile organic compounds using a colloidal crystal-based chemical sensor for environmental applications. *Sensors and Actuators. B, Chemical, 125*(2), 589–595. doi:10.1016/j.snb.2007.03.003

Germain, M. E., & Knapp, M. J. (2009). Optical explosives detection: From color changes to fluorescence turn-on. *Chemical Society Reviews, 38*(9), 2543–2555. doi:10.1039/b809631g PMID:19690735

Goldman, E. R., Medintz, I. L., Whitley, J. L., Hayhurst, A., Clapp, A. R., Uyeda, H. T., ... Mattoussi, H. (2005). A hybrid quantum dot-antibody fragment fluorescence resonance energy transfer-based TNT sensor. *Journal of the American Chemical Society, 127*(18), 6744–6751. doi:10.1021/ja0436771 PMID:15869297

Goos, F., & Lindberg-Hänchen, H. (1949). Neumessung des strahlversetzungseffektes bei totalreflexion. *Annalen der Physik*, *440*(3-5), 251–252. doi:10.1002/andp.19494400312

Gregory, J. W., Asai, K., Kameda, M., Liu, T., & Sullivan, J. P. (2008). A review of pressure-sensitive paint for high-speed and unsteady aerodynamics. *Proceedings of the Institution of Mechanical Engineers. Part G, Journal of Aerospace Engineering*, *222*(2), 249–290. doi:10.1243/09544100JAERO243

He, G., Yan, N., Yang, J., Wang, H., Ding, L., Yin, S., & Fang, Y. (2011). Pyrene-containing conjugated polymer-based fluorescent films for highly sensitive and selective sensing of TNT in aqueous medium. *Macromolecules*, *44*(12), 4759–4766. doi:10.1021/ma200953s

Homola, J. (2008). Surface plasmon resonance sensors for detection of chemical and biological species. *Chemical Reviews*, *108*(2), 462–493. doi:10.1021/cr068107d PMID:18229953

Hu, Z., Deibert, B. J., & Li, J. (2014). Luminescent metal–organic frameworks for chemical sensing and explosive detection. *Chemical Society Reviews*, *43*(16), 5815–5840. doi:10.1039/C4CS00010B PMID:24577142

Huber, C., Klimant, I., Krause, C., & Wolfbeis, O. S. (2001). Dual lifetime referencing as applied to a chloride optical sensor. *Analytical Chemistry*, *73*(9), 2097–2103. doi:10.1021/ac9914364 PMID:11354496

Jin, Z., Su, Y., & Duan, Y. (2000). An improved optical pH sensor based on polyaniline. *Sensors and Actuators. B, Chemical*, *71*(1-2), 118–122. doi:10.1016/S0925-4005(00)00597-9

Kannel, P. R., Lee, S., Lee, Y. S., Kanel, S. R., & Khan, S. P. (2007). Application of water quality indices and dissolved oxygen as indicators for river water classification and urban impact assessment. *Environmental Monitoring and Assessment*, *132*(1), 93–110. doi:10.100710661-006-9505-1 PMID:17279460

Kind, T., & Fiehn, O. (2007). Seven golden rules for heuristic filtering of molecular formulas obtained by accurate mass spectrometry. *BMC Bioinformatics*, *8*(1), 105. doi:10.1186/1471-2105-8-105 PMID:17389044

Lan, X., Huang, J., Han, Q., Wei, T., Gao, Z., Jiang, H., ... Xiao, H. (2012). Fiber ring laser interrogated zeolite-coated singlemode-multimode-singlemode structure for trace chemical detection. *Optics Letters*, *37*(11), 1998–2000. doi:10.1364/OL.37.001998 PMID:22660100

Lauwerys, R. R., & Hoet, P. (2001). *Industrial chemical exposure: guidelines for biological monitoring*. CRC Press.

Lee, J., Farha, O. K., Roberts, J., Scheidt, K. A., Nguyen, S. T., & Hupp, J. T. (2009). Metal-organic framework materials as catalysts. *Chemical Society Reviews*, *38*(5), 1450–1459. doi:10.1039/b807080f PMID:19384447

Lee, Y. S., Joo, B. S., Choi, N. J., Lim, J. O., Huh, J. S., & Lee, D. D. (2003). Visible optical sensing of ammonia based on polyaniline film. *Sensors and Actuators. B, Chemical*, *93*(1-3), 148–152. doi:10.1016/S0925-4005(03)00207-7

Li, B., Jiang, L., Wang, S., Zhou, L., Xiao, H., & Tsai, H. L. (2011). Ultra-abrupt tapered fiber Mach-Zehnder interferometer sensors. *Sensors (Basel)*, *11*(6), 5729–5739. doi:10.3390110605729 PMID:22163923

Lin, N., Jiang, L., Wang, S., Yuan, L., Xiao, H., Lu, Y., & Tsai, H. (2010). Ultrasensitive chemical sensors based on whispering gallery modes in a microsphere coated with zeolite. *Applied Optics*, *49*(33), 6463–6471. doi:10.1364/AO.49.006463 PMID:21102672

Liu, N., Hui, J., Sun, C., Dong, J., Zhang, L., & Xiao, H. (2006). Nanoporous zeolite thin film-based fiber intrinsic Fabry-Perot interferometric sensor for detection of dissolved organics in water. *Sensors (Basel)*, *6*(8), 835–847. doi:10.33906080835

Lou, J., Tong, L., & Ye, Z. (2005). Modeling of silica nanowires for optical sensing. *Optics Express*, *13*(6), 2135–2140. doi:10.1364/OPEX.13.002135 PMID:19495101

Luan, L., Evans, R. D., Jokerst, N. M., & Fair, R. B. (2008). Integrated optical sensor in a digital microfluidic platform. *IEEE Sensors Journal*, *8*(5), 628–635. doi:10.1109/JSEN.2008.918717

McDonagh, C., Burke, C. S., & MacCraith, B. D. (2008). Optical chemical sensors. *Chemical Reviews*, *108*(2), 400–422. doi:10.1021/cr068102g PMID:18229950

Meier, B., Werner, T., Klimant, I., & Wolfbeis, O. S. (1995). Novel oxygen sensor material based on a ruthenium bipyridyl complex encapsulated in zeolite Y: Dramatic differences in the efficiency of luminescence quenching by oxygen on going from surface-adsorbed to zeolite-encapsulated fluorophores. *Sensors and Actuators. B, Chemical*, *29*(1), 240–245. doi:10.1016/0925-4005(95)01689-9

Meng, C., Xiao, Y., Wang, P., Zhang, L., Liu, Y., & Tong, L. (2011). Quantum-dot-doped polymer nanofibers for optical sensing. *Advanced Materials*, *23*(33), 3770–3774. PMID:21766349

Mori, S., & Barth, H. G. (2013). *Size exclusion chromatography*. Springer Science & Business Media.

Murphy, C. J. (2002). Optical sensing with quantum dots. *Analytical Chemistry*, *74*(19), 520A–526 A. doi:10.1021/ac022124v PMID:12380801

Murrieta-Rico, F. N., Mercorelli, P., Sergiyenko, O. Y., Petranovskii, V., Hernandez-Balbuena, D., & Tyrsa, V. (2015). Mathematical modelling of molecular adsorption in zeolite coated frequency domain sensors. *IFAC-PapersOnLine*, *48*(1), 41–46. doi:10.1016/j.ifacol.2015.05.060

Pavia, D. L., Lampman, G. M., Kriz, G. S., & Vyvyan, J. A. (2008). *Introduction to spectroscopy*. Cengage Learning.

Payra, P., & Dutta, P. K. (2003). Development of a dissolved oxygen sensor using tris (bipyridyl) ruthenium (II) complexes entrapped in highly siliceous zeolites. *Microporous and Mesoporous Materials*, *64*(1), 109–118. doi:10.1016/j.micromeso.2003.06.002

Penza, M., Cassano, G., Aversa, P., Cusano, A., Cutolo, A., Giordano, M., & Nicolais, L. (2005). Carbon nanotube acoustic and optical sensors for volatile organic compound detection. *Nanotechnology*, *16*(11), 2536–2547. doi:10.1088/0957-4484/16/11/013

Pramanik, S., Zheng, C., Zhang, X., Emge, T. J., & Li, J. (2011). New microporous metal– organic framework demonstrating unique selectivity for detection of high explosives and aromatic compounds. *Journal of the American Chemical Society*, *133*(12), 4153–4155. doi:10.1021/ja106851d PMID:21384862

Remillard, J. T., Jones, J. R., Poindexter, B. D., Narula, C. K., & Weber, W. H. (1999). Demonstration of a high-temperature fiber-optic gas sensor made with a sol-gel process to incorporate a fluorescent indicator. *Applied Optics*, *38*(25), 5306–5309. doi:10.1364/AO.38.005306 PMID:18324032

Remillard, J. T., Poindexter, B. D., & Weber, W. H. (1997). Fluorescence characteristics of Cu-ZSM-5 zeolites in reactive gas mixtures: Mechanisms for a fiber-optic-based gas sensor. *Applied Optics*, *36*(16), 3699–3707. doi:10.1364/AO.36.003699 PMID:18253395

Resendiz, M. J., Noveron, J. C., Disteldorf, H., Fischer, S., & Stang, P. J. (2004). A self-assembled supramolecular optical sensor for Ni (II), Cd (II), and Cr (III). *Organic Letters*, *6*(5), 651–653. doi:10.1021/ol035587b PMID:14986941

Rodríguez-Iznaga, I., Rodríguez-Fuentes, G., & Petranovskii, V. (2018). Ammonium modified natural clinoptilolite to remove manganese, cobalt and nickel ions from wastewater: Favorable conditions to the modification and selectivity to the cations. *Microporous and Mesoporous Materials*, *255*, 200–210. doi:10.1016/j.micromeso.2017.07.034

Shirasu, M., & Touhara, K. (2011). The scent of disease: Volatile organic compounds of the human body related to disease and disorder. *Journal of Biochemistry*, *150*(3), 257–266. doi:10.1093/jb/mvr090 PMID:21771869

Shyamal, M., Maity, S., Mazumdar, P., Sahoo, G. P., Maity, R., & Misra, A. (2017). Synthesis of an efficient Pyrene based AIE active functional material for selective sensing of 2, 4, 6-trinitrophenol. *Journal of Photochemistry and Photobiology A Chemistry*, *342*, 1–14. doi:10.1016/j.jphotochem.2017.03.030

Smith, J. V. (1984). Definition of a zeolite. *Zeolites*, *4*(4), 309–310. doi:10.1016/0144-2449(84)90003-4

Smith, J. V. (1984). Definition of a zeolite. *Zeolites*, *4*(4), 309–310. doi:10.1016/0144-2449(84)90003-4

Sohrabnezhad, S., Pourahmad, A., & Sadjadi, M. A. (2007). New methylene blue incorporated in mordenite zeolite as humidity sensor material. *Materials Letters*, *61*(11), 2311–2314. doi:10.1016/j.matlet.2006.09.006

Striebel, C., Hoffmann, K., & Marlow, F. (1997). The microcrystal prism method for refractive index measurements on zeolite-based nanocomposites. *Microporous Materials*, *9*(1), 43–50. doi:10.1016/S0927-6513(96)00090-9

Sun, S., Hu, W., Gao, H., Qi, H., & Ding, L. (2017). Luminescence of ferrocene-modified pyrene derivatives for turn-on sensing of Cu 2+ and anions. *Spectrochimica Acta. Part A: Molecular and Biomolecular Spectroscopy*, *184*, 30–37. doi:10.1016/j.saa.2017.04.073 PMID:28477514

Tang, X., Provenzano, J., Xu, Z., Dong, J., Duan, H., & Xiao, H. (2011). Acidic ZSM-5 zeolite-coated long period fiber grating for optical sensing of ammonia. *Journal of Materials Chemistry*, *21*(1), 181–186. doi:10.1039/C0JM02523B

Tittl, A., Mai, P., Taubert, R., Dregely, D., Liu, N., & Giessen, H. (2011). Palladium-based plasmonic perfect absorber in the visible wavelength range and its application to hydrogen sensing. *Nano Letters*, *11*(10), 4366–4369. doi:10.1021/nl202489g PMID:21877697

Tschmelak, J., Proll, G., & Gauglitz, G. (2005). Optical biosensor for pharmaceuticals, antibiotics, hormones, endocrine disrupting chemicals and pesticides in water: Assay optimization process for estrone as example. *Talanta*, *65*(2), 313–323. doi:10.1016/j.talanta.2004.07.011 PMID:18969801

Turhan, H., Tukenmez, E., Karagoz, B., & Bicak, N. (2018). Highly fluorescent sensing of nitroaromatic explosives in aqueous media using pyrene-linked PBEMA microspheres. *Talanta*, *179*, 107–114. doi:10.1016/j.talanta.2017.10.061 PMID:29310209

Villatoro, J., & Monzón-Hernández, D. (2005). Fast detection of hydrogen with nano fiber tapers coated with ultra thin palladium layers. *Optics Express*, *13*(13), 5087–5092. doi:10.1364/OPEX.13.005087 PMID:19498497

Wild, W. J., & Giles, C. L. (1982). Goos-Hänchen shifts from absorbing media. *Physical Review A.*, *25*(4), 2099–2101. doi:10.1103/PhysRevA.25.2099

Xiao, H., Zhang, J., Dong, J., Luo, M., Lee, R., & Romero, V. (2005). Synthesis of MFI zeolite films on optical fibers for detection of chemical vapors. *Optics Letters*, *30*(11), 1270–1272. doi:10.1364/OL.30.001270 PMID:15981503

Yang, M., Sun, Y., Zhang, D., & Jiang, D. (2010). Using Pd/WO_3 composite thin films as sensing materials for optical fiber hydrogen sensors. *Sensors and Actuators. B, Chemical*, *143*(2), 750–753. doi:10.1016/j.snb.2009.10.017

Ye, J. W., Zhou, H. L., Liu, S. Y., Cheng, X. N., Lin, R. B., Qi, X. L., ... Chen, X. M. (2015). Encapsulating Pyrene in a Metal Organic Zeolite for Optical Sensing of Molecular Oxygen. *Chemistry of Materials*, *27*(24), 8255–8260. doi:10.1021/acs.chemmater.5b03955

Yoshino, T., Kurosawa, K., Itoh, K., & Ose, T. (1982). Fiber-optic Fabry-Perot interferometer and its sensor applications. *IEEE Transactions on Microwave Theory and Techniques*, *30*(10), 1612–1621. doi:10.1109/TMTT.1982.1131298

Zetie, K. P., Adams, S. F., & Tocknell, R. M. (2000). How does a Mach-Zehnder interferometer work? *Physics Education*, *35*(1), 46–48. doi:10.1088/0031-9120/35/1/308

Zhang, J., Luo, M., Xiao, H., & Dong, J. (2006). Interferometric study on the adsorption-dependent refractive index of silicalite thin films grown on optical fibers. *Chemistry of Materials*, *18*(1), 4–6. doi:10.1021/cm0525353

Zhang, J., Tang, X., Dong, J., Wei, T., & Xiao, H. (2008). Zeolite thin film-coated long period fiber grating sensor for measuring trace chemical. *Optics Express*, *16*(11), 8317–8323. doi:10.1364/OE.16.008317 PMID:18545545

Zhang, J., Tang, X., Dong, J., Wei, T., & Xiao, H. (2009). Zeolite thin film-coated long period fiber grating sensor for measuring trace organic vapors. *Sensors and Actuators. B, Chemical*, *135*(2), 420–425. doi:10.1016/j.snb.2008.09.033

Zhao, Z., Sevryugina, Y., Carpenter, M. A., Welch, D., & Xia, H. (2004). All-optical hydrogen-sensing materials based on tailored palladium alloy thin films. *Analytical Chemistry*, *76*(21), 6321–6326. doi:10.1021/ac0494883 PMID:15516124

ADDITIONAL READING

Baerlocher, C., McCusker, L. B., & Olson, D. H. (2007). *Atlas of zeolite framework types*. Elsevier.

Breck, D. W. (1984). *Zeolite molecular sieves: structure, chemistry and use*. Krieger.

Hernández, G. (1988). *Fabry-perot interferometers (No. 3)*. Cambridge University Press.

Krohn, D. A., MacDougall, T., & Mendez, A. (2000). *Fiber Optic Sensors: Fundamentals and Applications (Press Monograph PM247)*. SPIE.

Luxmoore, A. R. (1983). Optical transducers and techniques in engineering measurement. Elsevier: Applied Science.

Ogawa, T., & Kanemitsu, Y. (Eds.). (1995). *Optical properties of low-dimensional materials* (Vol. 2). World Scientific.

Righini, G. C., Tajani, A., & Cutolo, A. (2009). *An introduction to optoelectronic sensors* (Vol. 7). World Scientific. doi:10.1142/6987

Stöcker, M., Karge, H. G., Jansen, J. C., & Weitkamp, J. (Eds.). (1994). *Advanced zeolite science and applications* (Vol. 85). Elsevier. doi:10.1016/S0167-2991(08)60776-4

Wolfbeis, O. S. (2008). Fiber-optic chemical sensors and biosensors. *Analytical Chemistry*, *80*(12), 4269–4283. doi:10.1021/ac800473b PMID:18462008

Xu, R., Pang, W., Yu, J., Huo, Q., & Chen, J. (2009). *Chemistry of zeolites and related porous materials: synthesis and structure*. John Wiley and Sons.

KEY TERMS AND DEFINITIONS

Analyte: A substance whose chemical composition is identified and measured.

Cladding: The coating of a surface in a structure.

Coating: A thin layer of a material that covers a surface.

Photostability: Refers to the degree of stability in of a substance during the exposition to the light.

Quantum Yield: It related to the radiation-induced process, and it is known as the amount or number of times that a specific event happens per photon that is absorbed by the system. In this case, an event that happens is in terms of a chemical reaction.

Sensor: A device that detects changes in a parameter of interest.

Zeolite: A group of materials that can be found in nature or synthesized in the laboratory. Such materials are crystalline, with pores and channels of nanometric dimensions. They are primarily constituted by Si, Al, and Na.

Chapter 2
Analog Circuit of Light Detector for CMOS Image Sensor

Arjuna Marzuki
Universiti Sains Malaysia, Malaysia

Mohd Tafir Mustaffa
Universiti Sains Malaysia, Malaysia

Norlaili Mohd Noh
Universiti Sains Malaysia, Malaysia

Basir Saibon
Universiti Kuala Lumpur, Malaysia

ABSTRACT

Plant phenotyping studies represent a challenge in agriculture application. The studies normally employ CMOS optical and image sensor. One of the most difficult challenges in designing the CMOS sensor is the need to achieve good sensitivity while achieving low noise and low power simultaneously for the sensor. At low power, the CMOS amplifier in the sensor is normally having a lower gain, and it becomes even worse when the frequency of the interest is in the vicinity of flicker noise region. Using conventional topology such as folded cascode will result in the CMOS amplifier having high gain, but with the drawback of high power. Hence, there is a need for a new approach that improves the sensitivity of the CMOS sensor while achieving low power. The objective of this chapter is to update CMOS sensors and to introduce a modified light integrating circuit which is suitable for CMOS image sensor.

INTRODUCTION

This section will review current optical sensor, electrical noises, state of the art of circuit technique, technology, and the proposed CMOS analog circuit for the sensor.

DOI: 10.4018/978-1-5225-5751-7.ch002

Optical Sensor in Machine Vision for Agriculture Application

Optical crop sensors assess crop health conditions by shining light of specific wavelengths at crop leave and measuring the type and intensity of the light wavelengths reflected back to the sensors. Not all optical sensors use the same light wavelengths. Different color light waves can be used to measure different plant properties. Commercially available crop sensors use two or more of red, green, blue or near infrared (NIR) color light waves. Example of usage of a typical optical crop sensor is shown in Figure 1. The optical sensor employs two types of light sources which are Near Infra-Red and Red. Even though some of the optical sensors (Bragagnolo *et al.*, 2013) employ xenon transmitter as the light sources, the photodetector is the common device for the reflected light detection devices.

Figure 1. A typical of optical crop sensor application

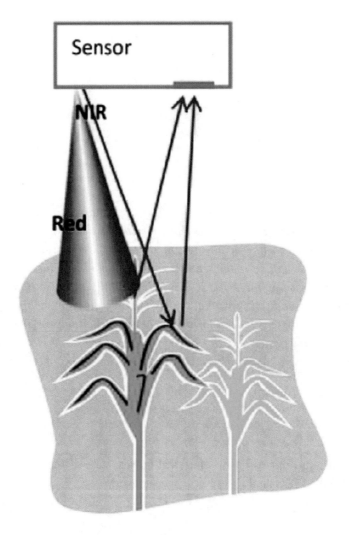

The leaf color indicates the level of nitrogen required by the plants. A dedicated chlorophyll meter is very expensive (Bragagnolo *et al.*, 2013) due to multiple light source (LEDs) requirements. The single wavelength based crop sensor is recently investigated (Intaravanne & Sumriddetchkajorn, 2015) which is based on android device. This work employs CMOS Image sensor chip (camera chip) as the photodetector. Even though it is simple, it cannot be used for in more advance horticulture or greenhouse application.

Figure 2 shows a state of the art of the electronic part (Front-End) of the optical crop sensor (Biggs *et al.*, 2002). Normally, the optical crop sensor used the constant light source, as this offer the simplest design. The amplifier is, therefore, a standard CMOS operational amplifier. The amplifier is used to amplify small voltage which is converted from light energy using the photodetector. The Front-End topology inherits low sensitivity due to the high dark current of the photodiode (when there is no light) and noisy CMOS operational amplifier. A very sensitive optical crop sensor indicates that it can detect dim light which means the light source power consumption can be reduced, thus leads to the potential of a portable system. The CMOS operational amplifier gain should be high when the sensitivity is low, this, however, will incur high power consumption to the CMOS operational amplifier. For low power CMOS operational amplifier, a sensitive photodetector is therefore required. It is also a requirement to have low noise CMOS operational amplifier in order not to affect the overall sensitivity.

An active multispectral imager whereby a monochrome Charge Coupled Device (CCD) camera (5 MPix) is mounted in a position two meters above the canopy surface inside a box with a LED light panel also inside the box illuminating the surface to produce nine spectral has been developed (Pajares *et al.*, 2016). However, it is uncertain whether the design can eliminate the unwanted reflected light.

Figure 2. Electronic parts and the detectors

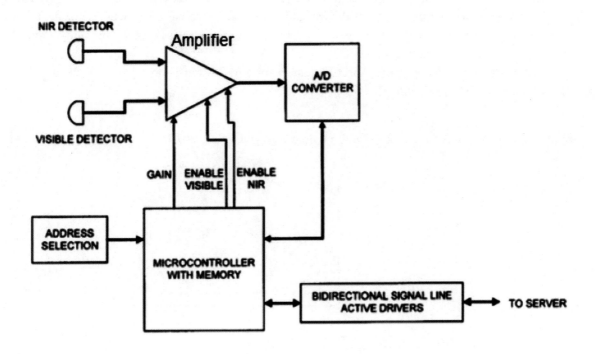

Electrical Noise

There are several type of electrical noises in an integrated circuit, for instance thermal noise, flicker noise, shot noise, burst noise, avalanche noise etc. (Gray *et al.*, 2009), but the main noise which dominates in low frequency is flicker noise.

Figure 3 shows the frequency spectrum of MOS noise voltage where thermal noise is the flat part of a circuit's intrinsic noise spectrum which also called Johnson noise and flicker noise, also known as 1/*f* noise is more intense at lower operating frequencies. Flicker noise that occurs practically in any electronic component which is caused by the traps near *Si/SiO*$_2$ interface that randomly capture and release carriers. The flicker noise of transistor M_x is given in Equation (1).

$$\overline{(V_{nfx}^{\,2})} = \frac{K_f}{(W \bullet L)_x C_{oxf}} \bullet \Delta f \tag{1}$$

Where K_f is the flicker noise coefficient, W is the channel width, L is the channel length, C_{ox} is the gate oxide capacitance per unit area, f is the frequency, and Δf is the bandwidth. The thermal noise that spread uniformly up to very high frequencies of transistor M_x is given in Equation (2).

$$\overline{(V_{ntx}^{\,2})} = 4kT \frac{2}{3} \frac{1}{g_{mx}} \bullet \Delta f \tag{2}$$

Where k is the Boltzmann's constant in joules per kelvin, T is the temperature in degrees Kelvin, and g_m is the transconductance parameter of the MOSFET device. K_f which is a process-dependant factor can be different for NMOS and PMOS devices. In the sub-micron process used in this study, the K_f for the PMOS device is less than that of NMOS device.

Therefore the input referred noise spectral density for a MOS transistor is given by (Jui-Lin *et al.*, 2010; Chan *et al.*, 2001) Equation (3).

Figure 3. Frequency spectrum of MOS noise voltage (Jui-Lin et al., 2010)

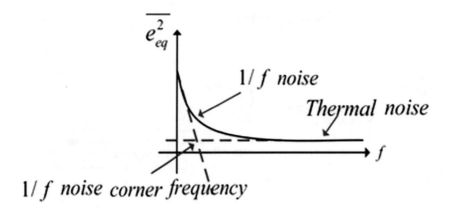

$$\overline{V_n^2} = \frac{8kT}{3g_m}\Delta f + \frac{K_f}{WLC_{oxf}}\bullet\Delta f \tag{3}$$

Equation (3) shows that flicker noise can be reduced by properly increasing the area of the transistor, which are the width and length of the transistor.

Since flicker noise component is usually larger than thermal noise component for frequencies below 10 kHz, the total noise for folded cascode circuit contribution after neglecting the thermal noise can be approximated as in Equation (4) (Chan *et al.*, 2001).

$$\overline{V_{nfc}^2} = 2\left[\overline{V_{n1}^{\,2}} + \overline{V_{nx}^{\,2}}\left(\frac{g_{mx}^2}{g_{m1}^2}\right)\right]\bullet\Delta f \tag{4}$$

The transconductance is related to Equation (5).

$$g_m^{\,2} = 2\mu_{n/p}C_{ox}\frac{W}{L}I_D \tag{5}$$

The above equations (Equation (4) and Equation (5)) justify that width and length of the transistors are the main parameter that will affect the noise contribution.

State of the Art of Analog Circuit of Optoelectronic Device

This section discusses the architecture or topology for the analog circuit design of the optoelectronic devices.

CMOS Color Sensor

A typical light or color sensor uses photodiodes and trans-impedance amplifier (TIA). The TIA is used to convert the photodiode current to voltage. This is shown in Figure 4. The resistor and capacitor at the output are used to filter out high frequency (unwanted) signal. The filter will affect the time response of the color sensor. The size of the feedback resistor, $R_{feedback}$ is also big and this consumes size or area of the silicon. The amplifier gain is normally huge and this will increase the current consumption.

CMOS Light to Frequency (LTF) Converter

Several digital color sensors employ light to frequency (LTF) technique and photodiodes which is similar to other studies (Ho *et al.*, 2011; Nahtigal & Strle, 2016). These frequency-based digital sensors require an advanced processor such as Digital Signal Processor (DSP) or Personal Computer (PC) to measure or calculate the frequency (Scalzi *et al.*, 2014; Lee *et al.*, 2006). Figure 5 shows an example of the sensor.

Figure 4. TIA-based color sensor

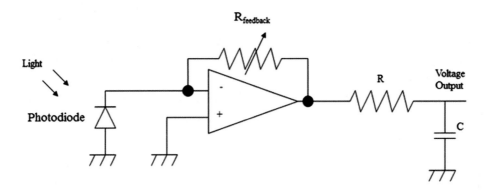

Figure 5. Light to frequency sensor

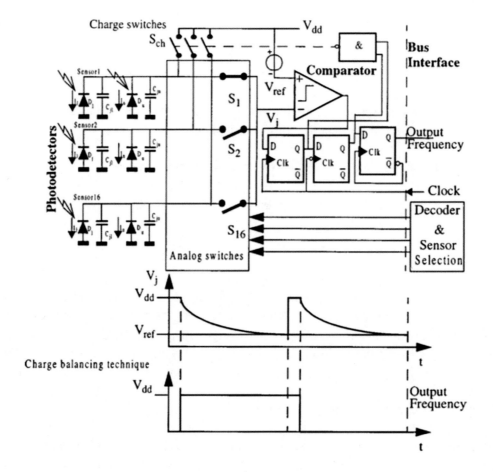

CMOS Image Sensor (CIS)

CIS utilizes smallest photodiodes/phototransistors compared to the color sensor and LTF converter. The data converter is one of the major sub-blocks in CIS, its function is to convert the analog signal into digital signal. This approach will improve signal robustness and provide for further processing. Analog to Digital Converter (ADC) is a typical circuit used as the data converter.

Pixel Level ADC CIS

Digital pixel sensor offers a wide dynamic range (Trépanier, Sawan, Audet, & Coulombe, 2002). Processing can also be done at the pixel level. The topology is similar to memory architecture. The major disadvantage is low fill factor.

Column-Level ADC CIS

This approach offers the best performance trade-off. A Correlated doubled sampling (CDS) circuit is normally used to reduce noises (Feng, 2014). This circuit is employed prior the column ADC.

Chip-Level ADC CIS

Chip-level ADC or sometimes called matrix level ADC. The ADC for this topology has to be very fast (David, 1999), this topology would also consume very high current. ADC type suitable for CIS topology is pipelined ADC. However, Successive Approximation Register (SAR) (Deguchi *et al.*, 2013) and Flash type ADC (Loinaz, 1998) has also been reported in this type of CIS design. Table 1 is a summary of ADC choice of location for CMOS Image Sensor (CIS) (Feng, 2014). Due to shutter requirement, column-level ADC is suitable for CIS architecture.

ADC Type

This section discusses major ADC architecture namely slope ADC, SAR ADC, flash ADC, pipelined ADC and Delta Sigma ADC. The dynamic range of ADC should be higher than the final CIS dynamic range.

Table 1. ADC Choice of Location in CIS.

Parameter	Pixel level ADC	Column-level ADC	Chip-level ADC
Power	Low (Trépanier *et al.*, 2002)	Medium (Takayanagi *et al.*, 2013)	High
Area	Fill factor low	Area medium	High
Speed (Frame Rate)	Highest	Medium	Limited by ADC
Noise	Elimination of temporal noise (Gamal, 2002)	Medium	Low
Global Shutter	Implementable	Implementable	Not Implementable

Slope ADC

Dual slope ADC type has greater noise immunity than the other type ADCs. While single slope ADC is sensitive to the switching error. However, a single slope ADC approach was still used in BSI application due to its small area (Suzuki, 2015). In Kleinfelder, Lim, Liu, & Gamal (2001), slope ADC is also employed.

SAR ADC

This is a popular ADC type for column-level ADC (Takayanagi *et al.*, 2013), also for chip level ADC (Deguchi *et al.*, 2013). A 3-bit SAR ADC has been developed for pixel level ADC (Zhao *et al.*, 2014). The conversion in this type of ADC always starts with MSB decision follow with LSB (Song, 2000). The SAR algorithm is important (to control Digital to Analog Converter (DAC)) output using progressing dividing the range by 2) in SAR ADC.

Flash ADC

This is the most straightforward way of designing an ADC. Another improvement of flash ADC is folding ADC. The Flash ADC probably loses importance in CMOS Image sensor since it is limited to 8 or 10 bit (Loinaz, 1998; Innocent, n.d). However, this ADC topology is suitable for chip level ADC.

Pipelined ADC

This ADC topology is suitable for chip level ADC (Hamami, Fleshel, Yadid-pecht, & Driver, 2004). The 1.5-bit stage is normally used in designing pipelined ADC (Abdul Aziz & Marzuki, 2010), thus requires only tri-level DAC (Song, 2000). Error correction is required to produce the correct output. Further improvement to pipelined ADC topology is algorithmic, cyclic or recursive ADC.

Delta Sigma ADC

Oversampled ADCs have the advantage of filtering temporal noise (Norsworthy, Schreier, & Temes, 1997). The idea is similar to synchronous analog voltage to frequency converter. This ADC has been employed at pixel (Mahmoodi & Joseph, 2008) level and column level (Chae et al., 2011). Table 2 shows the ADC performances.

In summary, the column-level ADC topology is a popular choice for CMOS image sensor due to a good trade-off between readout speed, silicon area and power consumption (Lyu, Yao, Nie, & Xu, 2014; Marzuki, Aziz, & Manaf, 2011).

Table 2. ADC Performances.

Topology	Latency	Speed	Accuracy	Area
Flash	Low	High	low	High
SAR	Low	Low-medium	Medium-high	Low
Delta-sigma	High	Low	High	Medium
Pipeline	High	Medium-high	Medium-high	Medium
Slope	low	low	High	low

Pixel Circuitry

This type of circuit is normally used in CIS. A low fixed pattern noise (LFPN) capacitive trans-impedance amplifier (CTIA) for active pixel CMOS image sensors (APS) with high switchable gain and low read noise is shown in Figure 6. The LFPN CTIA APS uses a switched capacitor voltage divider feedback circuit to achieve high *sensitivity*, low gain FPN, and low read noise. The circuit consists of a trans-conductance amplifier TA1, a photodiode, a network of feedback capacitors and switches (Cl, C2, Cf, Ml, M2), and a bit line select transistor M3. WORD is used to select each row of pixels, BIT is the output bus for each column in the sensor, RESET and GAIN are used to reset the pixel and control the pixel gain, and VREF is the pixel bias voltage.

Another circuit is shown in Figure 7. The integration capacitor, C_{int} is used as a feedback component. The photocurrent now is coming from C_{int} and V_{diode} remains constant throughout the integration period.

Technology

This section describes CMOS technology, Backside Illumination (BSI) technology and photo devices which applicable for sensor design.

General Comments on Technology or Process for CMOS Image Sensor

Generally, 4 types of processes are used, standard CMOS, Analog-Mixed Signal CMOS, Digital CMOS and CMOS Image Sensor process. The latter is the process developed specifically for CMOS Image Sensor. There are many foundries available for the development of CMOS Image Sensor. The most obvious difference between this process and other process is the availability of photo devices such as a pinned

Figure 6. CTIA (Fowler, Balicki, How, & Godfrey, 2001)

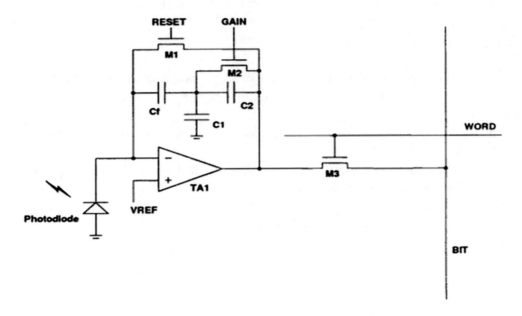

Figure 7. Pixel schematic (Goy, Courtois, Karam & Pressecq, 2001)

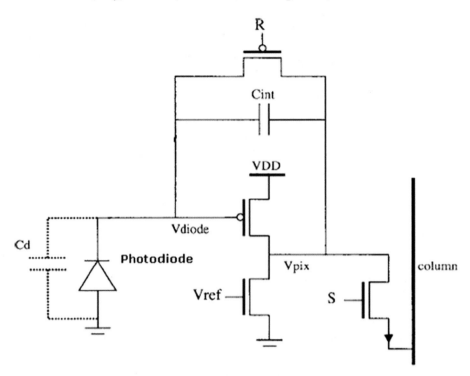

photodiode. Advantages of smaller dimension technology are a smaller pixel, high spatial resolution, and lower power consumption. A technology lower than 100 nm, requires a modification to fabrication process (not following the digital roadmap) and pixel architecture (Wong, 1997).

Fundamental parameters such as leakage current (will affect the sensitivity to the light) and operation voltage (will affect dynamic range, i.e., the saturation, pinned photodiode most likely not going to work at low voltage (Wong, 1997) are very important when a process is selected for CIS development. Because of this limitation, new circuit technique is introduced.

1.   An old circuit such as standard pixel circuit cannot be used when using 0.1 microns and lower (Bigas et al., 2006). This is due to topology requires high voltage; because the maximum supply voltage is now lower.
2.   Calibration circuit and cancellation circuit are normally employed to reduce noises.

In order to increase the resolution into multi-megapixel and hundreds of frame rate, lower dimension technology is normally chosen. Evidently, it has been reported that 0.13 micron (Takayanagi *et al.*, 2013) and 0.18 micron (Xu *et al.*, 2014) are good enough to achieve good imaging performance.

These modifications of CMOS process has started at 0.25 microns and below to improve their imaging characteristics –As process scaling going to much lower than 0.25 microns and below several fundamental parameters are degraded namely photo responsibility and dark current. Therefore the modifications are focused to mitigate these parameters degradation (Gamal, 2002; Bigas *et al.*, 2006). System requirement (such as supply voltage, temperature) is also one of the criteria in selecting suitable process. Price of tool and development cost will also determine the process selection.

Backside Illumination (BSI)

BSI technology eliminates the need to push light through the layers of metal interconnections. With this, high Quantum efficiency (QE) can be achieved. However, this technology incurs additional costs due to extra process, such as stacking and Through Silicon Via (TSV). The pixel of 1.1 microns seems to be the tipping point advantage over Front side illumination (FSI) (Aptina, 2010). A work in Sukegawa (2013) uses BSI to improve the resolution through the pixel size. As the BSI is targeted for very small pixel, a process such as 90 nm is preferred, of course, this is a very expensive process, and thus this CMOS Image sensor is meant for the expensive application (e.g. high-end camera).

Photo Devices

The typical photodetector devices are photodiode and phototransistor. Typical photodiode devices are N+/Psub, P+/N_well, N_well/Psub and P+/N_well/Psub (back to back diode) (Ardeshirpour, 2004). Phototransistor devices are P+/n_well/Psub (vertical transistor), P+/N_well/P+ (Lateral transistor) and N_well/gate (tied phototransistor) (Ardeshirpour, 2004).

This standard photo device still requires micro-lens and color filter array (CFA). Quantum efficiency (QE) of photodiodes in standard CMOS is usually below 0.3 (Scheffer, Dierickx, & Meynants, 1997).

The devices which are normally developed for modified CMOS process are a photogate, pinned photodiode, and an amorphous silicon diode. These devices will improve the sensitivity of CIS. Pinned photodiode which has low dark current offers good imaging characteristic for CIS (Lulé, 2000).

A Proposed CMOS Crop Sensor and Its Application

The proposed sensor will utilize light integrating concept with the modulated light source. The proposed application will improve the sensitivity of the sensor because, with the modulated light source, the background or noise detected by the detectors can be eliminated. The concept is shown in Figure 8. The device controller, which is a PWM-based LED driver with LEDs (such as Near Infra-Red and Visible LEDs) provides PWM light to a subject/crop. The reflected PWM light is later detected by the photo-sensor array which converts the light to current. A photocurrent to voltage block is used to convert the current to voltage for ADC. This block is synchronized with LEDs via the device controller. Therefore, PWM ON/OFF time is known or can be used in the photocurrent to voltage block. Hence during the OFF time/duration, any detected light is due to other light sources. The proposed concept requires some sort of color calibration in order to improve the accuracy of the color reading.

AN IMPROVED ANALOG CIRCUIT FOR CIS

The objective of this section is to discuss a simple analog mixed-signal circuit front-end circuit prior to ADC which suitable for CMOS Image sensor when the light source is PWM-based LEDs. The standard TIA has a stability problem. This is shown in Figure 9. Due to pulsating current, it can be seen that the ripple happens during 'ON' time. This indicates instability condition. In other words, the TIA could oscillate or becomes an oscillator. The huge gain of opamp also contributes to the instability of the cir-

Figure 8. Proposed CMOS sensor and its application

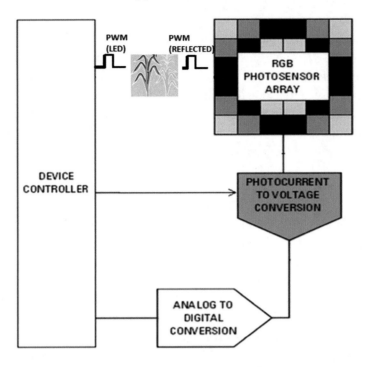

Figure 9. Stability issue of the conventional color sensor

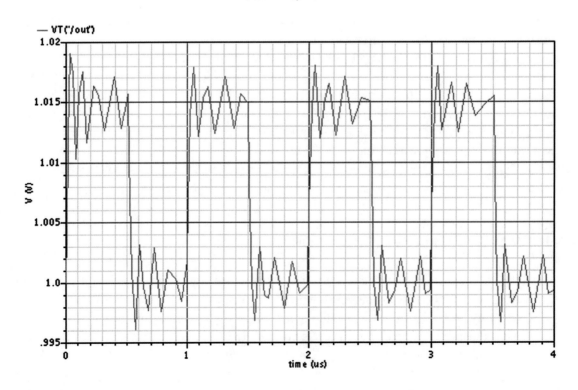

cuit/TIA. As mentioned in the previous section, the opamp consumes high current and this will affect the overall power consumption.

The proposed work could have a similar idea to any CTIA based image sensor (Xui, 2014). However, the conventional CTIA is not capable to eliminate the 'noise' or reflected light due to the unknown light source. This is because the idea of the light/color detection using PWM-based LEDs for agriculture application is new.

The new approach is based on light integrating technique. A Standard 0.18 µm analog technology was chosen for this front-end circuit while a standard Cadence Analog design flow was used in the designing of these circuitries. A calibration is required to improve the detection/color accuracy.

Design

This section discusses the design aspect. It covers concept, Integrator and Capacitor array, self-biased folded cascode amplifier, differential and sample/hold (S/H) buffer amplifier.

Concept Overview

The design objective is to integrate 1. Analog function (including gain stage functions), 2. RGB sensor and 3. Analog to digital converter function, into single chip solution. Figure 10 shows a basic block diagram of a proposed sensor with an ADC. REFERENCE block which consists of bandgap and IBIAS circuits are used to provide the voltage references and current references for ADC block. TEST_MUX is used for testing purposes. It employs buffer amplifiers and switches (NMOS type). However, only the analog circuit portion will be discussed in this chapter. The sub-blocks will be explained in next sub-sections. The basic concept was derived from previous work (Marzuki *et al.*, 2016). The proposed control signals for the sensor are shown in Figure 11. From Equation (6).

$$V_{in} = V_{prech\arg e} - \frac{I_{photodiode} \times T_{int\,egration}}{C_{int\,egration}} \tag{6}$$

When the light incident on the photodiode and the integration phase (see Figure 11) is activated, then the voltage V_{in} is inversely proportional to the light intensity. A differential amplifier is then used to produce differential voltages for differential input ADC, these values are later sampled by the S/H amplifier. The sampled values are held for analog to digital conversion. At the same time, the $C_{integration}$ is charged back to $V_{precharge}$ value. A pipeline ADC with 8-bit resolution is used for analog to digital conversion. The ADC can receive ± 1.2 V, i.e. differential voltages with the nominal voltage (common mode voltage or VCM) of 1.2 V. For output of 0 DEC, the differential voltages is -1.2 V (e.g. in p = 0. 6 V, in m = 1.8 V) while for output 256 DEC (2^8), the differential voltage is 1.2 V. Equation (7) describes the relationship of voltages and ADC output. Both Equation (6) and (7) have shown the concept capability of integrating the three mentioned functions into single silicon.

$$ADCOutput(DEC) = \left(\frac{(inp - inm) + VCM}{2VCM} \right) \times 256 \tag{7}$$

Figure 10. Block Diagram of proposed analog circuit

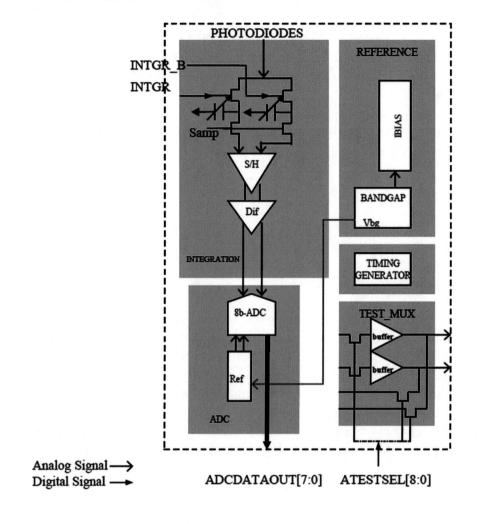

Integrator

Figure 12 shows the integrator circuit with a capacitor of 10 pF. The value of the capacitance can be changed through the capselect. The capacitor block is pre-charged to ~1.8V ($V_{precharge}$) when 'ctrln' signal is high. The 'ctrln' signal is low when integrating is selected. During the integration period, the pre-charged capacitor voltage starts to decrease as described by the Equation (6). The capacitor array together with photodiode sizes can be used to adjust the required output voltage. Transmission gates (TG) is used as the switch in the integrator.

Figure 13 shows the complete design of front-end analog circuit of the integrator circuit. Figure 14 shows integrator output simulation results. When all capacitors are selected (10 pF) the photodiode current (I) is 100 nA (assumed), the slope or dV/dT is agreeing on the Equation (8). As in the integrator, TG is used as switches in this circuit. The inverters are used to create the inverted control signals.

$$I = C \frac{dV}{dT} \tag{8}$$

Figure 11. Control signals

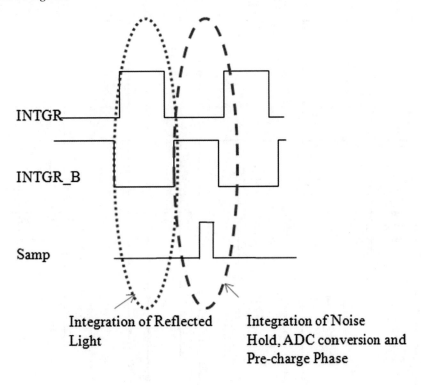

Integration of Reflected
Light

Integration of Noise
Hold, ADC conversion and
Pre-charge Phase

Figure 12. Integrator

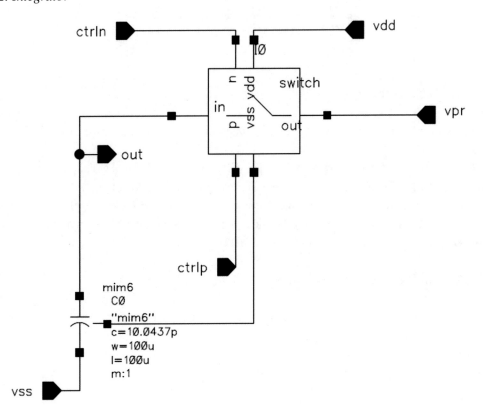

Figure 13. Complete integrator

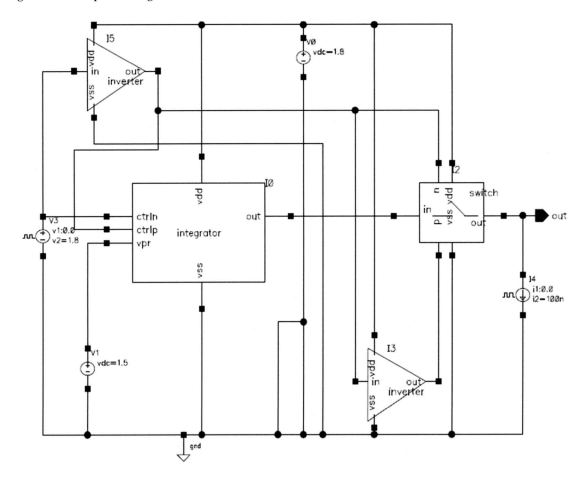

Sample and Hold Amplifier

An S/H buffer amplifier as shown in Figure 15 consists of an operational amplifier, 1 capacitor, and 2 switches.

Two S/H buffer amplifiers are used prior to the differential amplifier. Two capacitors with a value of 100 fF are used as the S/H capacitors. The value of the capacitor is chosen to achieve the best kT/C noise.

As the objective is low power consumption, the opamp in the S/H buffer amplifiers should consume low current. The opamp design is discussed in next section. The charge injection and feed-through issue should be reduced with dummy switch (Baker, 2010).

Self-Biased Folded Cascode Amplifier

The folded cascode op amp is characterized by various performances such as voltage gain, unity gain bandwidth, phase margin, total noise level and power consumption. These measurements are fixed by the design parameters, for instance, transistor sizing, bias currents, supply voltage, and other compo-nent values. The design parameters have been optimized to meet the objective feature while satisfying

Figure 14. Integrator output

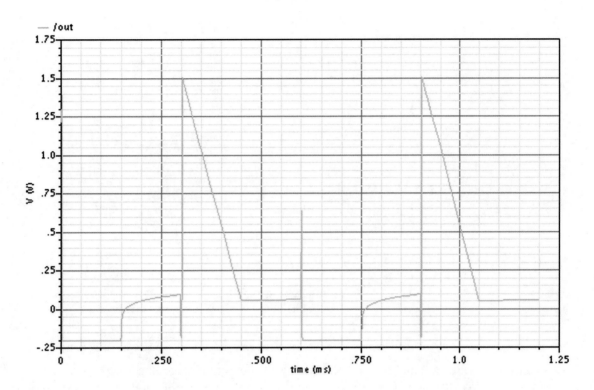

Figure 15. S/H buffer amplifier

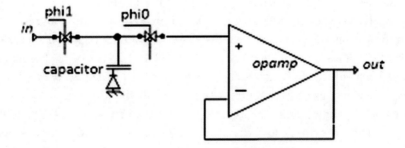

the design specifications and constraints. Figure 16 shows the design flow which clarifies a top-down synthesis methodology for CMOS folded cascode op amp architecture.

In order to determine the unknowns such as transistor sizes, drain current, and biasing current, it starts from stating the specifications target range. For instance, voltage gain and unity gain bandwidth are two main specifications that have to be made from the beginning in order to determine the unknowns like MOS device sizes $(W/L)_1$ and bias current I_{D1}. Model parameter extraction is important as to perceive the CMOS characteristics of the 180 nm SILTERRA technology.

Figure 16. Design flow of folded cascode OTA

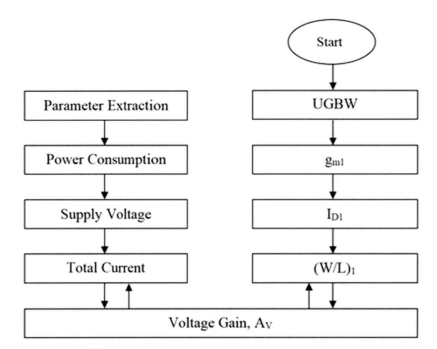

A MOS folded cascode amplifier uses cascoding in the output stage combined with an unusual implementation of the differential amplifier to obtain better input common mode has been used in this work. The folded cascode circuit basically consists of common source configuration, common gate configuration, cascaded mirror and constant current source. This folded cascode amplifier requires three biasing voltages to ensure all the transistors operate in saturation region.

Based on Figure 17, M_1 is connected in a common source configuration and M_{10} is connected in a common gate configuration. M_3 acts as a current source for both the common source configuration of M_1 and M_2. Since PMOS has lower flicker noise than NMOS, PMOS has been chosen for the differential input pairs M_1 and M_2 to reduce the flicker noise of the amplifier *(Mohamed et al., 2013; Nemirovsky et al., 2001)*. As M_4 and M_5 act as constant current source, small signal variations in the drain current of M_1 and M_2 are conducted primarily through M_{10} and M_{11} respectively. It is said to be folded as it reverses the direction of the signal flow back towards the current mirror. This reversal has two benefits when differential pair is being applied. One of them is it increases the output swing and so, it increases the common mode input range. The current mirror sends variation in the drain current of M_{10} to the output where it converts the differential signal into single ended output. Bias is realized by allowing the current in current sources M_4 and M_5 larger than $|I_{D3}|/2$.

Equation (9) shows the relations between I_{D10} and I_{D3} (Gray *et al.*, 2009).

$$I_{D10} = I_{D11} = I_{D4} - \left(\frac{\left|I_{D3}\right|}{2}\right) = I_{D5} - \left(\frac{\left|I_{D3}\right|}{2}\right) \tag{9}$$

Figure 17. Self-biased folded cascode amplifier

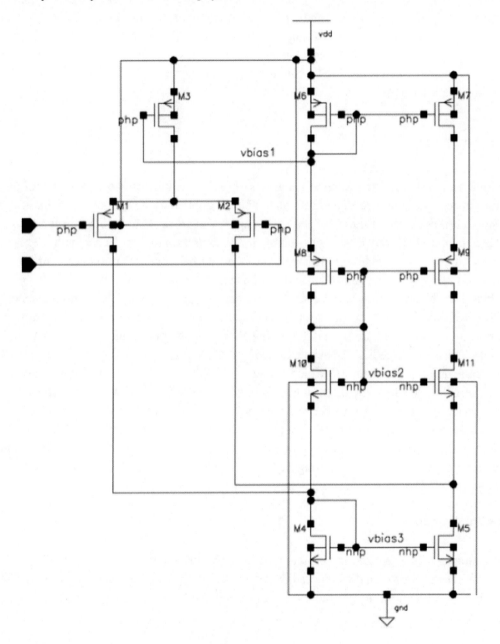

The small signal voltage gain of folded cascode op amp at low frequencies is shown in Equation (10).

$$A_v = G_m R_o \tag{10}$$

G_m is the transconductance and R_0 is the output resistance. The variation in the drain current of M_1 and M_2 contribute productively to the transconductance due to the appearance of the current mirror of M_6 to M_7. Thus, $G_m = g_{m1} = g_{m2}$.

In order to get R_0, both inputs, V_{in1}, V_{in2}, are connected to AC ground. The sources of M_1 and M_2 do not operate at AC ground even though the input voltages do not move in this case. The drain current of M_1 stays constant when the source of M_1 and M_2 connecting to AC ground. Besides, as M_6 and M_8 are diode connected, the Thevenin equivalent resistances at the gates of M_7 and M_9 are very small. Therefore, the gates of M_7 and M_9 are assumed to be connected to small signal ground. Thus, the calculation of R_0 is shown in Equation (11) (Gray *et al.*, 2009).

$$R_o = (R_{out|M11}) \| (R_{out|M9}) \qquad (11)$$

The folded cascode amplifier requires three sets of biasing voltages in contemplation of optimizing the performance of the amplifier. The first voltage bias is applied to the gate of transistor *M3*, which is the constant current source of the input stages. An approximate 1.3 V to 1.45 V of the voltage biasing is capable of enabling the transistor *M3* to operate in the saturation region. Hence, voltage bias 1 can be obtained by connecting the gate of transistor *M3* to the internal node where the drain of transistor *M6* and source of transistor *M8* are joined since the voltage at this node is 1.391 V. Next, voltage bias 2 and voltage bias 3 are biasing voltages to be applied to transistor pair *M10* and *M11* and transistor pair *M4* and *M5* respectively. The biasing voltages range to ensure both the aforementioned transistor pairs to be operated in saturation region are from 0.8 V to 0.9 V. for voltage bias 2 and around 0.4 V to 0.5 V for voltage bias 3. Thus, voltage bias 2 can be attained by connecting the transistor pair of *M10* and *M11* to the node at the drain of transistor *M8* which is 882.8 mV. Subsequently, voltage bias 3 can be realized from the internal node where the drain of transistor *M3* and drain of PMOS *M1* are attached which is 426.4 mV on the point of it is within the voltage range as stated earlier.

Assuredly, the overall circuit has become smaller since the additional space for extra bias circuitry is no longer needed. The self-biased folded cascode topology not only save a considerable amount of area, yet, noise contribution can be thoroughly reduced by reason of extra bias circuitry that susceptible to noise and crosstalk is avoided. Consequently, total power consumption is being reduced as well.

Differential Amplifier

The differential amplifier is used to eliminate the noise or unwanted reflected light while at the same time amplify the required signal if required. Figure 18 show the switched capacitor based differential amplifier which can be used to subtract two voltages. The switch capacitor resistor (SCR) is used to replace the resistor.

CMOS Implementation and Results

This section describes the CMOS implementation and design results. The main focus is on the INTEGRATION block. The rest of the blocks are the same as in previous work (Marzuki, 2016). This section also covers the model parameter extraction.

Figure 18. Circuit for voltages subtraction

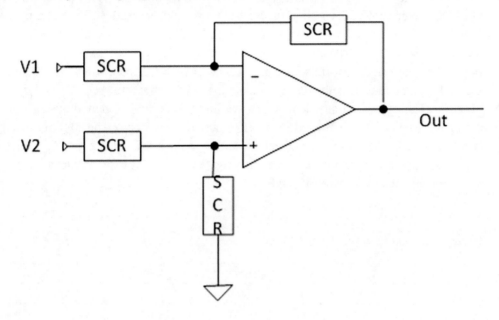

Model Parameter Extraction

Model parameter extraction is extremely crucial as the characteristics of the transistor of SILTERRA 180 nm technology must be extracted before any circuit design can be started. These parameters are extracted from the SILTERRA's BSIM Model.

The three parameters of a single PMOS and NMOS are transconductance parameter- $\mu_p C_{ox}$ and $\mu_n C_{ox}$, channel length modulation- λ_p and λ_n, as well as threshold voltage- $V_{T0,p}$ and $V_{T0,n}$. Equation (12), Equation (13) and Equation (14) are the transistor parameter equations.

$$\lambda = \frac{I_{D2} - I_{D1}}{I_{D1}V_{DS2} - I_{D2}V_{DS1}} \tag{12}$$

$$V_{T0} = \frac{V_{GS1} - V_{GS2}\sqrt{\dfrac{I_{D1}}{I_{D2}}}}{1 - \sqrt{\dfrac{I_{D1}}{I_{D2}}}} \tag{13}$$

$$K' = \frac{2I_D}{\dfrac{W}{L}(V_{GS} - V_{T0})^2(1 + \lambda V_{DS})} \tag{14}$$

The first parameter to be measured is λ, the channel length modulation. For the case of NMOS, I_D is plotted with DC sweep of V_{DS}. Two different V_{GS} values have been used, 0.5 V and 0.6 V whereby each curve represents a different V_{GS} values. Drain current I_D of two points on the saturation portion of a single curve which is at $V_{DS} = 1.0$V and 0.5 V are indicated in Table 3 and Table 4.

Here, four different transistor's widths have been tested or characterized, which is 200 nm, 900 nm, 2 μm and 3 μm. The main idea of collecting these data is to obtain the average $\mu_p C_{ox}$ and $\mu_n C_{ox}$ of SIL-TERRA 180 nm technology with transistors' width within few hundreds to few thousands nm. Lower width transistor is selected to be extracted in order to achieve low power consumption design since transistor with higher width draws more current. Note that the average $\mu_p C_{ox}$ and $\mu_n C_{ox}$ is not literally accurate as it varies with the changes of drain current or gate to source voltage. Figure 19 and Figure 20 show the I-V characteristic plot at given V_{GS} values.

Table 3. Characteristic and parameter of NMOS

NMOS									
L (nm)	180								Average
W (nm)	220		900		2000		3000		Average
V_{GS} (V)	0.5	0.6	0.5	0.6	0.5	0.6	0.5	0.6	-
V_{DS1} (V)	0.5								-
V_{DS2} (V)	1.0								-
I_{D1} (μA)	1.831	6.230	3.749	15.380	7.023	30.490	9.887	43.820	14.801
I_{D2} (μA)	2.207	7.150	4.602	17.920	8.662	35.690	12.220	51.390	17.480
λ_n (mV^{-1})	516.84	308.85	589.09	395.64	608.84	411.23	617.69	417.66	483.23
$V_{T0,n}$ (V)	0.3750		0.3973		0.4029		0.4048		0.3950
$\mu_n C_{ox}$ (μA/V^2)	152.4098	176.6011	109.8284	125.0056	102.7899	117.1811	100.0184	114.1603	124.7493

Table 4. Characteristic and parameter of PMOS

PMOS									
L (nm)	180								Average
W (nm)	220		900		2000		3000		Average
V_{SG} (V)	0.5	0.6	0.5	0.6	0.5	0.6	0.5	0.6	-
V_{SD1} (V)	0.5								0.5
V_{SD2} (V)	1.0								1.0
I_{D1} (μA)	0.3512	1.403	1.361	5.351	2.488	10.78	3.607	15.96	5.163
I_{D2} (μA)	0.4102	1.597	1.64	6.231	3.035	12.67	4.415	18.81	6.101
λ_p (mV^{-1})	403.83	320.93	515.71	393.65	563.63	425.20	577.35	434.78	454.39
$V_{T0,p}$ (V)	0.3972		0.3946		0.4041		0.4060		0.4005
$\mu_p C_{ox}$ (μA/V^2)	45.2465	48.1038	38.9591	42.3915	37.9900	41.6970	38.0138	41.7998	41.7782

Figure 19. NMOS I-V characteristic plot

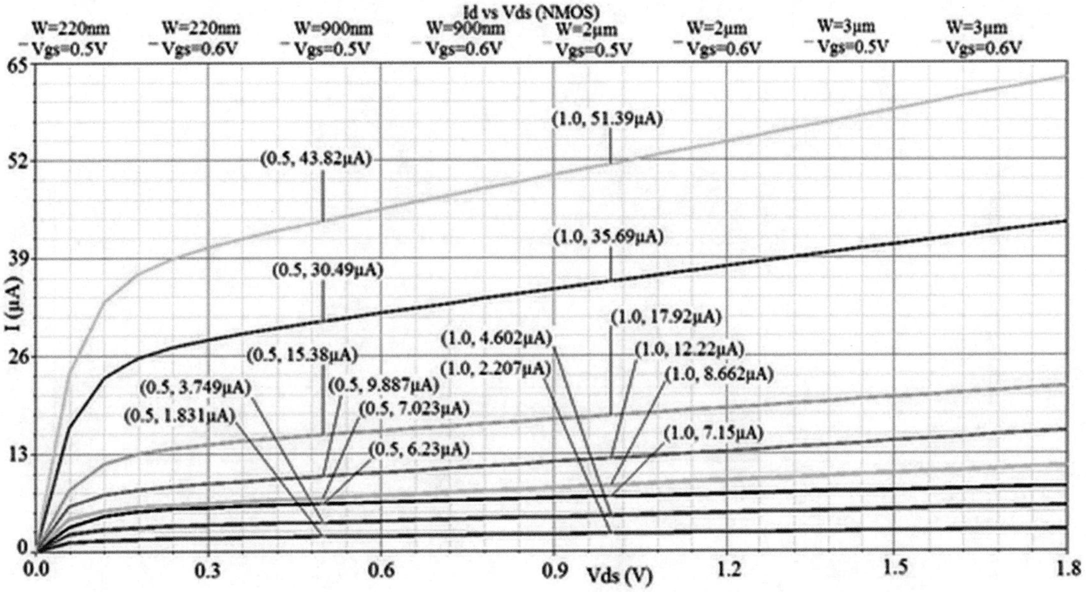

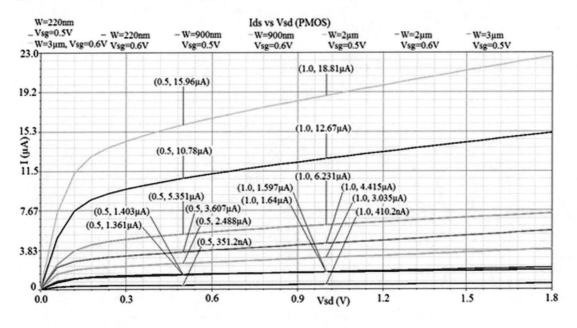

Figure 20. PMOS I-V characteristic plot

Unwanted Light Detection Results

Figure 21 shows the result when the photodiode current is set to the maximum to be 80 nA while the minimum current (which due to others reflected light) 20 nA. As it can be seen that, the output voltage does not remain constant between 450 μs to 600 μs. This is due to 20 nA current. Figure 22 shows the result when there is no 'noise' in the minimum current.

Figure 21. Integrator output with 'noise' current

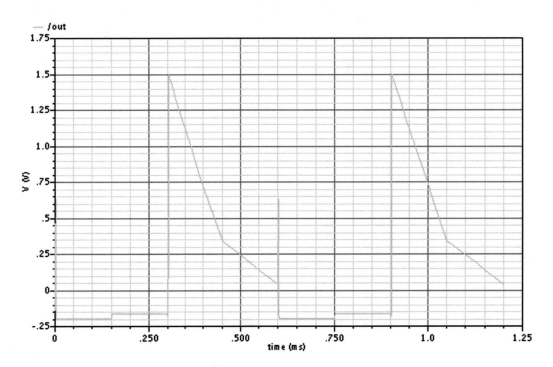

Figure 22. Integrator output without 'noise' current

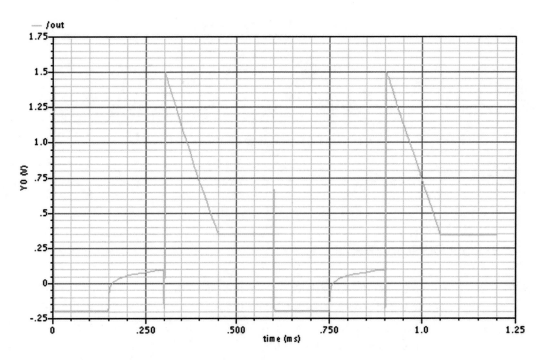

Complete CMOS Design

Figure 23 depicts the analog circuit for photodiodes/pixels. It is actually the CMOS implementation of INTEGRATION block (in Figure 10). The complete CMOS analog circuit is a combination of Figure 13 and 2 S/H buffer amplifiers. A buffer is required prior these S/H buffer amplifiers, this is to 'isolate' the integration capacitor and S/H capacitor. Figure 24 is the low power self-biased folded cascode amplifier. The amplifier consumes only 10 µA current. Figure 25 shows the layout of the analog circuit (Figure 23). The size of the layout is 200 µmx200 µm. Figure 26 shows the test bench for the analog circuit while

Figure 23. Complete CMOS analog circuit

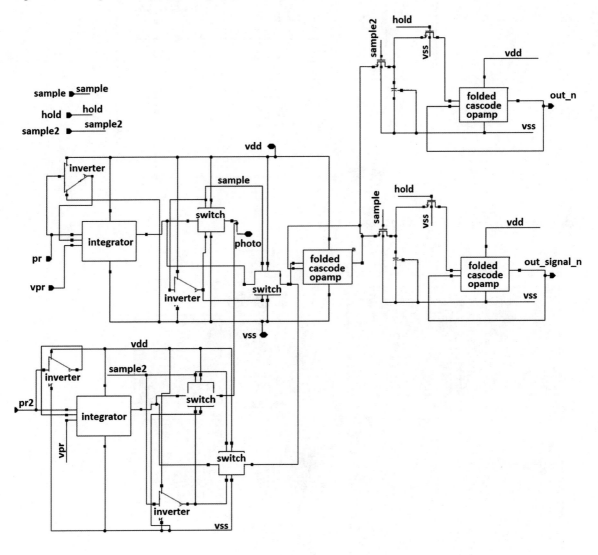

Figure 24. Self-biased folded cascode amplifier

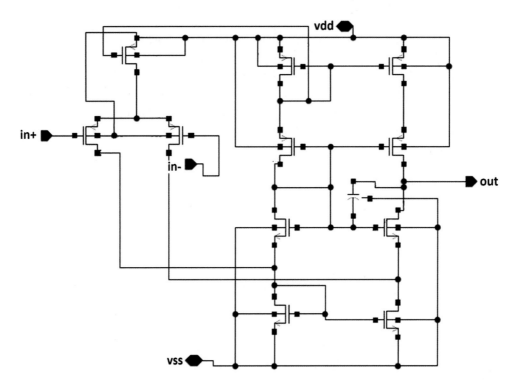

Figure 25. The layout of CMOS analog circuit

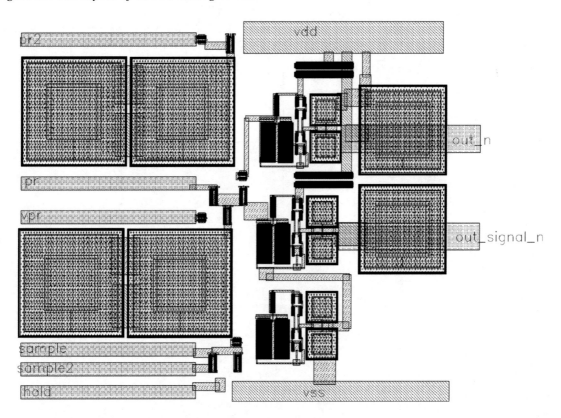

Figure 26. Test bench of CMOS analog circuit

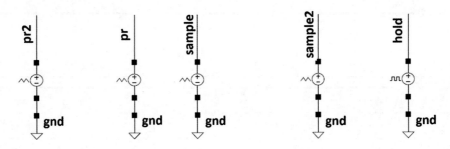

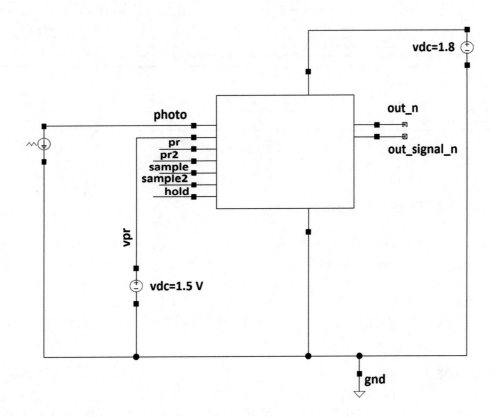

the results are depicted in Figure 27. 'sample' is the control signal for 1st integrator to sample voltage when the LEDs are ON. Whereas 'sample2' is the control signal for 2nd integrator to sample voltage when LEDs are OFF. So, the 2nd integrator captures the reflected light due to other light source. 'out_n' is the sampled voltage of the 2nd integrator which is held by an S/H buffer amplifier. 'out_signal_n' is the sampled voltage of the 1st integrator which is held by another S/H buffer amplifier. The difference of these two voltages is the real voltage due to LEDs.

Figure 27. Results of CMOS analog circuit

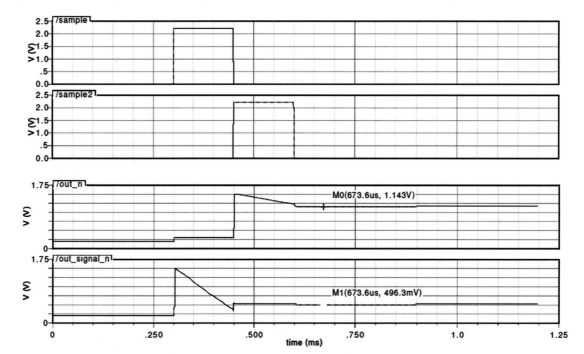

Corrected Light Detection Results

As discussed in the previous section, with appropriate sampling time and together with a differential amplifier, 'noise' due to unwanted reflected light can be reduced or eliminated. Figure 28 shows the result of the reduction. It can be seen that after the correction, the output voltage is 634 mV.

The discussed topology in Figure 10 only shows one pixel. This can be extended to more than one pixel for image sensor application. Figure 29 shows the 6x2 RGB pixel. It shows the level of green color.

CONCLUSION

There is at least three type of CMOS sensor which is used in vision for agriculture. The image sensor is probably most popular compared to TIA based color sensor or light to frequency color sensor. A new type of analog circuit for a color sensor which can be employed in the image sensor is proposed. The proposed analog circuit for the color sensor is based on light integrating concept. The analog circuits consist of 2 integrators and 2 S/H buffer amplifier. With the concept of 'controlled' modulated light source, a much simpler analog circuit has been achieved. With this concept, an unwanted 'noise' can be reduced or eliminated. The power consumption of the analog circuit is 54 μW. The low power consumption is achieved by using the self-biased folded amplifier as opamp in the S/H buffer amplifier. It is also predicted that, with a controlled light source, a normal or bare photodiode can be used to detect RGB light.

Figure 28. Integrator corrected output and the difference output

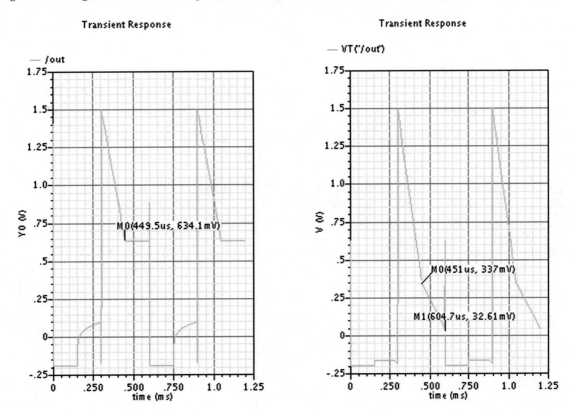

Figure 29. 6x2 RGB pixel

0,24,0	0,30,0
0,30,0	0,32,0
0,34,0	0,34,0
0,34,0	0,36,0
0,24,0	0,20,0
0,20,0	0,18,0

a) With 'noise'

0,64,0	0,70,0
0,70,0	0,72,0
0,74,0	0,74,0
0,74,0	0,76,0
0,64,0	0,60,0
0,60,0	0,58,0

b) corrected RGB

ACKNOWLEDGMENT

This work is partly supported by the Malaysia MOHE Grant, FRGS: 203/PELECT/6071348.

REFERENCES

Abdul Aziz, Z. A., & Marzuki, A. (2010). Residual Folding Technique adopting switched capacitor residue amplifiers and folded cascode amplifier with novel pmos isolation for high speed pipelined ADC applications. *3rd AUN/SEED-Net Regional Conference in Electrical and Electronics Engineering: International Conference on System on Chip Design Challenges (ICoSoC 2010)*, 14–17.

Aptina. (2010). *An Objective Look at FSI and BSI.* Aptina White Paper.

Ardeshirpour, Y., Deen, M. J., Shirani, S., West, M. S., & Ls, O. N. (2004). 2-D CMOS based image sensor system for fluorescent detection. In *Canadian Conference on Electrical and Computer Engineering* (pp. 1441–1444). IEEE.

Baker, R. J. (2010). *CMOS: Circuit Design, Layout, and Simulation* (3rd ed.). Wiley-IEEE Press. doi:10.1002/9780470891179

Bigas, M., Cabruja, E., Forest, J., & Salvi, J. (2006). Review of CMOS image sensors. *Microelectronics Journal, 37*(5), 433–451. doi:10.1016/j.mejo.2005.07.002

Biggs, G. L., Blackmer, T. M., Demetriades-Shah, T. H., Holland, K. H., Schepers, J. S., & Wurm, J. H. (2002). *Method and apparatus for real-time determination and application of nitrogen fertilizer using rapid, non-destructive crop canopy measurements.* Academic Press.

Bragagnolo, J., Amado, T. J. C., Nicoloso, R. D. S., Jasper, J., Kunz, J., & Teixeira, T. D. G. (2013). Optical crop sensor for variable-rate nitrogen fertilization in corn: I - plant nutrition and dry matter production. *Revista Brasileira de Ciência do Solo, 37*(5), 1288–1298. doi:10.1590/S0100-06832013000500018

Chae, Y., Cheon, J., Lim, S., Kwon, M., Yoo, K., Jung, W., & Han, G. (2011). A 2.1 M Pixels, 120 Frame/s CMOS Image Sensor ADC Architecture with Column-Parallel Delta Sigma ADC Architecture. *IEEE Journal of Solid-State Circuits, 46*(1), 236–247. doi:10.1109/JSSC.2010.2085910

Chan, P. K., Ng, L. S., Siek, L., & Lau, K. T. (2001). Designing CMOS folded-cascode operational amplifier with flicker noise minimisation. *Microelectronics Journal, 32*(1), 69–73. doi:10.1016/S0026-2692(00)00105-1

David, Y. (1999). *Digital pixel cmos image sensors.* Stanford University.

Deguchi, J., Tachibana, F., Morimoto, M., Chiba, M., Miyaba, T., & Tanaka, H. K. T. (2013). A 187.5μVrms -Read-Noise 51mW 1.4Mpixel CMOS Image Sensor with PMOSCAP Column CDS and 10b Self-Differential Offset-Cancelled Pipeline SAR-ADC. ISSCC 2013, 494–496.

El Gamal, A. (2002). Trends in CMOS Image Sensor Technology and Design. *Electron Devices Meeting*, 805–808. 10.1109/IEDM.2002.1175960

Feng, Z. (2014). *Méthode de simulation rapide de capteur d'image CMOS prenant en compte les paramètres d'extensibilité et de variabilité.* Ecole Centrale de Lyon.

Fowler, B., Balicki, J., How, D., & Godfrey, M. (2001). Low FPN High Gain Capacitive Transimpedance Amplifier for Low Noise CMOS Image Sensors. *Sensor and Camera Systems for Scientific, Industrial and Digital Photography Applications II, Proceedings of the SPIE*, 4306. 10.1117/12.426991

Goy, J., Courtois, B., Karam, J. M., & Pressecq, F. (2001). Design of an APS CMOS Image Sensor for Low Light Level Applications Using Standard CMOS Technology. *Analog Integrated Circuits and Signal Processing, 29*(1/2), 95–104. doi:10.1023/A:1011286415014

Gray, P. R., Hurst, P. J., Lewis, S. H., & Meyer, R. G. (2009). *Analysis and Design of Analog Integrated Circuits.* JohnWiley & Sons, Inc.

Hamami, S., Fleshel, L., Yadid-pecht, O., & Driver, R. (2004). CMOS Aps Imager Employing 3.3V 12 bit 6.3 ms/s pipelined ADC. *Proceedings of the 2004 International Symposium on Circuits and Systems.*

Innocent, M. (n.d.). *General introduction to CMOS image sensors.* Academic Press.

Intaravanne, Y., & Sumriddetchkajorn, S. (2015). Android-based rice leaf color analyzer for estimating the needed amount of nitrogen fertilizer. *Computers and Electronics in Agriculture, 116,* 228–233. doi:10.1016/j.compag.2015.07.005

Jui-Lin, L., Ting-You, L., Cheng-Fang, T., Yi-Te, L., & Rong-Jian, C. (2010). Design a low-noise operational amplifier with constant-gm. *Proceedings of SICE Annual Conference,* 322-326.

Kleinfelder, S., Lim, S., Liu, X., & El Gamal, A. (2001). A 10 000 Frames/s CMOS Digital Pixel Sensor. *IEEE Journal of Solid-State Circuits, 36*(12), 2049–2059. doi:10.1109/4.972156

Loinaz, M. J., Singh, K. J., Blanksby, A. J., Member, S., Inglis, D. A., Azadet, K., & Ackland, B. D. (1998). A 200-mW, 3.3-V, CMOS color Camera IC producing 352x288 24-b Video at 30 Frames/s. *IEEE Journal of Solid-State Circuits, 33*(12), 2092–2103. doi:10.1109/4.735552

Lulé, T., Benthien, S., Keller, H., Mütze, F., Rieve, P., Seibel, K., & Böhm, M. (2000). Sensitivity of CMOS Based Imagers and Scaling Perspectives. *IEEE Transactions on Electron Devices, 47*(11), 2110–2122. doi:10.1109/16.877173

Lyu, T., Yao, S., Nie, K., & Xu, J. (2014). A 12-bit high-speed column-parallel two-step single-slope analog-to-digital converter (ADC) for CMOS image sensors. *Sensors (Basel), 14*(11), 21603–21625. doi:10.3390141121603 PMID:25407903

Mahmoodi, A., & Joseph, D. (2008). Pixel-Level Delta-Sigma ADC with Optimized Area and Power for Vertically-Integrated Image Sensors. *51st Midwest Symposium on Circuits and Systems,* 41–44. 10.1109/MWSCAS.2008.4616731

Marzuki, A. (2016). CMOS Image Sensor: Analog and Mixed-Signal Circuit. In O. Sergiyenko & J. C. Rodriguez-Quiñonez (Eds.), Developing and Applying Optoelectronics in Machine Vision. Hershey, PA: IGI Global.

Marzuki, A., Aziz, Z. A. A., & Manaf, A. A. (2011, May). A review of CMOS analog circuits for image sensing application. In *2011 IEEE International Conference on Imaging Systems and Techniques (IST)* (pp. 180-184). IEEE. 10.1109/IST.2011.5962187

Mohamed, A. N., Ahmed, H. N., Elkhatib, M., & Shehata, K. A. (2013). A low power low noise capacitively coupled chopper instrumentation amplifier in 130 nm CMOS for portable biopotential acquisiton systems. In *International Conference on Computer Medical Applications,* 1-5. 10.1109/ICCMA.2013.6506168

Nemirovsky, Y., Brouk, I., & Jakobson, C. G. (2001). 1/f noise in CMOS transistors for analog applications. *IEEE Transactions on Electron Devices, 48*(5), 921–927. doi:10.1109/16.918240

Norsworthy, S. R., Schreier, R., & Temes, G. C. (Eds.). (1997). *Delta-sigma data converters: theory, design, and simulation* (Vol. 97). New York: IEEE press.

Pajares, G., García-Santillán, I., Campos, Y., Montalvo, M., Guerrero, J., Emmi, L., & Gonzalez-de-Santos, P. (2016). Machine-Vision Systems Selection for Agricultural Vehicles: A Guide. *Journal of Imaging*, *2*(4), 34. doi:10.3390/jimaging2040034

Scheffer, D., Dierickx, B., & Meynants, G. (1997). Random addressable 2048x2048 active pixel image sensor. *IEEE Transactions on Electron Devices*, *44*(10), 1716–1720. doi:10.1109/16.628827

Song, B. (2000). Nyquist-Rate ADC and DAC. In E. W. Chen (Ed.), VLSI Handbook. Academic Press.

Sukegawa, S., Umebayashi, T., Nakajima, T., Kawanobe, H., Koseki, K., Hirota, I., & Fukushima, N. (2013). A 1/4-inch 8Mpixel Back-Illuminated Stacked CMOS Image Sensor. In ISSCC 2013 (pp. 484–486). IEEE.

Suzuki, A., Shimamura, N., Kainuma, T., Kawazu, N., Okada, C., Oka, T., & Wakabayashi, H. (2015). A 1/1.7 inch 20Mpixel Back illuminated stacked CMOS Image sensor for new Imaging application. In ISSCC 2015 (pp. 110–112). IEEE.

Takayanagi, I., Yoshimura, N., Sato, T., Matsuo, S., Kawaguchi, T., Mori, K., & Nakamura, J. (2013). A 1-inch Optical Format, 80fps, 10. 8Mpixel CMOS Image Sensor Operating in a Pixel-to-ADC Pipelined Sequence Mode. *Proc. Int'l Image Sensor Workshop*, 325–328.

Trépanier, J., Sawan, M., Audet, Y., & Coulombe, J. (2002). A Wide Dynamic Range CMOS Digital Pixel Sensor. In *The 2002 45th Midwest Symposium on Circuits and Systems*. IEEE. 10.1109/MWS-CAS.2002.1186892

Wong, H. P. (1997). *CMOS Image sensors - Recent Advances and Device Scaling Considerations*. IEDM.

Xu, R., Ng, W. C., Yuan, J., Member, S., Yin, S., & Wei, S. (2014). A 1 / 2. 5 inch VGA 400 fps CMOS Image Sensor with High Sensitivity for Machine Vision. *IEEE Journal of Solid-State Circuits*, *49*(10), 2342–2351. doi:10.1109/JSSC.2014.2345018

Zhao, W., Wang, T., Pham, H., Hu-Guo, C., Dorokhov, A., & Hu, Y. (2014). Development of CMOS Pixel Sensors with digital pixel dedicated to future particle physics experiments. *Journal of Instrumentation: An IOP and SISSA Journal*, *9*(02), C02004–C02004. doi:10.1088/1748-0221/9/02/C02004

Chapter 3
Optoelectronic Differential Cloudy Triangulation Method for Measuring Geometry of Hot Moving Objects

Sergey Vladimirovich Dvoynishnikov
Novosibirsk State University, Russia

Vladimir Genrievich Meledin
Novosibirsk State University, Russia

ABSTRACT

The new differential cloudy triangulation method for quick and precise measurement of thickness of hot dynamic objects is proposed. The method is based on laser cloudy triangulation, used in the synchronous differential mode. The experimental model of the laser system for thickness measurements, which implements the proposed method, has been developed. Laboratory and industrial tests of the laser system have been performed. They confirm efficiency and possibilities of differential cloudy triangulation in metallurgy. The differential cloudy triangulation method may be used for geometry measurement of hot dynamic objects in a wide range of strong optical refractions in media.

INTRODUCTION

Machine vision-based optoelectronic methods are widely used in multiple fields of industrial applications. Development of new measuring industrial technologies is associated with the creation of measurement methods using the element base of modern optoelectronics and photonics (Meledin, 2011). Especially topical is an enhancement of methods for dynamic control of geometrical parameters in the industry (Degarmo, 2003). Presently, optical methods for measuring the geometry of complex shaped and fast-moving objects actively develop (Feng et al., 2011). High-precision measurement of moving products' thickness is a very important and complicated technological problem. Laser triangulation methods are being developed to solve this problem (Li et al., 2014; Sicard, Sirohi, 2013; Semenov et al., 2010; Penney,

DOI: 10.4018/978-1-5225-5751-7.ch003

Thomas, 1989; Batbakov et al., 2004; Plotnikov, 1995; Komissarov, 1990; Lee, 2001; Deponte, 2004; Gordon, Jacob, 1969; etc). They provide thickness measurement error of up to 10^{-4} in the differential scheme under laboratory conditions (Raymond, 1984).

There are various high-precision methods of thickness measurement, based on a differential laser triangulation (Dvoynishnikov et al., 2015). Triangulation measuring instruments are placed on different sides of the measured object. Information on thickness is obtained based on the difference in indications of the measuring instruments.

The error of the mentioned methods considerably increases at measuring the thickness of hot moving objects, for example, hot rolling in metallurgy. The error growth is caused, mainly, by the influence of phase inhomogeneity in thermal gradient medium on the structure of light beam tracers in the measuring scheme. Optical measurements are performed under intense air heating by hot surface of the measured object. The high temperature of the surface and high speed of motion form intensive non-stationary vortex flows near the object surface. These flows mix up with the convective motion of air masses, caused by an intense heat and mass transfer near a heated surface of the object. Air convection and motion of phase inhomogeneous in the medium form temperature phase gradients and change refraction index in the local medium. In such conditions, optical signals of laser triangulation of the measuring instruments are subjected to considerable non-stationary distortions. These distortions lead to significant and hard-to-predict errors of measurements (Dvoynishnikov et al., 2015). Therefore to improve the accuracy of laser triangulation methods, it is necessary to study and consider peculiarities of laser radiation propagation in the conditions of intensive temperature and phase distortions of optical signals.

Propagation of electromagnetic waves in the air was widely discussed. Most of these works aim at studying the propagation of electromagnetic waves in a large scale from a few hundred meters to hundreds of kilometers (Zuev et al., 1983; Bol'basova et al., 2009; Kravcov et al., 1983; Gurvich et al., 2014). In particular, a large number of studies are devoted to refractive distortions of laser radiation in a turbulent atmosphere (Pollinger et al., 2012; Aksenov et al., 2012; Nosov et al., 2009; Kandidov et al., 1998). The mentioned works touch upon the related areas of knowledge. However, direct use of the results of these studies in the analysis of propagation of laser radiation near hot dynamic objects is complicated. This is due to the specificity of phase and structural parameters of turbulent vortices in the air near the moving hot surfaces. The statistical distribution of these vortices' parameters is substantially different from the well-studied parameters of atmospheric vortices. This chapter presents a statistical analysis of propagation of laser radiation in the phase inhomogeneous media appropriated for laser measurements near hot dynamic objects.

A new optical method named laser cloudy triangulation is proposed. This method is used for the dynamic measurement of the geometry of hot moving objects. The method also works and ensures high accuracy and stability under time-dependent refractive distortions of optical signals. It also uses spatially structured illumination and a two-dimensional photodetector. The principle of work is based on analysis of the two-dimensional distribution of the structured illumination observed on the photodetector. This method allows measuring not only the distance to object's surface but also its 3D position.

For measurement thickness of hot moving objects, is proposed differential cloudy triangulation method. Measuring modules that implement laser cloudy triangulation method are installed on static massive base from the opposite sides of the measured object. Measuring modules are oriented in space to perform synchronous distance measurements to the object's surface. The thickness of the measured object is calculated as a difference in indications of the measuring modules with compensation of an angular misalignment of the object.

Proposed differential laser triangulation method was tested under industrial conditions at working steel plant. Measurements were carried out in the hot-rolled metallurgical shop of Kuzmina Metallurgical Plant, Novosibirsk. Performed industrial tests of differential laser triangulation methods for measuring the thickness of hot dynamic objects allowed to estimate measurement error by the proposed method. Obtained estimations are based on the comparison of the measurement results and data of direct measurements using hand micrometer. As a result of error estimation, it is shown that the actual error of the measured thickness is less than 20 μm. This corresponds to a relative error of 10^{-5}. Such accuracy level is progressive for optical measuring systems working in real conditions of hot metallurgical shops.

The chapter consists of 8 sections. Below first section provides an introduction and statement of the problem. Second section includes a theoretical analysis of the character of optical signals propagation in the presence of phase homogeneities in the medium. Third section is devoted to the description of the new cloudy triangulation method, which allows accurately measuring the distance to the surface of the hot object under phase distortions and temperature gradients. Fourth section describes the method of measuring the thickness of the hot dynamic object based on synchronous differential cloudy triangulation. Fifth section deals with calibration of the measurement system, which implements the proposed method. Sixth section focuses on the practical implementation of the thickness measuring method. Seventh section explains the calibration method, laboratory tests and measurement error estimation at laboratory conditions. Eight section shows the industrial tests and achieved characteristics of the measuring system, implementing the proposed measurement method.

OPTICAL SIGNALS PROPAGATION IN THE PRESENCE OF PHASE INHOMOGENEITIES IN THE MEDIUM

In the statistical analysis, authors use elements of the theory of wave propagation in random media (Ishimaru, 1999). It was supposed, that phase inhomogeneities were presented in the optical medium, located between the radiation source and the object, and between the object and the image forming system. These optical phase inhomogeneities are generated by moving air zones with temperature gradients, changing the refractive index of light in the medium. The resulting distortions of the optical signal generated by these inhomogeneities are stochastic. Further analysis is carried out for monochromatic signals corresponding to the laser triangulation system. The refractive index of air may be presented in the form seen in (1):

$$n\left(r,t\right) = n_0 + n_1\left(r,t\right) \tag{1}$$

Let refractive index has constant value n_0 and correction value n_1, which depends on time t, temperature gradient of air, and location in a scattering volume r. From Maxwell's equation (2):

$$\nabla^2 U + \frac{w^2 n^2}{c^2} U = 0 \tag{2}$$

Since $|n_1| \ll n_0$, field U may be presented as the sum of U_0 and small correction parameter U_1. U_0 agrees with a uniform refractive index of air in the optical measuring volume. U_1 takes into account the effect of perturbation of refractive index n_1. In this approximation, the wave equation takes the form (3):

$$\nabla^2 \left(U_0 + U_1 \right) + \frac{w^2 \left(n_0 + n_1 \right)^2}{c^2} \left(U_0 + U_1 \right) = 0 \tag{3}$$

Given that U_0 is the unperturbed solution, it must satisfy equation (1) for $n = n_0$ and $U = U_0$. Then equation (3) becomes:

$$\nabla^2 U_1 + \frac{w^2 n_0^2}{c^2} U_1 = -\frac{2 w^2 n_0 n_1}{c^2} U_0 \tag{4}$$

Solution of equation (4) is found as a convolution of Green's function for free space $\exp\left(j \frac{w n_0}{c} |r| \right) / |r|$.

As a result, it is obtained:

$$U_1(r) = \frac{1}{4\pi} \iiint \frac{e^{\frac{j w n_0}{c}|r-r'|}}{|r-r'|} \left[2 \frac{w^2 n_0^2}{c^2} n_1(r') U_0(r') \right] d^3 r' \tag{5}$$

Where integration is performed over the entire scattering volume. Field disturbance U_1 may be found by summing the set of spherical waves, generated at various points r' within the scattering volume V. The amplitude of the spherical wave generated at the point r' is proportionate to the product of the undisturbed incident radiation amplitude and the refractive index perturbation at that point. Let assume that the maximum transverse displacement within which the diffuser light enters the specified point is much less than the axial distance from the lens to the photodetector. Then, Fresnel approximation (Goodman, 1996) may be used in equation (5), which leads to the equation (6):

$$U_1(r) = \frac{w^2 n_0^2}{2\pi c^2} \iiint \frac{\exp\left\{ \frac{j w n_0}{c} \left[(z - z') + \frac{|\rho - \rho'|^2}{2(z - z')} \right] \right\}}{|z - z'|} n_1(r') U_0(r') d^3 r' \tag{6}$$

Where ρ and ρ' are the transverse displacement vectors r and r' from the z-axis. The resulting expression gives field disturbance U_1 as a superposition of a large number of independent contributions of various parts of inhomogeneous medium. In accordance with the central limit theorem the real and imaginary parts of U_1 comply with normal distribution. Full-wave intensity distribution may be presented as a sum of a constant complex value and a random complex value with normal distribution.

In (Goodman, 1985) it is shown that the marginal density function of the complex random variable amplitude presented as a sum of constant and complex random variables with normal distribution may be presented as Rice's density distribution:

$$p_A\left(a\right) = \frac{a}{\sigma^2} \exp\left(-\frac{a^2 + s^2}{2\sigma^2}\right) I_0\left(\frac{as}{\sigma^2}\right) \tag{7}$$

Where σ is the dispersion of a complex random variable U_1 with normal distribution, s is the constant amplitude of variable U_0, and I_0 is the modified Bessel function of zero order.

Since the deviation caused by the refractive index inhomogeneity is much smaller than the wave amplitude $(s \gg \sigma)$, the random variable tends to the Gaussian random variable:

$$p_A\left(a\right) \approx \frac{1}{\sqrt{2\pi}\sigma} \exp\left\{-\frac{(a - s)^2}{2\sigma^2}\right\} \tag{8}$$

Therefore, the presence of a random thermal lens in the medium provides stochastically random variations of the laser spot and normal distribution within larger time intervals. This enables precise measuring even under intense temperature and phase distortions of optical signals, which are specific to the measurement of hot dynamic objects. Consequently, random thermal gradients in the air provide stochastically random variations of the coordinate of the laser spot on the analyzed surface. These random variations are stochastic in nature and have a normal distribution within time intervals greater than the characteristic time parameters of convective and vortex processes (Walters, 1982), and the corresponding estimates are unbiased. Fulfillment of this condition allows implementing precise optical measurements of geometrical parameters under intense temperature and phase distortions of signals, which are peculiar to measuring of the hot dynamic objects.

METHOD OF CLOUDY TRIANGULATION

The method of cloudy triangulation is presented below. This method allows measuring distance to the surface of the dynamic objects under phase distortions of optical signals. Let suppose that the light spot with center at coordinates K is formed at the surface of the measured object. Due to the inhomogeneous refractive index of the optical medium, coordinates of point K will be described by the random variable with normal distribution:

K(t)=F(t,K0), (9)

Where *K0* determines initial position, orientation and internal parameters of the radiating system. At the rather long time of accumulation, coordinates of *K*, which characterizes the position of the spot in space, converge to a certain value:

$$\langle K \rangle = \frac{1}{T} \int F\left(t, K_0\right) dt \tag{10}$$

In measurements of hot dynamic objects, it is impossible to ensure long-term accumulation of data. Averaging over time or space ensemble is the only way to improve measurement accuracy with unbiased estimator under true conditions of the central limit theorem. Based on ergodic hypothesis, which was experimentally confirmed for measurement process in (Walters, 1982), we used averaging over spatial ensemble instead of time averaging to improve measurement accuracy. The proposed method of laser cloudy triangulation is based on the described approach. It allows performing quick and accurate measurements of the distance to the moving surface under conditions of significant optical distortions.

The optical source is a spatially structured illumination in the form of light rays. This illumination is dispersed on the object surface in the form of a set of light spots *{Ki}*, forming a spatially structured light cloud in the vicinity of the point with coordinates *<K>*. Emitters generating rays of the structured illumination are placed at a small distance from each other in comparison with geometrical dimensions of the measuring object. Light rays that agree with points K_i passing through the optical phase-inhomogeneous medium undergo random distortions. These deviations have the same statistical characteristics as for the point *K*. Consequently, the time dependence of coordinate K_i has a similar relationship with the point *K*:

$$K_i\left(t\right) = F\left(t, K_i^0\right). \tag{11}$$

Let light points on the photodetector form the known spatial dependence of intensity *Y (c1..ck)*, where *(c1..ck)* are parameters of distribution *Y*, describing distance to the object. Since K_i functions are normally distributed, parameters *(c1..ck)* may be found by the least square method. Here it is implicitly assumed that *F* distributions have Gaussian random deviation depending on time and spatial localization of the light beam. This approximation is quite possible according to the adopted ergodicity hypothesis of the dynamic system (Walters, 1982). It is assumed that distortions of the optical signal by phase-inhomogeneous medium have a random distribution of perturbations of the refractive index of the optical medium in time and space.

According to the least square method (Bretscher, 1995) parameters *(c1..ck)* may be found by minimizing the deviation of instantaneous values of K_i from the theoretical value *Y (s1..sk)*:

$$\frac{\partial\left(\sqrt{\sum\left(K_i - Y\left(x,y\right)\right)^2}\right)}{\partial c_j} = 0 \tag{12}$$

In cloudy triangulation method (Dvoynishnikov et al., 2015) the measured surface is probed by the set of light rays with predetermined spatial intensity distribution (two intersecting light clouds). Clouds of light points on the light-scattering surface of the object are formed as an image of two intersecting ellipses, strongly stretched along one of the axes. The receiving optical system detects scattered radiation in the form of two-dimensional spatially distorted projective distribution of the image of the points cloud (Figure 1). Regression spatial analysis of the distribution of these clouds on the registered image

Figure 1. Distribution of cloud of light points on the image in cloudy triangulation method.

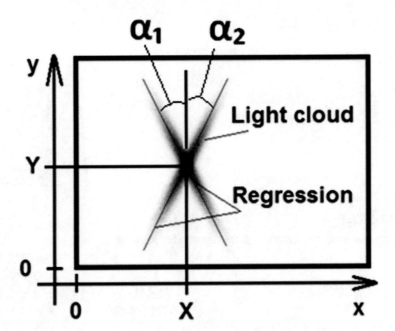

allows to determine the distance to the object surface (the X coordinate of the intersection of regression lines) and inclination in the space (angles of the regression lines α_1, α_2).

It is shown, that laser cloudy triangulation method works well for measuring the distance to the surface of hot dynamic objects and its angle of inclination in conditions of strong phase distortions of optical signals. The advantages of laser cloudy triangulation method are connected with using of a large set of data in each analyzed image and regression resistance to erroneous data.

THICKNESS MEASUREMENT METHOD FOR HOT DYNAMIC OBJECTS

Thickness measurement method based on differential cloudy triangulation is as follows: Measuring modules that implement laser cloudy triangulation method are installed on the opposite sides of the measured object. They are oriented in space to perform synchronous measurements at one given point of a reference plane (Figure 2). Object's thickness at the given point is calculated as a difference in indications of the measuring modules with compensation of an angular misalignment of the object. Such computations are performed using calibration data. Redundant data about the slope of the opposite sides of the flat object in a differential measurement allows estimating instantaneous measurement error or the object's curvature (Dvoynishnikov & Rakhmanov, 2013).

Distance from the laser light source is determined by the expression for classical triangulation point (Gruber et al., 1992):

Figure 2. Principle of thickness measurement by laser cloudy triangulation.

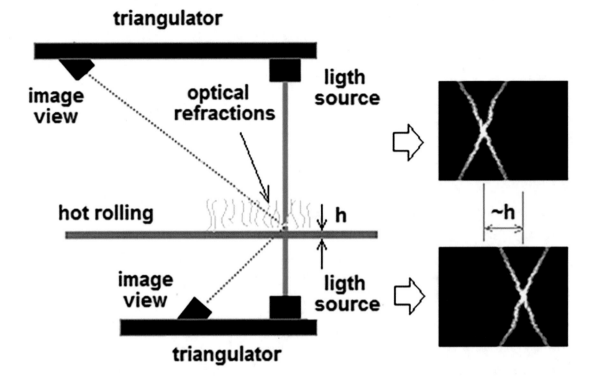

$$z = Z_0 + \frac{\Delta X * P}{f * \sin(\theta)} \tag{13}$$

Where ΔX is the characteristic parameter of spatial distribution function of the radiation intensity monitored by the receiver, $Z0$ is the parameter determining the initial position of the measured object, P is the distance from the focus of the receiving optical system to the measured object surface, f is the focal length of the receiving optical system, and θ is the triangulation angle. The thickness is calculated from the following expression:

$$h = L - Z_1 - Z_2 - q(\alpha_1, \alpha_2) \tag{14}$$

Where L is the distance between triangulators, z_1 is the distance to the top and bottom surface of the object measured by corresponding triangulators, q is the calibration correction for slope compensation of the measured object, and a_1, a_2 are the inclination angles of regression lines, determined on the registered image.

The presented thickness measurement method for hot dynamic objects based on synchronous differential cloudy triangulation is quite simple to implement, industrially applicable and may be used successfully in modern production. However, calibration of the measuring system based on proposed method is rather non-trivial. It is described in the following section.

CALIBRATION METHOD

Authors proposed 2-steps calibration method of a measuring system based on differential cloudy triangulation. The first step of calibration is aimed to determine parameters from equation (13) for each cloudy triangulator. The easiest way to implement this procedure is to place flat calibration sheet in several positions with known spatial coordinates and fixing of characteristic parameters of radiation intensity distribution on the received image. As a result functions F_1 and F_2, which allow determining the value of Z_1 and Z_2 in (14) will be obtained:

$$Z_1 = F_1(X), Z_2 = F_2(X) \qquad (15)$$

At the second step, the function q is determined experimentally. For this purpose authors arbitrarily moved calibration sheet with a known thickness in the measuring volume and simultaneously collected the calibration data array in the PC memory by special software. The software on PC collected series of measured values Z_1, Z_2 and parameters of radiation intensity distribution on received images. These parameters characterize spatial orientation of the sheet in the measured area. As a result of the described processes, the multidimensional array of calibration data will be accumulated. This array characterizes the dependence of q in accordance with the expression:

$$q(\alpha_1, \alpha_2) = L - Z_1 - Z_2 - H_0, \qquad (16)$$

Where *H0* is the thickness of calibration sheet.

The proposed calibration method allows making precise optical measurements of the thickness of hot dynamic objects using synchronous differential cloudy triangulation method considering all distortions including optical aberrations and transverse spatial modes of laser sources.

PRACTICAL IMPLEMENTATION

The experimental model of a laser system for thickness measurements of hot dynamic objects based on the proposed method of differential cloudy triangulation was developed (Figure 3). The laser system performs thickness measurements of hot metal directly behind the rolling mill in the hot-rolled shop of Kuzmin Metallurgical Plant, Novosibirsk. It was created to replace the outdated radiation-hazardous measurement system in the future.

The measuring system is mounted on a thermostable C-formed frame. At the top and bottom of the frame, the laser cloudy triangulators are placed. Triangulators synchronously emit spatially modulated illumination in the form of two crossing light clouds. After dissipation on the object surface, these light clouds form an image with two crossed extended areas on the photodetector (Figure 4). These extended areas may be analyzed as regression lines.

The horizontal coordinate of the crossing of regression lines of these clouds determines the distance to the object's surface. The inclination angles of regression lines in the plane of the photodetector determine the spatial orientation of the object's surface in the measurement area. Distance from top triangulator

Figure 3. Layout of main modules of thickness measurements system for hot dynamic objects: 1 – C-formed frame, 2 – guard case, 3 – optical receiver, 4 – optical source, 5 – triangulator base, 6 – synchronization module, 7 – power module.

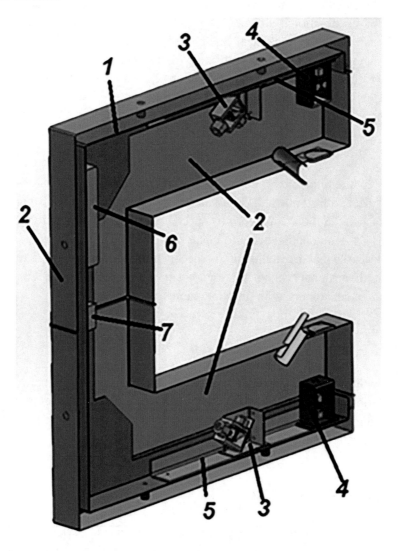

to the measured object surface is about 1 m. Distance from the bottom triangulator is about 0.5 m. Such distances are due to the mill configuration. In addition, the high temperature of rolled metal (1100 °C) and its high speed (5-10 m/s) do not allow positioning triangulators closer to hot metal. It may cause overheating of the measurement system or system failure in the event of an accident.

The active fluid thermostat is used to provide thermostability of measurement system and protect against high temperatures. It allows obtaining stable measurement results regardless of ambient temperature. Additionally, the guard case is insulated and has a multi-layered structure, which prevents rapid heating.

In the created laser measuring system the cloudy triangulators use powerful multimode semiconductor laser diode with 445 nm wavelength and 1-watt power. Each triangulator contains two laser emitters. They provide formation of spatial illumination in the form of two crossed light clouds. Optical receivers

Figure 4. An example of real image on the photodetector of laser cloudy triangulator at measuring hot metal rolling thickness

are based on digital photodetectors DMM 22BUC03-ML. They have hardware synchronization interface and can transmit the received images to a computer via USB interface. Synchronization of laser cloudy triangulators is provided by a specialized module controlled by a host PC (Dvoynishnikov & Rakhmanov, 2015).

The thickness measurement is performed as follows. The measurement is carried out by two cloud triangulators working in a differential scheme. The measurement process initiates by the synchronal pulse of generator integrated into synchronization module. On the pulse from the synchronization module, photodetectors open shutter and radiation modules generate light pulses that form spatially modulated illumination scattered on the surface of the measured object.

Part of the optical radiation scattered on the surface of the measured object passes through a set of optical filters and falls onto the photodetector matrix. Generated images are received as a digital image in the data processing module. The data processing module processes the images and calculates the measured thickness. The obtained value of thickness is stored as primary processing results. On the next processing step, low-frequency filtering based on the FIR digital filter is applied to the primary processing results of thickness measurement. The filter parameters are selected in accordance with the physical properties of the measured object. Obtained thickness measurement results are stored in a database and output to information monitors in real time.

The low-frequency data filtering is applied based on results of priori statistical analysis of the measured thickness values. In the case of measuring the thickness of hot rolled products directly behind the rolling mill, it is possible to successfully apply low-frequency filtering of the measured hot-rolling thickness data.

Let the primary results of the thickness measurements are set of discrete measurements *x(t)*. The most common and simple low-pass filter is a filter of moving average. If filter of moving average with window size *N* is used, then filtered measurement results *y(t)* can be calculated:

$$y\left(n\right) = \frac{1}{N}\sum_{k=0}^{N} x\left(n-k\right) \tag{17}$$

The frequency response of the moving average filter is expressed by the function *sin(x)/x*. Increasing the size of the averaging window narrows the main lobe, but does not significantly reduce the amplitude of the side lobes in the frequency response curve. Naturally, such filter is not suitable in the case of need to perform large attenuation of the frequency response in the attenuation band.

The filter, which has a smooth monotonous frequency response curve, for example, a Butterworth filter, is most preferable:

$$G^2\left(\omega\right) = \left|H\left(j\omega\right)\right|^2 = \frac{G_0^2}{1 + \left(\dfrac{\omega}{\omega_c}\right)^{2n}} \tag{18}$$

Where n — filter order, ω_c — cutoff frequency (frequency, when amplitude attenuation is equal to -3 dB) G_0 — a gain of the constant component of the signal (gain at zero frequency)

Let the applied FIR filter have coefficients *h(n)*. The final signal will be calculated as a convolution of the original signal with the filter coefficients:

$$y\left(n\right) = x\left(k\right) * x\left(n\right) = \sum_{k=0}^{N-1} x\left(k\right) \cdot x\left(n-k\right). \tag{19}$$

In frequency domain:

$$Y\left(w\right) = H\left(w\right) \cdot X\left(w\right) \tag{20}$$

Where

$$Y\left(w\right) = \sum_{n=0}^{N-1} y\left(n\right) \exp\left(-j\frac{2\pi}{N} nk\right) \tag{21}$$

$$X\left(w\right) = \sum_{n=0}^{N-1} x\left(n\right) \exp\left(-j\frac{2\pi}{N} nk\right) \tag{22}$$

$$H(w) = \sum_{n=0}^{N-1} h(n) \exp\left(-j\frac{2\pi}{N}nk\right) \tag{23}$$

In the case where the FIR filter frequency response curve is known, its coefficients can be calculated using the inverse Discrete Furrier transform (DFT):

$$x(t) = \frac{1}{N} \sum_{k=0}^{N-1} H(k) \exp\left(-j\frac{2\pi}{N}tk\right) \tag{24}$$

Thus, to reduce the measurement error by low-frequency filtering of the primary measurement results it is necessary to perform low-frequency data filtering under the condition of having a priori information about the statistic parameters of the measured value.

To filter the measurement data of the hot-metal rolling thickness on the basis of statistical observations of the technological process, a filter with a cutoff frequency of 2 Hz was calculated. As a result of direct measurements of rolled sheets, it was found that the frequency of characteristic fluctuations in the thickness of rolled metal does not exceed 1 Hz. Thus, the selected filter will not significantly distort the measured value.

Image processing in the cloudy triangulation method is performed to determine the regression parameters of the spatially modulated illumination image. Regressions allow us to restore the central lines of light clouds in a set of experimental data. In addition, the data processing algorithm must provide real-time image processing, so the processing time must be shorter than the measurement period. This requirement imposes the need to optimize image processing algorithms in the cloudy triangulation method.

The analyzed image is the distribution of two intersecting clouds of light points. Based on the analysis of these clouds in the plane of the photodetector, numerical regression functions corresponding to the central lines of the light clouds should be constructed. Knowing these functions the position of the desired point of intersection of the central lines of light clouds on the processed image corresponding to the coordinate of the surface of the object can be determined.

Thus, the output of the image processing algorithm should be coordinates of the intersection of the central lines and the angles of their tilt in the intersection area. Next, an image processing algorithm is described for the case where the regression lines are determined by linear functions.

The most obvious method of processing this kind of image is the application of the Radon transform. This transformation is integral and minimizes the error in determining the coordinates of the intersection of lines. The image on the photodetector of the cloudy triangulator is transformed according to the equation:

$$R(s,\alpha) = \int I\left(s\cdot\cos(\alpha) - z\cdot sin(\alpha), s\cdot sin(\alpha) + z\cdot cos(\alpha)\right)dz \tag{25}$$

Next local maximums on the image are determined. They completely characterize sought lines. However, because of the algorithmic complexity of implemented Radon transform, it was decided to develop a faster algorithm.

The proposed algorithm is based on the following sequence of operations: a rough estimate of the location of cloudy points forming the desired ellipses of their boundaries; image segmentation by masking, the definition of regression line equations; determining the point of their intersection and the slopes of the regression lines.

The spatial illumination is modulated in such a way that clouds of light points on the image appear to be located symmetrically with respect to two intersecting central lines L_1 and L_2. Then the points with the coordinates (x_1, y_1) and (x_4, y_2) belong to the line L_1, and the points (x_2, y_1), (x_3, y_2) - L_2. Hence equations of lines with respect to two points belonging to these lines are defined:

$$L_1 : y = a_1 x + b_1 \tag{26}$$

$$L_2 : y = a_2 x + b_2 \tag{26}$$

$$a_1 = \frac{y_1 - y_2}{x_1 - x_4}, \; b_1 = \frac{x_1 y_2 - y_1 x_4}{x_1 - x_4} \tag{27}$$

$$a_2 = \frac{y_1 - y_2}{x_2 - x_3}, \; b_1 = \frac{x_2 y_2 - y_1 x_3}{x_2 - x_3} \tag{28}$$

In the vicinity of the detected lines, masks are formed in such a way that the distance to the point belonging to the mask does not exceed the value characterizing the width of the cloud of points and exceeds the distance to the second line:

$$M_2\{(x,y)\} = \left\{ \begin{array}{l} \left|(x,y), y = a_1 x + b_1\right| < \varepsilon \\ \left|(x,y), y = a_2 x + b_2\right| > \varepsilon \end{array} \right\}, \tag{29}$$

$$M_2\{(x,y)\} = \left\{ \begin{array}{l} \left|(x,y), y = a_1 x + b_1\right| > \varepsilon \\ \left|(x,y), y = a_2 x + b_2\right| < \varepsilon \end{array} \right\}, \tag{30}$$

Next, the functions of the central lines L_1 and L_2 are determined by the linear regression method. Here, N_1 and N_2 are the number of points in the sets M_1 and M_2, respectively.

$$L_1 : y = a_1' x + b_1' \tag{31}$$

$$L_2 : y = a_2' x + b_2' \tag{32}$$

$$a_1' = \frac{\dfrac{\sum_{M1} x^2}{N1} \cdot \dfrac{\sum_{M1} y}{N1} - \dfrac{\sum_{M1} xy}{N1} \cdot \dfrac{\sum_{M1} x}{N1}}{\dfrac{\sum_{M1} x^2}{N1} - \left(\dfrac{\sum_{M1} x}{N1}\right)^2} \qquad (33)$$

$$b_1' = \frac{\dfrac{\sum_{M1} xy}{N1} - \dfrac{\sum_{M1} x}{N1} \cdot \dfrac{\sum_{M1} y}{N1}}{\dfrac{\sum_{M1} x^2}{N1} - \left(\dfrac{\sum_{M1} x}{N1}\right)^2} \qquad (34)$$

$$a_2' = \frac{\dfrac{\sum_{M2} x^2}{N2} \cdot \dfrac{\sum_{M2} y}{N2} - \dfrac{\sum_{M2} xy}{N2} \cdot \dfrac{\sum_{M2} x}{N2}}{\dfrac{\sum_{M2} x^2}{N2} - \left(\dfrac{\sum_{M2} x}{N2}\right)^2} \qquad (35)$$

$$b_2' = \frac{\dfrac{\sum_{M2} xy}{N2} - \dfrac{\sum_{M2} x}{N2} \cdot \dfrac{\sum_{M2} y}{N2}}{\dfrac{\sum_{M2} x^2}{N2} - \left(\dfrac{\sum_{M2} x}{N2}\right)^2} \qquad (36)$$

The last step of the algorithm is the calculation of the characteristic parameters X, α_1, α_2:

$$X = \frac{b_1' - b_2'}{a_2' - a_1'} \qquad (37)$$

$$\alpha_1 = \mathrm{atan}\left(a_1'\right) \qquad (38)$$

$$\alpha_2 = \mathrm{atan}\left(a_2'\right) \qquad (39)$$

The proposed algorithm for processing images of a cloudy triangulator works much faster than the algorithm based on Radon transform and allows real-time cloudy triangulation measurements with a measurement frequency up to 100 Hz on ordinary computers. To minimize the measurement error in the cloudy triangulator measurements, it is proposed to apply multilevel spatial, temporal and spectral filtering of information signals. Synchronization of the radiation pulse and the exposure of the photodetector provides a minimum of background illumination on the photodetector. Spatial amplitude-modulated illumination makes it possible to perform spatial regression analysis at the stage of digital data processing.

Optical spectral filtration (ultraviolet filtering, infrared filtering and narrowband filtering by interference filters) provides narrowband recording in the wavelength range of optical radiation. As a result, the spectral component of the background signal that has a spectral composition close to the emission spectrum of the optical source falls on the photodetector. Digital signal is filtered by the photodetector. Amplitude filtering and threshold discrimination can effectively eliminate the influence of background illumination and allow to perform regression analysis with the best accuracy. Thus, using the multilevel filtering of information signals, it is possible to measure geometric parameters by the cloudy triangulation method with a record-low error.

During the operation of the measuring system, the measurement system software is executed on two computers: a computer measuring module located directly on the optic-electronic module trunk and a server located in the special room. In the measurement process, the software of the measurement module receives data from photodetectors of cloudy triangulators, the temperature values of the measuring system at several critical points, the temperature value of the metal rolling. The software of measurement module processes received data in real time, calculates the thickness, stores the measurement results in the archive and generates a data file with a special format for visualizing measurement results.

The server software reads data from a file with special format generated by the measurement module software and visualizes measurement results on the monitor in real time. Three monitors are connected to the server (one is connected directly to the system unit of the server and two are connected through a network interface based on "thin client" technology).

The internal software architecture can be divided into the following modules with different functionality: data processing, interaction with a photodetector, collection of diagnostic parameters, calibration, image processing module, configuration parameters, visualization, saving results and a module for storing diagnostic parameters:

- The module of interaction with photodetectors implements communication interface and provides a synchronous recording of images from the upper and lower cloudy triangulators;
- The image processing module implements a fast algorithm for processing the data of cloudy triangulators. It provides the determination of coordinates and spatial distortions of the observed spatially modulated illumination;
- The module of collecting diagnostics information implements communication interface with diagnostics unit of the measuring system, which collects diagnostic information about the state of the measuring system;
- The calibration module contains calibration data, which allows calculating the thickness based on the data received by the photodetectors;
- The configuration parameters module provides access to the configuration data of the measuring system;
- The visualization module implements communication interface with the visualization subsystem of the measuring system and transmits information about measured thickness and about the internal state of the measuring system in real time;
- The module for storing measurement results and diagnostic data ensures storing information for further analysis.

The subsystem of measurement results visualization is designed to display results of measuring the thickness of hot metal rolling in real time at different sections of the process chain. Visualization of the results is carried using specialized software, which displays graphic images with thickness data on monitors.

The subsystem of measurement results visualization consists of a server, two thin hardware clients (a specialized device that is fundamentally different from the computer: the hardware thin client does not have a hard disk and uses a specialized local OS whose main task is to organize the session with the server) and three monitors. A monitor is connected to each thin client and server which visualize measurement results. Thin clients interact with the server and actually transmit graphical window of the program running on the server to connected monitor.

The server reads the measurement results from the file storage of the data processing module in real-time mode. The measurement results are processed and visualized to users on monitors.

The interaction of thin clients with the server and the server with the data processing module is implemented based on the computer network Ethernet IP4 protocol. Interaction of all devices is carried out through the network switch.

The data processing module is located in close proximity to the optoelectronic measuring module and is implemented using an industrial computer and specialized software.

The computer receives a video signal from cloudy triangulation cameras through the USB protocol and processes digital image data using software algorithm. First, the information of the video stream of each cloudy triangulator is converted into distance data from the given triangulator to the nearest metal rolling surface. Then the distance data from the two cloudy triangulators is converted into the thickness information of the hot metal rolling.

Processing results are sent via Ethernet connection to the application server on information monitors.

In addition, data collecting and processing module interact with the adjustment board by registering and adjusting the trim factors for the thickness calculations.

Data collecting and processing module communicates with the diagnostic system control unit, receives, processes and archives temperature data at various points in the optoelectronic module, power voltages data on internal electrical circuits, data about the temperature of hot metal rolling using built-in optical pyrometer.

To ensure a low measurement error of hot metal rolling thickness by the method of cloudy triangulation, it is necessary to provide a phase-inhomogeneous medium in the region of the traces of the optical signals of the triangulators with required statistical characteristics. As shown above, the cloudy triangulation method effectively operates in a phase-inhomogeneous medium that distorts optical beams by adding random deviations with zero mathematical expectation.

The solution of this serious scientific and technical problem in the work is connected with the organization of controlled phase inhomogeneity of the medium in the areas of propagation of optical signals. For this purpose, a specially prepared airflow with a given intermittency is introduced into the optical signal paths. This flow is formed by the fan so that the dimensions of the phase inhomogeneities are small in comparison with the spatial scale of the optical paths, and the structure, due to intensive mixing, becomes on the average spatially homogeneous, excluding systematic influences on the light rays. Fluctuations of indications are effectively smoothed by averaging over a large data ensemble, obtained in the cloud triangulation method.

Figure 5. Calibrating functions of top (solid line) and bottom (dotted line) triangulators.

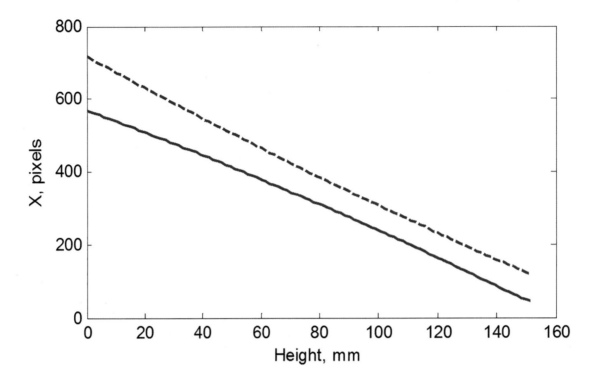

MEASUREMENT SYSTEM CALIBRATION

The measurement system was calibrated before industrial tests. The calibration method is described in a section below. Functions F_1 and F_2 (15) are obtained at the first stage of calibration. The procedure is implemented by placing the flat calibration metal sheet in several positions with known spatial coordinates. Characteristic parameters of the radiation intensity distribution on the received image are recorded at each position of the metal sheet. As a result functions F_1 and F_2 are obtained (Figure 5).

Function q is determined at the second stage of calibration. This function compensates distortions caused by a spatial slope of the metal sheet considering the integral action of all optical distortions and aberrations (Figure 6). The function is obtained by multiple direct measurements of the calibration metal sheet, randomly oriented in space, according to (16). Inclination angles of regression lines on the image of the light source are used as parameters, which characterize spatial slope of the measured metal sheet. Obviously, these angles uniquely determine the spatial slope of the measured metal sheet.

The thickness of 2 mm thick flat metal sheet was measured to estimate the measurement error under laboratory conditions. A considerable amount of data and distribution of measurement results were obtained. Average value of the measured thickness was 2.0017 mm, and the standard deviation was 0.0021 mm (Figure 7). The results agree with the relative error of thickness measurement 10^{-6}.

In the conditions of the hot metallurgical shop, the measuring results depended on the temperature of triangulation modules. Temperature meters were installed at the optical triangulation modules of the measuring system to monitor the state during operation. Figure 8 shows dynamics of the temperature

Figure 6. Calculated function q, which compensates distortions caused by spatial inclination of the measured object

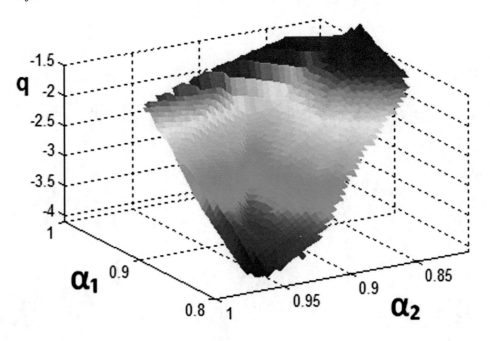

Figure 7. Measurement result for 2 mm thick calibration sheet: Standard deviation is 0.0021 mm.

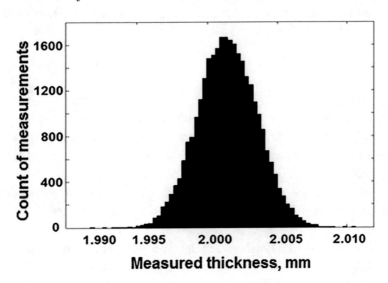

Figure 8. Dynamics of the temperature of the upper triangulation module for 16 days (green line shows work time of the shop)

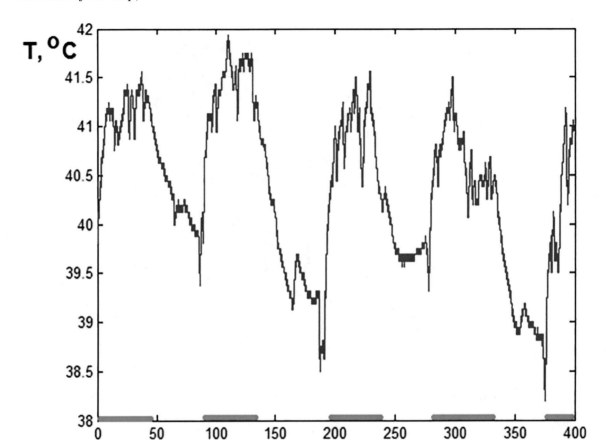

of the upper triangulation module for 16 days. The data were obtained in the hot-rolling shop of the Kuznina Metallurgical Plant, Novosibirsk. The presented graph in Figure 8 shows that the temperature of the triangulation modules varies by several degrees during operation despite thermostabilization of the measuring system. It can affect the measurement error negatively.

Figures 9 and 10 show the measurement results of an immovable 3 mm calibration sheet and the temperature of the upper triangulation meter. The measurements were carried out for 15 hours after shop work time.

Graphs show a fairly good correlation between the temperature and the measured thickness. Figure 11 shows the dependence of the measured thickness on the temperature of the triangulation module.

The distribution of the measured thickness has a spread at a level of 40 μm (Figure 12). Active compensation of dependence of the thickness measurement results on the temperature is allowed to significantly reduce the spread of the measured thickness. The distribution of measured thickness obtained after adding a linear compensating function corresponding to the dependence of the thickness on temperature is shown in Figure 13.

Figure 9. Measurement result of the thickness of an immovable 3 mm calibration sheet

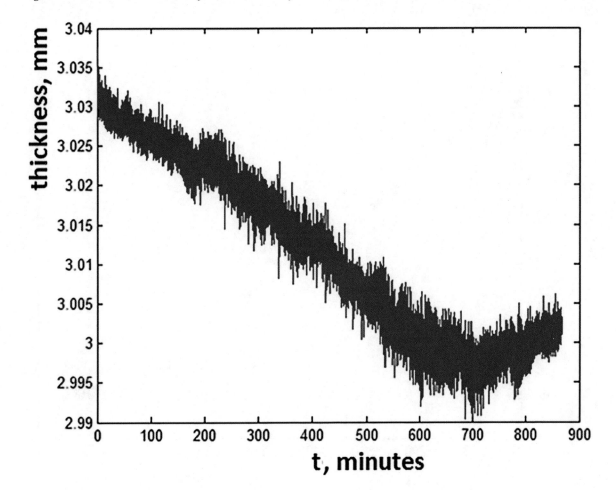

As a result, the method of active compensation of the temperature dependence of the triangulation module ensured a reduction of the measurement error by almost 3 times.

The estimation of measurement error was obtained in the absence of strong refractive distortion of optical signals under laboratory conditions. This error estimation of the proposed differential cloudy triangulation method is valid for thickness measurements of flat cold objects. The results of thickness measurements for hot-rolled metal in industrial environments are described below.

INDUSTRIAL TESTING

The laser system for measuring the thickness of hot dynamic objects was tested under industrial conditions at the working steel plant. Measurements were carried out in the hot-rolled metallurgical shop of Kuzmin Metallurgical Plant, Novosibirsk (Figure 14). The rolling mill was configured up to produce steel sheet with 2.85 mm thickness, branded 323863/02 2kp.

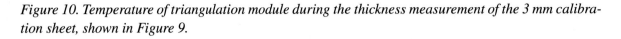

Figure 10. Temperature of triangulation module during the thickness measurement of the 3 mm calibration sheet, shown in Figure 9.

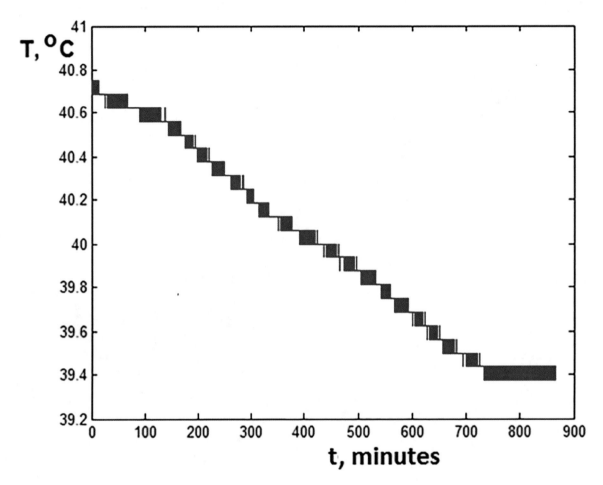

Graphs of Z_1 and Z_2 values obtained during hot rolling measurement are shown on Fig.15. The measurement time was about 20 sec. The figure shows that for the first 7 sec the metal sheet is strongly shaken and shifted in the measuring area. This is due to the fact that metal sheet is clamped only by the shafts of the finishing group when moves out of the finishing rolling of the mill. Then the rolling sheet is supplied to the reeling section and fixed by the guide rollers. These rollers clamp the sheet and limit its further oscillations.

As an example, Figure 16 shows one of the measured thickness profiles and result of control measurements taken at a later stage of metalworking of the rolling sheet. The graph shows that the deviation of thickness of the rolling sheet does not exceed 0.15 mm of the predetermined thickness. The thickness is in negative tolerance practically over the entire rolling sheet. This mode of rolling processing is the most preferable for the metallurgical industry. It allows to save metal, increases cost-effectiveness and makes production economically profitable.

Figure 11. Dependence of the measured thickness on the temperature of the upper triangulation module

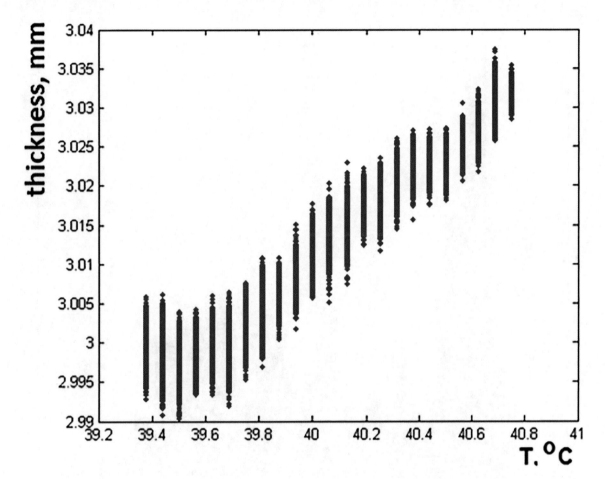

Figure 17 shows the trajectory of hot-rolled metal in the area of thickness measurement and its measured thickness. The green color in the graph signals shows that the measured thickness is identical to the reference thickness; blue color shows that the measured thickness is less than the reference, and red indicates that the measured thickness is greater than the reference.

The performed industrial tests of a laser thickness measurement system for hot dynamic objects allowed to estimate the measurement error of the differential cloudy triangulation method. Based on the comparison of the measurement results and the data of direct measurements it is shown that the actual error of the measured thickness is less than 20 μm. This corresponds to a relative error of 10^{-5}. This accuracy level is progressive for measuring systems, working in real conditions of hot metallurgical shops. Obtained results confirm perspectives of the performed research and big potential of the proposed method.

Figure 12. Distribution of the measured thickness of 3 mm calibration sheet

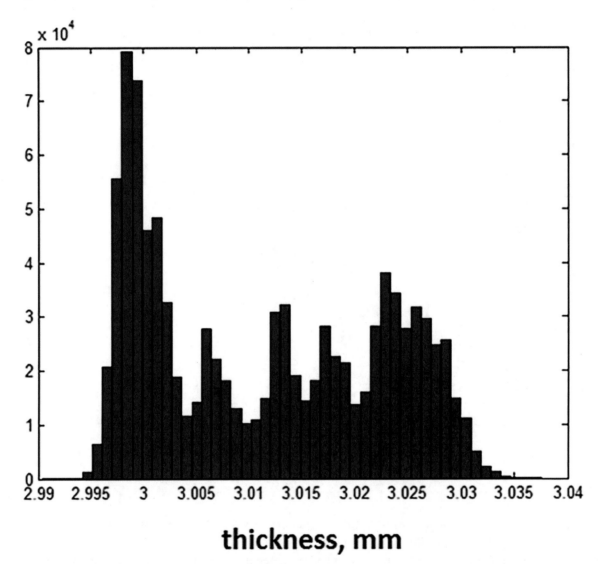

CONCLUSION

The new differential cloudy triangulation method for quick and precise measurement of thickness of hot dynamic objects is proposed. The method is based on laser cloudy triangulation, used in the synchronous differential mode. The experimental model of the laser system for thickness measurements, which implements the proposed method, has been developed. Laboratory tests of the laser system have been performed. It is shown that the relative error of measurement is about 10^{-6} under laboratory conditions. Industrial tests of the measuring system have been performed. They confirm efficiency and possibilities

Figure 13. Distribution of the measured thickness of 3mm calibration sheet after adding a linear compensating function

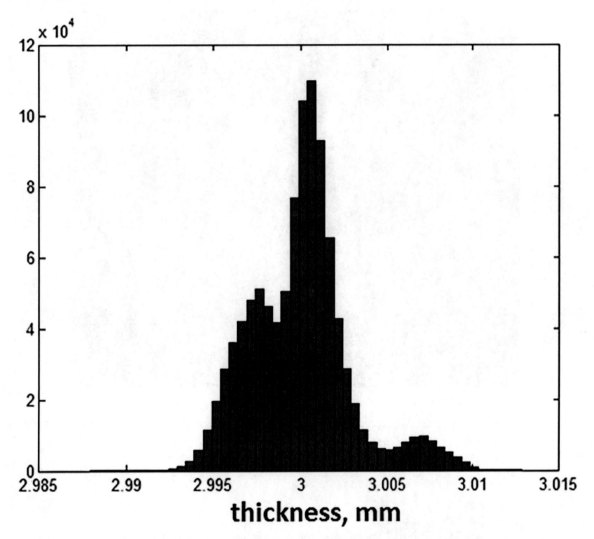

of differential cloudy triangulation in metallurgy. It has been shown that the relative measurement error of hot dynamic object thickness at real production corresponds to 10^{-5}. This level of error is progressive for laser measuring systems, operating in working conditions of hot metal shops. The differential cloudy triangulation method may be used for geometry measurement of hot dynamic objects in a wide range of strong optical refractions in media. The obtained results are particularly promising for the use in geometry measurements of hot metal rolling under existing steel production.

Figure 14. Laser system for measuring the thickness of hot dynamic objects tested under industrial conditions at the working steel plant.

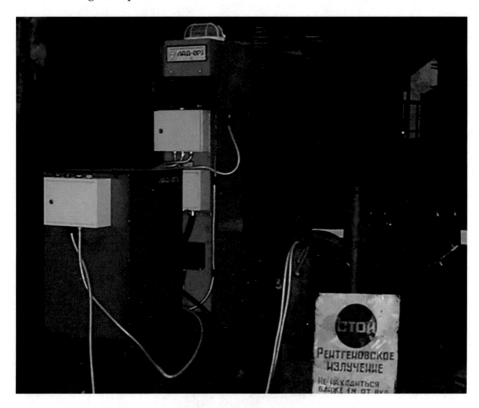

Figure 15. Indications of laser cloudy triangulators in thickness measurements of hot metal rolling (solid line – upper triangulator, dashed line – bottom triangulator)

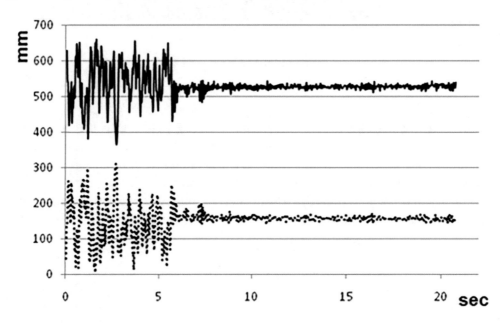

Figure 16. Measured thickness of metal rolling (solid line) and control measurements of thickness at various points (round markers)

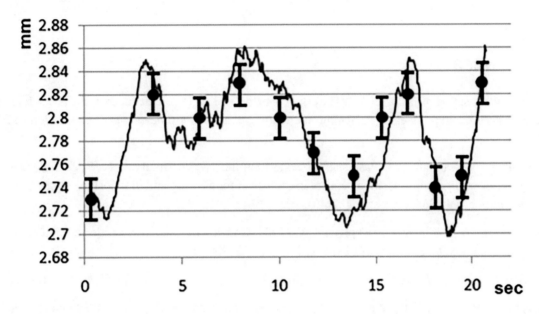

Figure 17. Trajectory of hot-metal rolling in the area of thickness measurement and its measured thickness

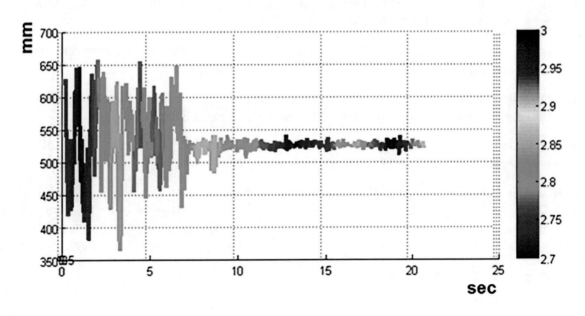

ACKNOWLEDGMENT

This research was supported by Russian Science Foundation (project No. 14-29-00093).

REFERENCES

Aksenov, V. p., & Pogutsa, Ch. (2012). The effect of an optical vortex on the random shifts of a Laguerre-Gaussian laser beam propagating in the turbulent atmosphere. *Atmospheric and Oceanic Optics*, *25*(7), 561–565.

Baybakov, A. N., Ladigin, B. I., Pastushenko, A. I., Plotnikov, S. V., Tukubaev, N. T., & Yunoshev, S. P. (2004). Laser triangulation position sensors in industrial monitoring and diagnostics. *Optoelectronics, Instrumentation and Data Processing*, *2*(40), 105–113.

Bol'basova, L. A., Lukin, V. P., & Nosov, V. V. (2009). Image jitter of a laser guide star in a monostatic formation scheme. *Optics and Spectroscopy*, *107*(6), 993–999. doi:10.1134/S0030400X09120236

Bretscher, O. (1995). *Linear Algebra With Applications* (3rd ed.). Upper Saddle River, NJ: Prentice Hall.

Degarmo, E. P., Black, J. T., & Kohser, R. A. (2003). Materials and Processes in Manufacturing (9th ed.). Wiley.

Deponte, Ed. (2004). *European Patent Nº EP1619466A1*. Munich, Germany: European Patent Office.

Dvojnishnikov, S. V., Bakakin, V. G., Glavnij, V. G., Kabardin, I. K., & Meledin, V. G. (2013). *RU Patent Nº 2537522*. Moscow: FIPS.

Dvoynishnikov, S., & Rakhmanov, V. (2015). Power installations geometrical parameters optical control method steady against thermal indignations. *EPJ Web of Conferences*, *82*, 01035-3.

Dvoynishnikov, S., & Rakhmanov, V. (2015). Power installations geometrical parameters optical control method steady against thermal indignations. *EPJ Web of Conferences*, *82*, 01035-3.

Dvoynishnikov, S. V., Bakakin, G. V., Glavnyj, V. G., Kabardin, I. K., & Meledin, V. G. (2015). *RU Patent Nº 1826697*. Moscow: FIPS.

Dvoynishnikov, S. V., Rakhmanov, V. V., Meledin, V. G., Kulikov, D. V., Anikin, Yu. A., & Kabardin, I. K. (2015). Experimental Assessment of the Applicability of Laser Triangulators for Measurements of the Thickness of Hot Rolled Product. *Measurement Techniques*, *57*(12), 1378–1385. doi:10.100711018-015-0638-x

Goodman, J. W. (1985). *Statistical Optics*. Wiley.

Goodman, J. W. (1996). *Introduction to Fourier Optics*. McGraw-Hill.

Gordon, K., & Jacob, S. (1969). *U.S. Patent Nº 3565531*. Washington, DC: U.S. Patent and Trademark Office.

Gruber, M., & Hausler, G. (1992). Simple, robust and accurate phase-measuring triangulation. *Optik (Stuttgart), 3*, 118–122.

Gurvich, A. S., Gorbunov, M. E., Fedorova, O. V., Fortus, M. I., Kirchengast, G., Proschek, V., & Tereszchuk, K. A. (2014). Spatiotemporal structure of a laser beam at a path length of 144 km: Comparative analysis of spatial and temporal spectra. *Applied Optics, 53*(12), 2625–2631. doi:10.1364/AO.53.002625 PMID:24787588

Hu, F. (2011). An automatic measuring method and system using a light curtain for the thread profile of a ballscrew. *Measurement Science & Technology, 22*(8), 085106. doi:10.1088/0957-0233/22/8/085106

Ishimaru, A. (1999). *Wave Propagation and Scattering in Random Media.* Wiley. doi:10.1109/9780470547045

Kandidov, V. P., Tamarov, M. P., & Shlenov, S. A. (1998). The influence of the outer scale of atmospheric turbulence on the dispersion of the center of gravity of the laser beam. *Atmospheric and Oceanic Optics, 12*(1), 27–33.

Komissarov, A. G. (1990). *RU Patent Nº 1826697.* Moscow: FIPS.

Kravcov, Y. A., Feizulin, Z. I., & Vinogradov, A. G. (1983). *Passage of radio waves through Earth atmosphere.* Moscow: Radio i svyaz.

Lee, D. (2001). *U.S. Patent Nº 20030007161.* Washington, DC: U.S. Patent and Trademark Office.

Li, R., Fan, K., Miao, J., Huang, Q., Tao, S., & Gong, E. (2014). An analogue contact probe using a compact 3D optical sensor for micro-nano coordinate measuring machines. *Measurement Science & Technology, 25*(9), 094008. doi:10.1088/0957-0233/25/9/094008

Meledin, V. (2011). Optoelectronic Measurements in Science and Innovative Industrial Technologies. *Optoelectronic Devices and Properties, 18*, 373–399.

Nosov, V. V., Lukin, V. P., Nosov, E. V., Torgaev, A. V., Grigoriev, V. M., & Kovadlo, P. G. (2009). *Coherent structures in the turbulent atmosphere. In Mathematical models of nonlinear phenomena* (pp. 120–154). Nova Science Publishers.

Penney, C. M., & Thomas, B. (1989). High performance laser triangulation ranging. *Journal of Laser Applications, 1*(2), 51–58. doi:10.2351/1.4745229

Plotnikov, S. V. (1995). Comparison of methods for signal processing in the triangulation measurement systems. *Optoelectronics, Instrumentation and Data Processing, 6*, 58–63.

Pollinger, F., Hieta, T., Vainio, M., Doloca, N. R., Abou-Zeid, A., Meiners-Hagen, K., & Merimaa, M. (2012). Effective humidity in length measurements: Comparison of three approaches. *Measurement Science & Technology, 23*(2), 025503. doi:10.1088/0957-0233/23/2/025503

Raymond, M. (1984). *Laser Remote Sensing: Fundamentals and Applications.* Wiley.

Semenov, D. V., Sidorov, I. S., Nippolainen, E., & Kamshilin, A. A. (2010). Speckle-based sensor system for real-time distance and thickness monitoring of fast moving objects. *Measurement Science & Technology, 21*(4), 045304. doi:10.1088/0957-0233/21/4/045304

Sicard, J., & Sirohi, J. (2013). Measurement of the deformation of an extremely flexible rotor blade using digital image correlation. *Measurement Science & Technology, 24*(6), 065203. doi:10.1088/0957-0233/24/6/065203

Walters, P. (1982). *An Introduction to Ergodic Theory*. Berlin: Springer-Verlag. doi:10.1007/978-1-4612-5775-2

Zuev, V. V., Zuev, V. E., Makushkin, Y. S., Marichev, V. N., & Mitsel, A. A. (1983). Laser sounding of atmospheric humidity: Experiment. *Applied Optics, 22*(23), 3742–3746. doi:10.1364/AO.22.003742 PMID:18200259

Chapter 4
Surface Measurement Techniques in Machine Vision:
Operation, Applications, and Trends

Oscar Real Real
*Universidad Autónoma de Baja California,
Mexico*

Daniel Hernández-Balbuena
*Universidad Autónoma de Baja California,
Mexico*

Moises J. Castro-Toscano
*Universidad Autónoma de Baja California,
Mexico*

Moises Rivas-Lopez
*Universidad Autónoma de Baja California,
Mexico*

Julio Cesar Rodríguez-Quiñonez
*Universidad Autónoma de Baja California,
Mexico*

Wendy Flores-Fuentes
*Universidad Autónoma de Baja California,
Mexico*

Oleg Serginyenko
*Universidad Autónoma de Baja California,
Mexico*

Lars Lindner
*Universidad Autónoma de Baja California,
Mexico*

ABSTRACT

Surface measurement systems (SMS) allow accurate measurements of surface geometry for three-dimensional computational models creation. There are cases where contact avoidance is needed; these techniques are known as non-contact surface measurement techniques. To perform non-contact surface measurements there are different operating modes and technologies, such as lasers, digital cameras, and integration of both. Each SMS is classified by its operation mode to get the data, so it can be divided into three basic groups: point-based techniques, line-based techniques, and area-based techniques. This chapter provides useful information about the different types of non-contact surface measurement techniques, theory, basic equations, system implementation, actual research topics, engineering applications, and future trends. This chapter is particularly valuable for students, teachers, and researchers that want to implement a vision system and need an introduction to all available options in order to use the most convenient for their purpose.

DOI: 10.4018/978-1-5225-5751-7.ch004

INTRODUCTION

Currently there are numerous applications in the area of research and engineering that implements Surface Measurement Systems (SMS) to perform accurate measurements of surface geometry. SMS works essentially by measuring a set of points across an object's surface, from which can be constructed a representation based on triangulation or surface patches (images). Accurate geometry measures are used in the creation of three-dimensional computational models for different applications, such as autonomous navigation in robotic tasks, in topography for direct survey to determine accurately the terrestrial position of points, distances and angles using leveling instruments, on structural health monitoring to obtain a full database of the structure and get a full assembly diagnosis, and mostly in manufacturing industry in the creation of computer-aided design (CAD) to design or reverse engineer parts. The SMS usually work with meticulous processes especially in autonomous navigation to detect obstacles and avoid collisions or when designed parts must fit on existing equipment, minimizing the possibility of error in the geometry measures. It is often that SMS have contact measuring techniques like micrometers or coordinate measuring machines, however sometimes the object of interest cannot have direct contact with the instruments to obtain the geometry information by different factors: its size, the difficulty to reach it or the nature of the material. For these types of task there are numerous Non-Contact surface measurements techniques that are capable of performing an accurate surface geometry measurement. Some areas where the Non-Contact surface measurements techniques applies are, the archaeology, where the objects are delicate and must have less contact to avoid irreparable damage, other areas include the medicine, where three-dimensional recordings of human bodies surface are used to diagnose orthopedically diseases in a non-invasive way. It also has applications on electronic industry, in the quality control for delicate parts of integrated circuits or computer devices. To carry out the Non-Contact surface measurements there are different operation modes and technologies, such as lasers, digital cameras and integration of both. Each SMS can be classified by its operation mode to get the data, so it can be divided into three basic groups: Point based techniques, Line based techniques and Area based techniques.

This chapter provides useful information to students, teachers and researchers that want to learn about the different types of non-contact surface measurement techniques, how they are used in actual research topics, applications, future trends and the integration of one or more non-contact SMS to enhance the performance.

BACKGROUND

Surface measurement systems in its common definition is the possibility of a machine (by sensing means and computer mathematic processing consecutively) to obtain information about surrounding environment for further analytical treatment. According to the common definition, any complete surface measurement system or machine vision system is a combination of two components: technical means (hardware) and information processing mathematics and algorithm (software), (Rivas-Lopez, Sergiyenko, & Tyrsa, 2008).

Surface measurement systems are related to many fields, techniques developed from different areas are used for recovering information of objects like image processing, where its algorithms are useful in early stages of surface measurement systems, curve and surface representations and other techniques are used from computer graphics, also, computer graphics uses many techniques from surface measurement systems to enter models into the computer for creating realistic images. Other area is pattern recognition, where many statistical techniques have been developed for classification of patterns, artificial intelligence is used to analyze scenes by computing a symbolic representation of the scene contents after the information has been processed to obtain features (Cuevas-Jimenez, 2006).

In this chapter a compilation of non-contact measurement techniques are presented, they are divided according to their mode of operation, also, the working principle, basic equations, system implementation and future trends are shown and how all of these techniques can be implemented together to improve the performance of the vision system.

OPERATION MODES

Point Based Techniques

Point based techniques perform point by point measurements, this means that for each cycle a single measurement is obtained, this technique is often used to provide a complete set of data. Depending on the technique, these can be used in several areas like, autonomous navigation, driver assistance systems, shape recognition, structural health monitoring, etc. The advantages of using these techniques is that precise measurements can be obtained with a broader field of view in real time. The disadvantages of this techniques are the measuring performance over edges on the poor obtained information about the object texture.

Time of Flight

Time of flight consists on a laser emitting source, two photoreceptors and a signal processing unit. This method uses time, amplitude or frequency as a variable to calculate the distance between a laser source and its reflection over surface as shown in Figure 1.

Figure 1. Time of flight principle

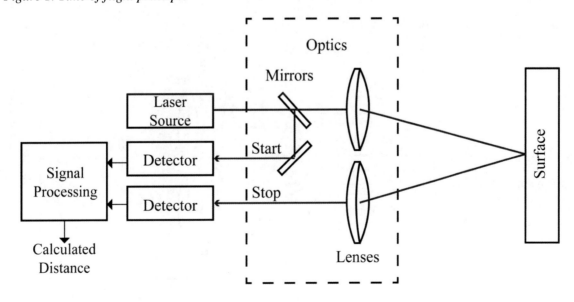

There are different ways to calculate the distance in this method, using pulse, amplitude or frequency modulation (Kilpela, 2002). This technique has several application in different areas like autonomous navigation systems, advanced driver assistance systems, motion analysis, and structure design, among others.

Pulse Modulation Method

The pulse modulation method functioning principle is as shown in Figure 2, where the pulsed laser source emits a pulsed laser beam which is reflected back to the start detector and the transmitter lens, which focuses the laser beam over the surface to scan, then the stop detector receives the signal and passes through the amplifier to the timing discriminator, which function is to keep the timing event at the same point regardless of the incoming pulse amplitude, at last the time interval measurement calculates the distance using the difference between the start signal and the stop signal (Palojärvi, Määttä, & Kostamovaara, 1997).

To calculate the distance, the time taken for a light pulse to travel back reflected from the surface to scan is measured (Bradshaw, 1999). Then the time is multiplied for the light speed to calculate the distance traveled by the laser light and divided by two, due to the distance from the laser source and the surface to scan is half the traveled distance by the laser beam as in equation 1, where C is the light speed constant

(3×10^8), T is time (s) and D is the distance (m).

$$D = (C * T) / 2 \qquad (1)$$

Amplitude Modulation Method

The amplitude modulation method also called laser radar or LIDAR, uses a phase shift given by the amplitude modulated laser beam, as it travels from the laser source to the surface and back (Payne, 1973). This system is capable of performing continuous measurements, but the data is quite noisy with

Figure 2. Time of flight pulse modulation

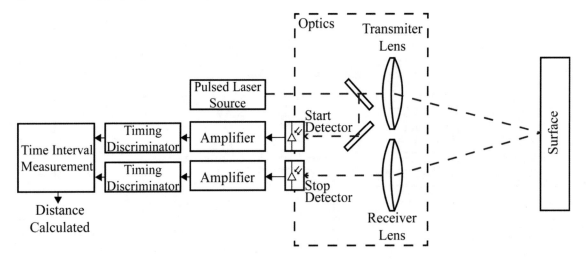

an appreciable number of errors on measurements. The distance is calculated by measuring the phase shift between the sine modulated transmitted and reflected beams (Figure 3).

To calculate the distance in the phase shift mode, equation 2 is used, where D is inferred from the phase shift (equation 3) between the photoelectric current and the modulated emitted signal, c is the light speed and Δ_t is the time of flight (Markus-Christian, Thierry Bosch, Risto, & Marc, 2001).

$$D = \frac{1}{2} c \frac{\Delta_\varphi}{2\pi} f_{rf} \qquad (2)$$

$$\Delta_\varphi = 2\pi f_{rf} \Delta_t \qquad (3)$$

Frequency Modulation Method

In the frequency modulation method, the modulation of the signal is produced by varying the laser cavity length, then the returning laser beam is mixed on the detector with the frequency-swept transmitter beam acting as a local oscillator and finally the frequency of the signal at the detector is related to the target distance (Hulme, Collins, Constant, & Pinson, 1981).

Some advantages of time of flight methods is that in comparison with stereo vision or triangulation systems, the whole systems is compact, due to the laser source is next to the receiver, whereas the other systems need a certain base line, also, no mechanical moving parts are used, this task uses a small amount of processing power due to the direct process to extract the distance information in contrast to stereo vision systems, where complex correlation algorithms are implemented. This methods can be used in real-time applications, since, they are able to measure distances within a complete scene in a single shot. On the other side, this methods have disadvantages when using photoelectric sensors that use visible or near infra-red light, the measurements can be affected by the background light generated by the sun or artificial lighting. Also, if several time of flight vision systems are to be used simultaneously, the systems may disturb each other's measurements. There are different methods to deal with this problem, like time multiplexing or use different modulation frequencies.

Figure 3. Amplitude modulation

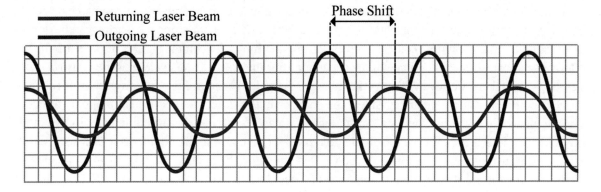

Construction Considerations

Usually, the construction of time of flight vision systems are divided in 2 parts: electronics and optics. The electronics part is composed by a pulsed semiconductor for the pulsed laser source, which peak power vary in the range of 5 watts to 100 watts. The required power relies on the desired measurement distance, optical losses and the reflection coefficient of the target. For the laser source a single heterostructure or a double heterostructure type of laser can be used with a peak power level that varies from 2 watts to 30 watts. The circuit used to generate the pulses, includes a capacitor, which is discharged to the laser with a high power switch, which commonly is an avalanche transistor. The detector is composed by a photodiode and an amplifier, which is also composed by a preamplifier and a postamplifier. For the photodiode there are 3 different options: PIN, avalanche and metal-semiconductor-metal. The selection of it relies on the application needs, for example, the response time of the avalanche photodiode is 50-200 times higher than PIN photodiodes, but require a higher voltage (50-200 V) and the avalanche multiplication produces extra noise. PIN silicon photodiodes have almost the same bandwidth as silicon avalanche photodiodes with a frequency of 1-2 GHz, but with silicon metal-semiconductor-metal over 100 GHz frequencies can be measured at a wavelength of 400nm. The amplifier is composed by a transimpedance type preamplifier and a post-amplifier. For the transimpedance amplifier a simple structure is commonly used, due to the need of a wide bandwidth and a small delay to have enough phase margin. There are three options for the input stages: the common emitter, which is the most usual, the common collector or common base circuits. This input stages are affected by the Miller effect, which affects the input impedance and frequency response of the amplifier, due to apparent scaling of the impedance connected from the input to output of the amplifier (Miller, 1920). In the common emitter the Miller effect of the collector-base capacitance is reduced by a cascade transistor. If in the input transistor the dominating pole is defined by the Miller-capacitance, the photodiode's capacitance may vary in a wide range without significantly affecting the bandwidth of the amplifier. For the common base stage, the Miller effect if minimized by adding another bias voltage supply. The common base stage advantage is that it has a small input resistance, which reduces the importance of input capacitance and permits the use of feedback current, which increases the bandwidth. On the other side the common collector stage has the disadvantage of small gain, which results in the need of low noise for the next stage. The timing discriminator task is to convert the analog timing pulses to logic pulses and separate the timing pulses from the noise pulses. The time interval measurement consists on an accurate time measuring unit for short time intervals. There are different methods to measure time, they can be analog, digital or a combination of both. For the digital measure of time, a number a clock pulses are counted between the signals of the start and the stop detector to calculate de time. The analog time measuring can be done by discharging a capacitor with a constant current during the time between the signals of the start and stop detectors and measuring the voltage of the capacitor after the discharging cycle. For the construction of the optical part there are several things that must be taken into account, like the properties of the target surface and the measurement angle, due to the changes in the reflection coefficient that may affect the measurement. There are two options for the lenses, it can be single-axis or double axis. In the single-axis, only one lens is used for the transmitter and receiver, whereas in the double-axis 2 lenses are used, one for the transmitter and one for the receiver. Depending on the application it is recommended to use single-axis for short distances (maximum 5 meters) and double-axis for long distances (5 to 30 meters). For the beam splitter it is normally used a mirror based on a metal film that is used to redirect part of the laser beam to the start detector and the other part to the surface to scan (Kilpela, 2002).

Laser Triangulation

Laser triangulation consists on a positioning laser, a photoelectric sensor and a fixed distance between both. The emitting source focus the laser beam over the surface to scan, then it is detected by the photoelectric sensor and forms a triangle between the positioning laser, the fixed point over the surface to scan and the photoelectric sensor. This vision systems are applied in various ways for 3D measurements or shape recognition, like vision assisted assembly, autonomous navigation, structural health monitoring, micro surface inspections and precise automated surgery.

Static Laser Triangulation

Depending on the type of photoelectric sensor the laser triangulation can be divided on static or dynamic. The static laser triangulation is carried out by the laser source, which is adjusted to a known angle, the laser beam, which is reflected on the surface and detected by the array of sensors or the ccd camera (Figure 4).

In Figure 5 the necessary variables to calculate the distance are shown, the red line represents the laser beam ($\overrightarrow{Li}$), the blue line represents the vector $\overrightarrow{Cxy}$ between the optical center of the camera and point A, finally the vector ($\vec{d}$) between the laser source and the optical center of the camera.

On equation (4) this geometrical relation is expressed in spherical coordinates, where φ is the angle between the laser beam plane and the plane Y-Z, and θ is the angle between the vector $\overrightarrow{Li}$ and the Y axis, as in equations (5), (6) and (7) (Acosta, Garcia, & Aponte, 2006).

$$\overrightarrow{Li} = \vec{d} + \overrightarrow{Cxy} \tag{4}$$

$$\left|Li\right|\sin\left(\varphi\right) = d_x + Cxy_x \tag{5}$$

$$\left|Li\right|\cos\left(\varphi\right)\cos\left(\theta\right) = d_y + Cxy_y \tag{6}$$

Figure 4. Static laser triangulation principle

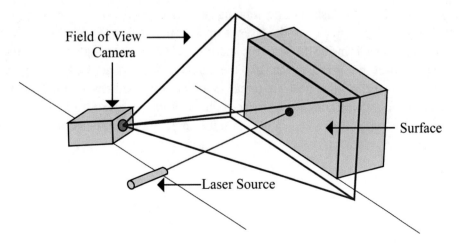

Figure 5. Static laser triangulation variables

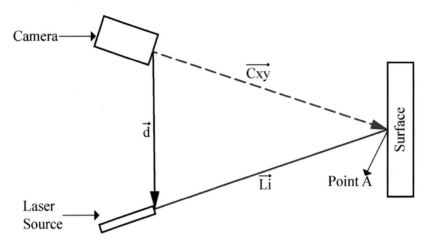

$$\left|Li\right|\cos\left(\varphi\right)\sin\left(\theta\right)=d_z + Cxy_z \tag{7}$$

A disadvantage of this method is that the field of view is limited by the specifications of the camera. To get a broader field of view, a method known as dynamic laser triangulation is used, where the camera is substituted by a scanning aperture.

The construction of static laser triangulation vision systems is divided in four parts: Data acquisition system, mechanical system, control system and the software. For the data acquisition system a laser generator and a ccd based camera (usually a webcam) are used. The mechanical system is built using solid structure and a motor for the rotational movement of the system. The control system works with an input, which is provided by a computer, the outputs are the signals for the motor, the activation of the data acquisition system and feedback to the computer. The software obtains the data from the ccd based camera, provides the inputs to the control system, calculates de distance with the triangulation algorithm and works as a human-machine interface to visualize the model and export the obtained data (Acosta, Garcia, & Aponte, 2006).

Dynamic Laser Triangulation

Dynamic laser triangulation consists on a positioning laser, a photoelectric sensor, and a fixed distance between both. The emitting source focus the laser beam over the surface to scan, then it is detected by a photoelectric sensor (which is inside the scanning aperture) and forms a triangle between the positioning laser, the fixed point over the surface to scan and the scanning aperture as shown in Figure 6.

Once the triangle is formed it is possible to calculate the distance between the laser vision system and the surface to scan implementing the sine and cosine laws using equations 8, 9, 10 and 11 (Basaca et al., 2010), since the angle of the laser source (*Cij*), the angle of the photoelectric sensor (*Bij*) and the distances between them is known.

Figure 6. Dynamic laser triangulation

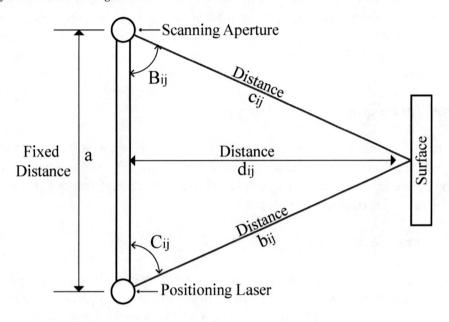

$$x_{ij} = a \frac{sinB_{ij}sinC_{ij}cos\sum_{j=1}^{j}\beta_j}{sin\left[180° - \left(B_{ij}\right) + C_{ij}\right]} \tag{8}$$

$$y_{ij} = a\left(\frac{1}{2} - \frac{sinB_{ij}cosC_{ij}}{sin\left[180° - \left(B_{ij}\right) + C_{ij}\right]}\right) \quad at\, B_{ij} \leq 90° \tag{9}$$

$$y_{ij} = -a\left(\frac{1}{2} + \frac{sinB_{ij}cosC_{ij}}{sin\left[180° - \left(B_{ij}\right) + C_{ij}\right]}\right) \quad at\, B_{ij} \geq 90° \tag{10}$$

$$Z_{ij} = a \frac{sinB_{ij}sinC_{ij}sin\sum_{j=1}^{j}\beta_j}{sin\left[180° - \left(B_{ij} + C_{ij}\right)\right]} \tag{11}$$

Laser triangulation techniques have great precision, but limited range and depth variation. On the other side the opposite happens with time of flight techniques where they have low precision but great range and depth variation (França, Gazziro, Ide, & Saito, 2005).

The construction of this vision system consists of two main parts: the positioning laser and the scanning aperture. Usually both parts are built using aluminum for the structure, but in most recent works a scanning aperture was designed using PLA and a 3D printer (Real-Moreno et al., 2017). The positioning laser consists on a laser source, a 45° mirror and a dc or stepper motor to position the laser over the surface to scan. The scanning aperture is the most complex and important part of the vision system, it is composed by a dc or stepper motor, a 45° mirror, a zero sensor, which indicates when the mirror is in the position of 0°, a photoelectric sensor and a pair of lenses to focus the laser beam on the photoelectric sensor .

Line Based Techniques

Line based techniques perform measurements using a laser line across the surface, these techniques are faster than point based techniques but require more data processing and both techniques require a stable surface during the scan. These techniques can be applied in reverse engineering, mechanical design, artistic reconstruction, surface recognition, etc.

Line Laser Triangulation

Laser line triangulation works by the same principle as point based technique and the construction of it is almost the same, the difference between them is that in line laser triangulation the laser source is a laser line (Figure 7). Since this method works by the same principle, the equations used to calculate the distance are the same (4, 5, 6 and 7).

It is a fast technique with a broad range of applications in reverse engineering, simulation and mechanical design, it can be found in numerous applications and academic researches. Besides its engineering applications, this technique is also applied in artistic reconstruction and to maintain a record of real artistic objects. As in previous techniques, this also is applied to autonomous navigation for environment modeling and project recognition (Acosta, Garcia, & Aponte, 2006).

To obtain the 3D model from this vision system, six modules are used as shown in the block diagram from Figure 8. The camera is the module used for the input image acquisition, the laser source is composed

Figure 7. Line laser triangulation

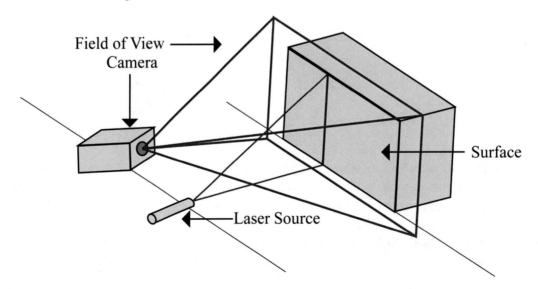

Figure 8. Line laser triangulation block diagram

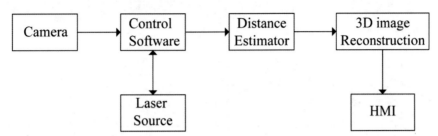

by a laser line generator, the motors and the encoders to position the laser over the surface, the control software receives the video frames and corrects the distortion caused by the intrinsic camera parameters, the distance estimator module uses explicit camera calibration methods for 3D coordinate mapping and implicit approach to distortion correction, also, it applies filters for noise elimination, edge detection and image digitization, the output of this module is a cloud point image, the 3D image reconstruction module uses this cloud point image as its input, then using image alignment and image fusion it reconstructs the main object, finally, the HMI (Human Machine Interface) module it's a graphic user interface that allows setting and modifying the parameters for the scan (França, Gazziro, Ide, & Saito, 2005).

Area Based Techniques

Area Based Techniques is a classic topic in computer vision. The goal is to obtain a three-dimensional description from one or more images, i.e. a recovery shape from a 2D image. The images in which the pixel value is a distance function of the corresponding point in the scene from the sensor are called range images. There are two commonly used principles to obtain the range image, radar and triangulation (Jain, Kasturi, & Schunk, 1995). The three-dimensional description can be expressed in different forms; Depth that considers the relative distance from the camera to a surface point or the surface height above the x-y plane. Surface Normal (*nx, ny, nz*), which is the orientation of a vector perpendicular to the tangent plane on the object surface. Surface gradient is the expression on percentage of change on the depth measures in the x and y directions and Surface slant is related to the surface normal as ($lsin\phi cos\theta$, $lsin\phi sin\theta$, $lcos\phi$), where l is the magnitude of the surface normal (Zhang, Tsai, Cryer, & Shah, 1999). In essence area based techniques obtain the data using a grid across the surface, these techniques usually requires multiple images of the surface to scan, resulting on high computational cost, which means that they are not usually in real time, unlike other Surface Measurement Techniques. Area based techniques present variations in their methodology to obtain a three-dimensional description, these variations are Shape from Shading, Depth from focus/defocus, Stereophotogrammetry and Phase Measurement Profilometry.

Shape From Shading

Shape from shading (SFS) determine the surface through a camera, under different illumination conditions, analyzing the reflectance of the illuminated surface and set surface from a height field. "Shape" has different meaning in the literature, but in shape from shading, shape is the area outline invariant under the scene illumination change from various pictures of that scene (Koenderink, van Doorn, Chalupa, & Werner, 2003). Shape from shading recovers the shape of an object from a gradual variation

Figure 9. Image plane coordinate system to camera coordinates

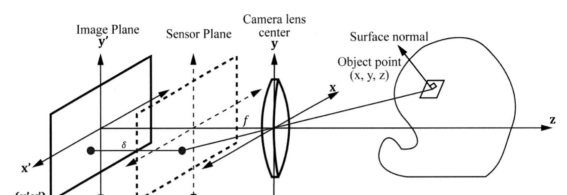

of shading in one or more images. The aim is to recover the light source and the surface shape at each pixel in the image, using a gray level image or Lambertian model, in which the gray rate at the image pixel depends on the light source direction and the surface normal (Richter & Roth, 2015). In order to be useful, the surface reflectance and scene illumination must be in the coordinate system of the image plane, this orientation must be formulated in camera coordinates, as the Figure 9 shows (Jain, Kasturi, & Schunk, 1995).

Figure 10 consider the surface as a Gaussian sphere aligned with the optical axis of the lens, also is regarded that the plane is tangent to the point (Jain, Kasturi, & Schunk, 1995). Therefore the surface normal of the image plane is also the corresponding point surface normal on the sphere.

Consider a point at position (*x, y, z*), the point is at a distance *z* from the camera lens plus the distance from the image plane to the camera lens center, and the parallel projection is used to map the point onto the image plane. To estimate the orientation of a surface patch in the image scene, consider a point nearby the image plane at position (*x* + δ*x*, *y* + δ*y*). The depth of the point is (*z* + δ*z*), (Jain, Kasturi, & Schunk, 1995) considerate the depth a function of image plane coordinate:

$$z = z\left(x, y\right)$$

(12)

Figure 10. Image plane coordinate system considering Gaussian sphere surface

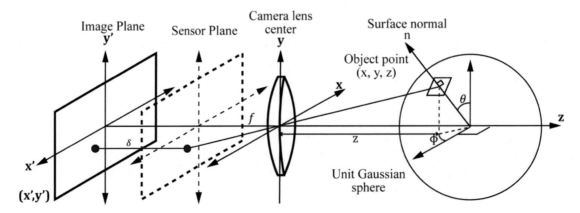

And the change in depth δz of the point relate to the function $z(x, y)$ about the point (x, y) is described by:

$$\delta z \approx \frac{\partial z}{\partial x}\delta x + \frac{\partial z}{\partial y}\delta y \tag{13}$$

The partial derivatives of z respect to x and y are related to the tangent plane inclination on the scene surface at the point (x, y, z). The gradient surface at (x, y, z) can be describe by the vector (p, q).

$$p = \frac{\partial z}{\partial x} \ \ and \ \ q = \frac{\partial z}{\partial y} \tag{14}$$

So the surface normal from the patch is related to the gradient by:

$$\boldsymbol{n} = \left(p, q, 1\right) \tag{15}$$

Understood as the amount of displacement in x and y corresponding to a unit change in depth z. Dividing the surface normal by its length is obtained the unit surface normal:

$$\hat{\boldsymbol{n}} = \frac{\boldsymbol{n}}{|\boldsymbol{n}|} = \frac{\left(p, q, 1\right)}{\sqrt{1 + p^2 + q^2}} \tag{16}$$

Any function or property of the surface can be referenced in terms of image plane coordinates; in particular, the surface orientation as function of x and y, since p and q have been developed to specify the orientation of any surface by two functions:

$$p = p\left(x, y\right) \ \ and \ \ q = q\left(x, y\right) \tag{17}$$

The pixel intensity as function of the surface orientation of the scene is captured in the reflectance map. The reflectance map is characterized by the combination of scene illumination, surface reflectance and the representation of surface orientation in viewer coordinates (Durou, Falcone, & Sagona, 2008). So, recovering the surface shape from the image intensity change is the relation between the image irradiance $E(x, y)$ and the orientation (p, q) of the surface point:

$$E\left(x, y\right) = R\left(p, q\right) \tag{18}$$

Where, $R(p, q)$ is the surface reflectance map; the goal is to recover the surface shape by calculating the surface orientation (p, q) of each point in the image (x, y). Also this technique can be divided into four groups by it approach to get shape: minimization approaches, propagation approaches, local approaches and linear approaches (Zhang, Tsai, Cryer, & Shah, 1999). Minimization approaches recover the surface gradient; since the surface point has two unknowns for the surface gradient and gray pixel value the system is undetermined. To solve this, two constraints are introduced: brightness constraint and smoothness constraint. The brightness constraint require that the shape reconstructed have the same brightness as the image at each surface point, while smoothness constraint guarantee a smooth surface reconstruction obtaining the shape information by minimizing an energy function (Zhang, Tsai, Cryer, & Shah, 1999). Propagation approaches is essentially a characteristic strip method. The characteristic strip is a line along the image, which the depth and surface orientation are computed if quantities are known at the starting point of the line. The construction begins around the neighborhood of a singular point using a spherical approximation. The direction of the characteristic strip is obtained as the direction of intensity gradients. So, spread the shape information from a set of surface points to the whole image. The local approach recover the shape information from the intensity and its first and second derivatives. Also is assumed that the surface is locally spherical at each point. Therefore local approaches derive shape based on the assumption of surface type. Linear approaches compute the solution based on the linearization of the reflectance map and solve for shape. The basic idea is a linear approximation of the reflectance function in terms of the surface gradient and a numerical method applied (Fourier transform or discrete approximation) to approach a solution for depth at each point (Durou, Falcone, & Sagona, 2008).

The basic principles of shape from shading are simple, nevertheless, there are difficulties in the practical implementation. In practice, the reflectance properties are not always known and the scene illumination is not controlled. Therefore to carry out the shape from shading is necessary to specify a variety of parameters, for example, all pictures should be from the same point of view and different images

Figure 11. Shape from shading

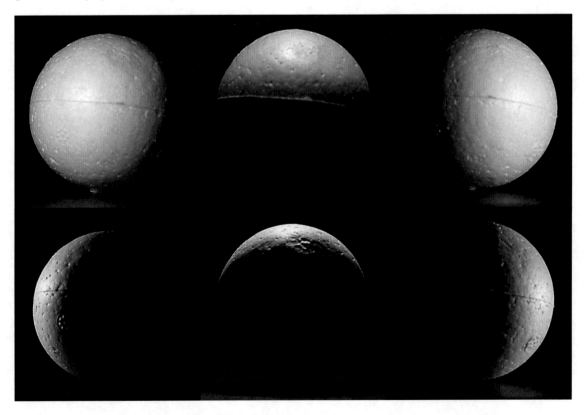

must be taken by the same camera i.e. the different parameters must remain the same during the photo shoot (Koenderink, van Doorn, Chalupa, & Werner, 2003). The shape from shading main advantage is the simplicity of the equipment (CCD camera and a light source) and is a well-established method (Zhang, Tsai, Cryer, & Shah, 1999), however, it has been demonstrated that some impossibly shaded images exist, which could not been shaded under the assumption of uniform reflectance properties and lighting, resulting on incorrect images solution.

Depth From Focus/Defocus

Depth information from focusing and defocusing have been long noticed as important sources of depth information for human and surface measurement systems. Depth from focus/defocus are techniques where the geometry of a surface is determined by analyzing the focal properties of the surface images. By moving the object or varying the focal properties of the camera, the focus of the image will increase, reach the high peak and fall off again. As with many techniques this one relies on the surface exhibiting a high level of detectable texture. Obtaining depth information by controlling camera parameters is an important task in surface measurement systems, due to its passive and monocular characteristics. Compared to stereo vision methods, focus/defocus methods does not have the correspondence and occlusion problems (Deng, Yang, Lin, & Tang, 2007), therefore these are valuable methods as an alternative to stereo vision for depth recovery. These 2 methods are depth from focus, which obtains the distance to the surface by taking various images with different focal length with a better focus from each one, and depth from defocus, where a small number of images is used under different lens parameters to determine depth at all points in the scene (Xiong & Shafer, 1993).

While other surface measurement systems obtain the 3D information of a surface using pinhole cameras, depth from focus/defocus use real aperture cameras. These cameras have a short depth of field, which makes the images to be focused only on a small 3D slice of the scene, as seen in Figure 12, where the images are shown with different depth of field.

Figure 12. Objects with different focal length

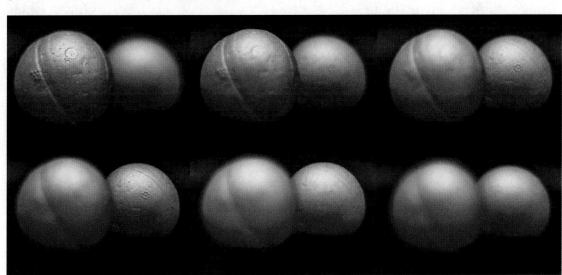

In Figure 13 can be observed the basic image formation geometry. The light rays radiated from the object P and intercepted by the lens are refracted to converge at point Q in the image plane. The relation between the object distance *o*, focal distance *f* and the image distance *i* is given by the Gaussian lens law shown in equation 19.

$$\frac{1}{o} + \frac{1}{i} = \frac{1}{f}$$

(19)

All points on the object plane are projected on the image plane, resulting on a clear or focused image *If(x, y)* to be formed on the image plane.

The goal of depth from focus/defocus is to calculate the focus measure at point (i, j) using a focus measure operator, a few focus measure operators have been proposed (Krotkov, 1988), in this case it is presented as the sum of modified Laplacian values in a "small" window around (i, j) that are greater than a threshold value as shown in equation (20).

$$F\left(i,j\right) = \sum_{x=i-N}^{i+N} \sum_{y=j-N}^{j+N} ML\left(x,y\right) \ for \ ML\left(x,y\right) \geq T_1$$

(20)

Where N is the parameter used to determine the window size used to calculate the focus measure and ML is the discrete approximation to the modified Laplacian (Nayar & Nakagawa, 1990).

In literature exists a large variety of approximation models to calculate the focus measurement. Also, it exists a number of real-time systems in depth from defocus. Depth from defocus proved to be effective for small distances and it has been compared to stereo vision, provided that the optical system and the scanned object are properly re-scaled.

Figure 13. Formation of images

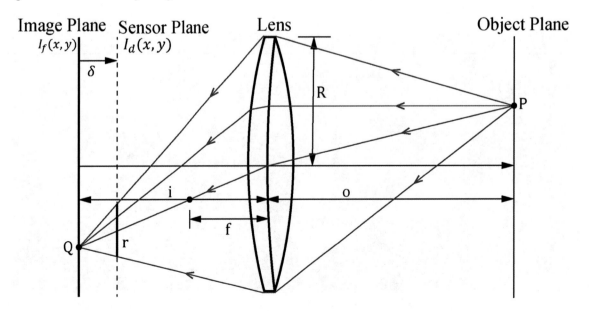

Stereophotogrammetry

The stereophotogrammetry or stereo vision works by the principles of triangulation, a point in an image corresponds to a line of points in a three-dimension space (Rodriguez-Quiñonez et al., 2017). By taking images in known position and intersecting their vectors, the point in three-dimension is found. This makes multiple images of the same scene taken from different position generating problems to find the correspondences between the different images.

The simplest model of stereophotogrammetry is two identical cameras separated only in the x direction by a baseline b, with their image planes coplanar from each other. A point over the surface to scan is viewed by two cameras at different positions on the image plane. The displacement between the locations of these two points in the image plane is called disparity. The plane passing through the camera centers and the surface point in the scene is called the epipolar plane. The epipolar line is defined by the intersection of the epipolar plane with the image plane. As it is shown in Figure 14, every point on the image plane will lie on the same row on the other image plane. Usually, the formulations of binocular stereo algorithms assume zero vertical disparity (Jain, Kasturi, & Schunk, 1995).

Considering the same epipolar plane or zero vertical disparity, Figure 15 show a scene object point *P*, which is observed at left camera lens point *pl* and the right camera lens point *pr* image planes. Assuming the coordinate system origin coincides with the lens center, the triangle from point *P*, *M* and *Cl* (*PMCl*) is similar to triangle *plLCl*, (Jain, Kasturi, & Schunk, 1995) described by:

$$\frac{x}{z} = \frac{x_l'}{f} \tag{21}$$

Figure 14. Stereophotogrammetry principle

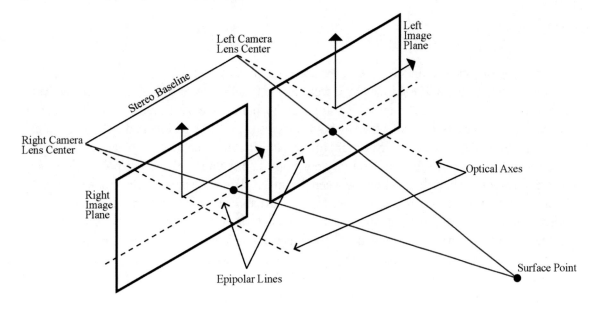

Figure 15. Stereophotogrammetry variables

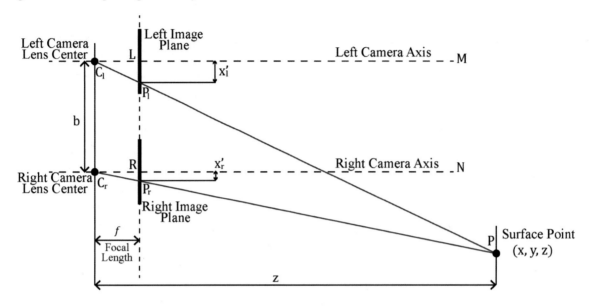

For *PNCr* and *prRCr*:

$$\frac{x - b}{z} = \frac{x_r^{'}}{f} \tag{22}$$

Combining (21) and (22) the distances *z* is:

$$z = \frac{bf}{\left(x_l^{'} - x_r^{'}\right)} \tag{23}$$

By knowing the image point disparities, the scene point's depth may be recovered. The depth accuracy for a given scene point is enhanced by increasing the baseline distance b, thus increasing the baseline distance, all scene points fraction that are seen by both cameras decreases (GÖRan, 1989). In consequence, even the region that is seen by both cameras is probably to appear different due to distortion introduced by the perspective projection, making it difficult to identify conjugate pairs.

Even when the two cameras are in arbitrary position and orientation, the image points correspond to a scene point along the intersection lines between the image planes and the epipolar plane (Jain, Kasturi, & Schunk, 1995). Some recent research works focus on dynamically controlling position and orientation parameters to facilitate the image analysis. On image analysis systems the process of dynamic control of camera parameters and its movements is known as active vision systems. The conjugate pair detection in stereo images, has been a challenging research issue known as the correspondence problem, that can be stated as: for each point in left image, find the right image corresponding point (Jain, Kasturi, & Schunk, 1995). To determine that points, it is necessary to measure the point similarity, due to the matching points that should be distinctly different from its surrounding pixels.

Phase Measurement Profilometry

Phase measurement profilometry is evolved from laser interferometry and it is one of the most used methods for optical 3D measurement (Zuo-chun, Hua-wei, & Yan-zhou, 2011), it is based on the principle of laser triangulation using a light pattern. This pattern can be a single light spot, a stripe or some complex light pattern as seen in Figure 16, used over surface to scan and observe it from an angle by a camera, then the depth information is extracted by the amount of deviation that the reflected light pattern.

Once the system is calibrated, the 3D coordinates of the surface to scan can be calculated by using the projector-to-world matrix *mwp* and camera-to-projector matrix *mwc*. Before the scan, the projector coordinates are known based on the pattern being projected. The camera coordinates can be obtained from the captured camera images. To calculate the 3D coordinates of the surface to scan, equation 24 is used (Yalla & Hassebrook, 2005).

$$P = \begin{bmatrix} X^W & Y^W & Z^W \end{bmatrix} = C^{-1} D \tag{24}$$

Where matrices C and D are:

$$C = \begin{bmatrix} m_{11}^{wc} - m_{31}^{wc} x^c & m_{22}^{wc} - m_{32}^{wc} x^c & m_{23}^{wc} - m_{33}^{wc} x^c \\ m_{21}^{wc} - m_{31}^{wc} y^c & m_{22}^{wc} - m_{32}^{wc} y^c & m_{23}^{wc} - m_{33}^{wc} y^c \\ m_{21}^{wp} - m_{31}^{wp} y^p & m_{22}^{wp} - m_{32}^{wp} y^p & m_{23}^{wp} - m_{33}^{wp} y^p \end{bmatrix} \tag{25}$$

$$D = \begin{bmatrix} m_{34}^{wc} x^c - m_{14}^{wc} \\ m_{34}^{wc} y^c - m_{24}^{wc} \\ m_{34}^{wp} y^p - m_{24}^{wp} \end{bmatrix} \tag{26}$$

Where *mwc* and *mwp* are given by the camera parameter matrix and the projector parameter matrix respectively.

Phase measurement profilometry's advantage is that while starting the surface scan, point to point operation can be used, this means that the phase value of a point is not affected by light intensity values of adjacent points. This technique can measure the entire surface of the object at a time, without additional one or two dimensional scanning.

SOLUTIONS AND RECOMMENDATIONS

Non-contact SMS perform meticulous processes, especially in practical implementations. The non-contact SMS has been used as a practical solution with different goals and purposes, according to its operation mode, working principles, advantages and disadvantages. Thereby, the following recommendations are used: to perform real time measurements of surface geometry without high texture information, the time of flight techniques fulfill the task goals. Task that requests great precision and real time measurements,

Figure 16. Phase measurement profileometry principle

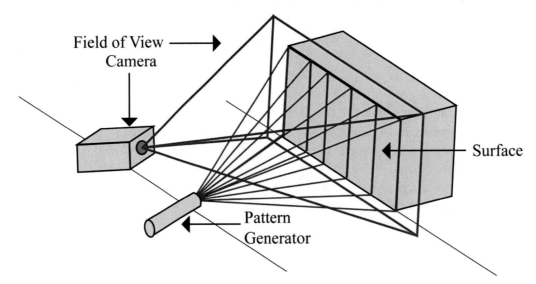

laser triangulation serves the purpose, however, the limited range and depth variation cause poor surface geometry measurements for large objects. Area based techniques perform correctly the detection of high texture information, shade edges, outlines areas and accurate surface geometry measurements, however, they require multiple images resulting on high computational cost and non-real time measurements. To select the proper technique it is necessary to considerate the variables of the environment where the system will be required in order to have the best performance.

From previously mentioned techniques, can be identified several strengths and disadvantages, however, they can be integrated together into a system to compensate the disadvantages of a technique with the advantages of another technique and get the most of each technique, resulting in an increased accuracy measure and a larger field of view. As in Moghadam, Sardha, and Feng, (2008) in Nanyang Technological University from Singapore, where they present a path planning and mapping for autonomous navigation systems using a stereo vision to provide 3D structural data of complex objects and a 2D LIDAR to generate a separate cost map. Then a grid-based occupancy map approach is used to fuse the complementary information provided by the 2D LIDAR and the stereo vision. Also in the work of Li, Ruichek, and Capelle (2013) they present a novel extrinsic calibration algorithm between a binocular stereo vision system and a 2D LIDAR proposing a method that solves the problem based on 3D reconstruction of the chessboard and geometric constraints between views from the stereo vision and the LIDAR. On the other hand, the electronics industry has developed numerous commercial applications like ©Pix4D where a non-contact surface measurement system is implemented to create a professional mapping tool using UAVs. Google also has incorporated this measurement techniques to their autonomous vehicle prototype called WAYMO, where the vehicle has sensors and software that are designed to detect pedestrians, cyclist, vehicles, road work, etc. Moreover, Tesla Company, have already released an autonomous vehicle with full self-driving hardware composed by eight surround cameras, twelve ultrasonic sensors and a forward-facing radar. DJI Technology Company recently released a new UAV product called spark, which uses deep learning gesture recognition to control spark's movement and collision avoidance, which incorporates a 3D sensing system, IMU, camera, vision positioning system, 24 computing cores and gps/glonass.

FUTURE RESEARCH DIRECTIONS

Point Based Techniques

Time of Flight

Pulse Modulation Method

Even when the accuracy of this method has been proved, it still can be improved, like in Jahromi, Jansson, and Kostamovaara (2015), where they present a pulsed time of flight with a 2D SPAD-TDC receiver. This receiver is a single CMOS chip consisting of a 9x9 CMOS SPAD array and a 10-channel 10ps precision TDC2 and they have low timing jitter, photon detection probability of 2-35% (depending on the wavelength) and does not require analog signal processing (Kostamovaara et al., 2015). The main contribution of the work mentioned before is the high integration level of the receiver, the demonstration of the use of a large area detector array. With this configuration they have been able to achieve a single shot precision of ~170ps and a measurement range of tens of meters with non-cooperative targets have been achieved with an 18mm receiver aperture.

Amplitude Modulation Method

From the time of flight methods the most commonly used is the amplitude modulation method. In Zhou et al., (2017) the researchers present a real-time field programmable gate arrays platform that calculates the phase shift for a time of flight camera that works by amplitude modulation with low resource and power consumption that makes it suitable for embedded applications. Their experiments have shown that their platform can acquire ranging images at a frame rate of 131 frames per second with a range error of 5.1mm at 1.2m of distance.

Laser Triangulation

Static Laser Triangulation

Static laser triangulation technique has great precision, but limited range and depth variation in comparison with dynamic laser triangulation and time of flight techniques. Also, this technique works by the same principle as line laser triangulation and the construction of it is similar, the difference is in the laser source, where the point laser is substituted by a line laser, making the scanning process faster, which is the reason that researchers are actually working with line laser triangulation instead of static laser triangulation.

Dynamic Laser Triangulation

Dynamic laser triangulation has various applications, one of them is in the medical area, like in Rodriguez-Quiñonez et al., (2014), where a 3D medical scanner that works by the dynamic triangulation principle is presented. In this paper the method of operation, medical applications, orthopedically diseases and the most common types of skin to employ the system in proper way are analyzed. Also, three mathematical methods to eliminate errors in natural experimental data sets where used: Polak-Ribiere, Quasi-Newton and Levenberg-Marquardt. They were used to filter strong errors in the results of 3D-coordinates measurement. The three methods were tested and since Levenberg-Marquardt method shown a better performance it was applied to resolution improvement in their rotational 3D body shape scanner.

Line Based Techniques

Line Laser Triangulation

In Fu, Corradi, Menciassi, and Dario, (2011) the researchers present a miniaturized line laser triangulation scanner that was designed for use in a miniature mobile robot of $10 \times 10 \times 10\, cm^3$, the purpose of their work was to demonstrate the possibility of extracting basic information from the robot surroundings using small, simple, low-power, and low-cost demanding devices. The design was made for indoor applications and they reached a measurement range of 80 mm, with 1mm resolution, up to 600 mm, with 12 mm resolution.

Area Based Techniques

Shape From Shading

In the work Tao et al., (2017), the researchers present a principled algorithm for dense depth estimation, which combines defocus and correspondence metrics. Also, their analysis is extended to the additional shading information, to refine fine details of the surface. Their work shows that combining depth estimation algorithms in multiple scenarios is outperformed, compared to previous state-of-the-art light-field algorithms. For the shape from shading part, they focus on general scenes and unknown lighting, without requiring geometry or light priors. The authors propose and provide quantitative validation for a new shape estimation framework that use just single-capture passive light-field image. Their shape estimation algorithm is suitable for passive point-and-shoot acquisitions from consumer light-field cameras.

Depth From Focus/Defocus

Depth from focus and defocus techniques proved to be an effective surface measurement for small distances on laboratory scenes. In Suwajanakorn, Hernandez, and Seitz, (2015) the first depth from focus method capable of handling images from mobile phones and other hand-held cameras is introduced. Also, they present a novel uncalibrated depth from focus problem and propose a new focal stack aligning algorithm to account for scene parallax. The approach of the authors demonstrate high quality results on a range of challenging cases.

Stereophotogrammetry

Stereophotogrammetry or stereo vision has many applications, one of them in the autonomous navigation area, like in Song et al. (2017), where a real-time and robust line detection and forward collision warning technique based on stereo cameras is presented. This technique obtains the obstacles image through stereo matching and UV-disparity segmentation algorithm, then a top view of the road is generated by fusing the obstacles image and the original image. Then, the extreme points are calculated as the detected lanes according to the traffic lanes model. The host lane is selected among all the detected lanes and the nearest obstacle in this host lane is detected by the forward collision warning. Their results prove that their real-time robust line detection and forward collision warning technique can work efficiently in autonomous driving or driver assistant system.

Phase Measurement Profileometry

In the work Liu and Li (2014), the researchers propose a whole-field 3D shape measurement algorithm based on phase-shifting projected fringe profilometry. This algorithm casts the 3D reconstruction problem into a phase-to-point mapping task. A rigorous phase-to-height mathematical model is derived by considering the practical case of tilt angle in the fringe projection process. Their model achieves higher accuracy compared to conventional stereo vision models that require camera and projector calibration, this is because the projector in the system is not required to be calibrated. The practical experimental results performed by the authors achieved reasonably accuracy.

CONCLUSION

The precision of surface geometry measurements for three-dimensional computational models creation is an important task in industry and research for the creation of computer-aided design (CAD); therefore non-contact surface measurement systems gives a solution to the complex problem of performing accurate surface geometry measurements, this techniques can be automated, avoid direct contact whit the object and have a wide workspace range in comparison with contact techniques. This advantages turn non-contact surface measurement techniques into a relevant topic in research and engineering areas, where the performance of these techniques is continuously updated and improved.

This chapter provided useful information about the different types of non-contact surface measurement techniques, theory, basic equations, system implementation, actual research topics, applications and future trends. Also, introduces some recommendations for practical implementations of non-contact surface measurement systems and examples of solutions for general cases with its advantages and disadvantages. Non-contact surface measurements is a topic in continuous research by different investigation areas and industrial processes, turning it into a high-search topic nowadays; making this chapter useful for students, teachers, researchers and someone that wants to implement a vision system and need an introduction to all the options available and use the most convenient for their purpose, due to the technological and surface recognition progress.

REFERENCES

Acosta, D., Garcia, O., & Aponte, J. (2006). Laser Triangulation for shape acquisition in a 3D Scanner Plus Scan. In *Electronics, Robotics and Automotive Mechanics Conference* (pp. 14-19). IEEE.

Ahola, R., & Millylä, R. (1986). A Time-of-Flight Laser Receiver for Moving Objects. *IEEE Transactions on Instrumentation and Measurement, IM-35*(2), 216–221. doi:10.1109/TIM.1986.6499093

Basaca, L., Rodruiguez, J., Sergiyenko, O. Y., Tyrsa, V. V., Hernadez, W., Nieto-Hipolito, J. I., & Starostenko, O. (2010). Resolution improvement of Dynamic Triangulation method for 3D Vision System in Robot Navigation task. In *IECON 2010-36th Annual Conference on IEEE Industrial Electronics Society* (págs. 2886-2891). IEEE.

Bradshaw, G. (1999). *Non-Contact Surface Geometry Measurements Techniques*. Dublin: Trinity College Dublin, Department of Computer Science.

Cuevas-Jimenez, E. V. (2006). *Intelligent Robotic Vision* (Doctoral dissertation). Freie Universität Berlin.

Deng, Y., Yang, Q., Lin, X., & Tang, X. (2007). Stereo Correspondence with Occlusion Handling in a Symmetric Patch-Based Graph-Cuts Model. *IEEE Transactions on Pattern Analysis and Machine Intelligence, 29*(6), 1068–1079. doi:10.1109/TPAMI.2007.1043 PMID:17431303

Durou, J. D., Falcone, M., & Sagona, M. (2008). Numerical methods for shape-from-shading: A new survey with benchmarks. *Computer Vision and Image Understanding, 109*(1), 22–43. doi:10.1016/j.cviu.2007.09.003

França, J. G., Gazziro, M. A., Ide, A. N., & Saito, J. H. (2005). A 3D scanning system based on laser triangulation and variable field of view. In *Image Processing, 2005. ICPI 2005. IEEE International Conference on* (pp. I-425). IEEE.

Fu, G., Corradi, P., Menciassi, A., & Dario, P. (2011). An Integrated Triangulation Laser Scanner for Obstacle Detection of Miniature Mobile Robots in Indoor Environment. *IEEE/ASME Transactions on Mechatronics, 16*(4), 778–783. doi:10.1109/TMECH.2010.2084582

Göran, S. (1989). Roentgen stereophotogrammetry: A method for the study of the kinematics of the skeletal system. *Acta Orthopaedica Scandinavica*, 1–51. PMID:2686344

Hulme, K. F., Collins, B. S., Constant, G. D., & Pinson, J. T. (1981). A CO_2 laser rangefinder using heterodyne detection and chirp pulse compression. *Optical and Quantum Electronics, 13*(1), 35–45. doi:10.1007/BF00620028

Jahromi, S., Jansson, J. P., & Kostamovaara, J. (2015). Pulsed TOF Laser Rangefinding with a 2D SPAD-TDC Receiver. In SENSORS (pp. 1-4). IEEE.

Jain, R., Kasturi, R., & Schunk, B. G. (1995). *Machine Vision*. New York: McGraw-Hill.

Kilpela, A. (2002). *Pulsed time-of-flight laser range finder techniques for fast high-precision measurement applications*. Academic Press.

Koenderink, J. J., van Doorn, A. J., Chalupa, L. M., & Werner, J. S. (2003). Shape and Shading. *Visual Neuroscience*, 1090–1105.

Kostamovaara, J., Huikari, J., Hallman, L., Nissinen, L., Nissinen, J., Rapakko, H., ... Ryvkin, B. (2015). On Laser Ranging Based on High-Speed/Energy Laser Diode Pulses and Single-Photon Detection Techniques. *IEEE Photonics Journal*, 7(2), 1–15. doi:10.1109/JPHOT.2015.2402129

Krotkov, E. (1988). Focusing. *International Journal of Computer Vision*, 223-237.

Li, Y., Ruichek, Y., & Capelle, C. (2013). Optimal Extrinsic Calibration Between a Stereoscopic System and a LIDAR. *IEEE Transactions on Instrumentation and Measurement*, 62(8), 2258–2269. doi:10.1109/TIM.2013.2258241

Liu, J. Y., & Li, Y. F. (2014). A Whole-Field 3D Shape Measurement Algorithm Based on Phase-Shifting Projected Fringe Profilometry. In *Information and Automation (ICIA), 2014 IEEE International Conference* (pp. 495-500). IEEE.

Markus-Christian, A., Thierry Bosch, M. L., Risto, M., & Marc, R. (2001). Laser ranging: A critical review of usual techniques for distance measurement. *Optical Engineering (Redondo Beach, Calif.)*, 10–19.

Miller, J. M. (1920). Dependence of the input impedance of a three-electrode vacuum tube upon the load in the plate circuit. *Scientific Papers of the Bureau of Standards*, 367-385.

Moghadam, P., Sardha, W., & Feng, D. J. (2008). Improving Path Planning and Mapping Based on Stereo Vision and Lidar. In *Control, Automation, Robotics and Vision, 2008. ICARV 2008, 10th International Conference* (pp. 384-389). IEEE.

Nayar, S. K., & Nakagawa, Y. (1990). Shape from Focus: An Effective Approach for Rough Surfaces. In *IEEE International Conference* (pp. 218-225). IEEE. 10.1109/ROBOT.1990.125976

Palojärvi, P., Määttä, K., & Kostamovaara, J. (1997). Integrated Time-of-Flight Laser Radar. *IEEE Transactions on Instrumentation and Measurement*, 969–999.

Payne, J. (1973). An Optical Distance Measuring Instrument. *The Review of Scientific Instruments*, 44(3), 304–306. doi:10.1063/1.1686113

Real-Moreno, O., Rodriguez-Quiñonez, J. C., Sergiyenko, O., Basaca-Preciado, L. C., Hernandez-Balbuena, D., Rivas-Lopez, M., & Flores-Fuentes, W. (2017). Accuracy Improvement in 3D Laser Scanner Based on Dynamic Triangulation for Autonomous Navigation System. In *Industrial Electronics (ISIE), 2017 IEEE 26th International Symposium on* (pp. 1602-1608). IEEE.

Richter, S. R., & Roth, S. (2015). Discriminative Shape from Shading in Uncalibrated Illumination. In *Proceedings of the IEEE Conference on Computer Vision and Pattern Recognition* (pp. 1128-1136). IEEE. 10.1109/CVPR.2015.7298716

Rivas-Lopez, M., Sergiyenko, O., & Tyrsa, V. (2008). Machine Vision: Approaches and Limitations. In Computer Vision. InTech.

Rodriguez-Quiñonez, J. C., Sergiyenko, O., Flores-Fuentes, W., Rivas-Lopez, M., Hernandez-Balbuena, D., Rascon, R., & Mercorelli, P. (2017). Improve a 3D distance measurement accuracy in stereo vision systems using optimization methods' approach. *Opto-Electronics Review*, *25*(1), 24–32. doi:10.1016/j.opelre.2017.03.001

Rodriguez-Quiñonez, J. C., Sergiyenko, O. Y., Basaca-Preciado, L. C., Tyrsa, V. V., Gurko, A. G., Podrygalo, M. A., . . . Hernandez-Balbuena, D. (2014). Optical monitoring of scoliosis by 3D medical laser scanner. *Optics and Laser in Engineering*, 175-186.

Sergiyenko, O., Tyrsa, V., Hernandez-Balbuena, D., Lopez, M., Lopez, I., & Cruz, L. (2008). Precise Optical Scanning for practical multi-applications. In *Industrial Electronics, 2008. IECON 2008. 34th Annual Conference of IEEE* (pp. 1656-1661). IEEE.

Song, W., Fu, M., Yang, Y., Wang, M., Wang, X., & Kornhauser, A. (2017). Real-Time Lane Detection and Forward Collision Warning System Based on Stereo Vision. In *Intelligent Vehicles Symposium (IV), 2017 IEEE* (pp. 493-498). IEEE. 10.1109/IVS.2017.7995766

Suwajanakorn, S., Hernandez, C., & Seitz, S. M. (2015). Depth from Focus with Your Mobile. In *IEEE Conference on Computer Vision and Pattern Recognition* (pp. 3497-3506). IEEE.

Tao, M. W., Srinivasan, P. P., Hadap, S., Rusinkiewicz, S., Malik, J., & Ramamoorthi, R. (2017). Shape Estimation from Shading, Defocus, and Correspondance Using Light-Field Angular Coherence. *IEEE Transactions on Pattern Analysis and Machine Intelligence*, *39*(3), 546–560. doi:10.1109/TPAMI.2016.2554121 PMID:27101598

Xiong, Y., & Shafer, S. A. (1993). Depth from focusing and defocusing. In *IEEE Computer Society Conference* (pp. 68-73). IEEE.

Yalla, V. G., & Hassebrook, L. G. (2005). Very High Resolution 3-D Surface Scanning Using Multi-frequency Phase Measuring Profilometry. *SPIE*, 44-53.

Zhang, R., Tsai, P.-S., Cryer, J. E., & Shah, M. (1999). Shape from Shading: A Survey. *IEEE Transactions on Pattern Analysis and Machine Intelligence*, *21*(8), 690–706. doi:10.1109/34.784284

Zhou, W., Lyu, C., Jiang, X., Zhou, W., Li, P., Chen, H., ... Liu, Y.-H. (2017). Efficient and Fast Implementation of Embedded Time-of-Flight RAnging System Based on FPGAs. *IEEE Sensors Journal*, *17*(18), 5862–5870. doi:10.1109/JSEN.2017.2728724

Zuo-chun, S., Hua-wei, L., & Yan-zhou, Z. (2011). Calibration of Measurement System Based on Phase Measurement Profilometry. In *Optoelectronics and Microelectronics Technology (AISOMT), 2011 Academic International Symposium* (pp. 200-203). IEEE.

Section 2
Machine Vision for Modeling the Human Approach

With the beginning of the deep learning era, the rate of advancement in machine vision for modeling the human approach has been astonishingly increased. Classical computer vision techniques have been enhanced and some of them replaced by machine learning techniques. In this section, Chapter 5 makes use of these techniques to perform a localization and identification task over a set of EEG Electrodes images. Chapter 6 describes a 3D vision system based on logic programming for analysis, recognition, and understanding of human anomalous behavior detection. Chapter 7 describes the design of a system based on classical computer vision techniques enhanced with artificial intelligence algorithms for moving object classification under illumination changes. Chapter 8 resumes the machine vision trends and its application on science and industry.

Chapter 5
Leveraging Models of Human Reasoning to Identify EEG Electrodes in Images With Neural Networks

Alark Joshi
University of San Francisco, USA

Phan Luu
Philips Neuro, USA

Don M. Tucker
Philips Neuro, USA

Steven Duane Shofner
Philips Neuro, USA

ABSTRACT

Humans have very little trouble recognizing discrete objects within a scene, but performing the same tasks using classical computer vision techniques can be counterintuitive. Humans, equipped with a visual cortex, perform much of this work below the level of consciousness, and by the time a human is conscious of a visual stimulus, the signal has already been processed by lower order brain regions and segmented into semantic regions. Convolutional neural networks are modeled loosely on the structure of the human visual cortex and when trained with data produced by human actors are capable of emulating its performance. By black-boxing the low-level image analysis tasks in this way, the authors model solutions to problems in terms of the workflows of expert human operators, leveraging both the work performed pre-consciously and the higher-order algorithmic solutions employed to solve problems.

DOI: 10.4018/978-1-5225-5751-7.ch005

INTRODUCTION

Until very recently, approaches to computer version have been dominated by the application of what the authors refer to as classical computer vision techniques. These include morphological operations, thresholding, feature extractors, Hough transforms, and edge detectors among many others. As researchers and practitioners have gained experience and insights into the problem, slow and steady progress has been realized, but with the advent of the era of deep learning the rate of advancement has been astonishing.

In "The Unreasonable Effectiveness of Data", the authors argue that "we should stop acting as if our goal is to author extremely elegant theories, and instead embrace complexity and make use of the best ally we have: the unreasonable effectiveness of data." (Halevy, Norvig, & Pereira, 2015). Image processing is to shape analysis and feature extraction what natural languages are to context-free grammars, and defining a theory or formula to capture either represents an intractable problem. At the same time, humans are capable of performing these tasks to a reasonable degree of accuracy with little effort. What Halevy, Norvig, and Pereira (2015) suggest is that within certain domains no elegant theories can reasonably model the complexity of the problem space, and that data along with machine learning has proven to be an effective means of doing this.

Training a deep convolutional neural network with human labeled data is essentially designing an algorithm capable of emulating the performance of the human labeler(s). In this chapter, we outline an approach that combines this emulation derived from the labeled data with the algorithmic approach the same users apply in performing a localization and identification task over a set of images.

BACKGROUND

Philips Neuro manufactures complete dense array electroencephalography (EEG) systems. EEG, as a technology, has existed since 1875 (Swartz, 1998) and is much older than many other technologies used for non-invasive neural imaging, such as magnetic resonance imaging (MRI) or functional magnetic resonance imaging (fMRI), but offers some unique advantages such as high temporal resolution and movement tolerance.

Epileptic seizures result from excessive synchronization in the firing of neurons in the brain (often starting in a small highly localized region). The high temporal resolution (often 1,000 samples per second or more) makes EEG ideal for the diagnosis of epilepsy, as the associated spikes are most easily characterized by the millisecond resolution enabled by EEG. In some cases, epileptic seizures are not controlled by pharmacological means and a determination is made that a brain resection should be performed to remove the tissue that serves as the source of the problem. When resectioning is deemed appropriate source localization of the recovered electrical signals can assist in identifying the region of the brain responsible. In order to perform source localization using the recovered electrical signals, the sensor locations must be registered to the surface of a model of the head.

There are a number of other situations where the ability to localize recovered signals is highly desirable. There is some evidence that neuronal activation can be depressed by low frequency pulsed direct current stimulation (Groppa et al., 2010). This may offer a non-invasive means for the treatment of neurological disorders like Epilepsy or Parkinson's. For these disorders, it may be possible to use source localization of signals in EEG to identify the region of origin of the unwanted activity and then, given the reciprocity principle (Fernández-Corazza, Turovets, Luu, Anderson, & Tucker, 2016) (which holds that electrical current will follow the same pathways when injected into the scalp as it follows to reach the scalp from a specific brain region), target the same region with transcranial injected current to modulate the activation or plasticity of the region.

In both cases, an accurate registration of the sensors to the scalp surface is needed. Several means of achieving this exist with their own advantages and disadvantages.

RF systems can derive 3D position by measuring an RF signal emission at a number of receivers. These systems require a technician measure the placement of sensors while the subject remains completely still and is sensitive to the effects caused by other materials in the vicinity of the receivers. Laser scanners can derive very precise models quickly, but pose a risk to eyes and produce a dense mesh that would need to be further analyzed to derive sensor positions. Photogrammetry is a technique that allows the subject to be imaged quickly, poses no risk to the eyes, and produces a permanent record of net application, but requires advanced image analysis to identify and localize the sensors.

Despite the challenges, the advantages of photogrammetry for this purpose lead to the development of the Geodesic Photogrammetry System (GPS): 11 cameras arranged in a geodesic dome structure are triggered simultaneously and the captured images are analyzed offline to derive the 3D positions of the sensors placed on the head of the patient. Though classical computer vision techniques have been applied to identifying these sensors, significant user intervention is still required to adjust placement and identification of the sensors.

Because of the permanence of this record it is also possible the extract a significant amount of data that can be used to train a neural network to perform the tasks and achieve a higher level of automation.

Neural Networks

The class of machine learning algorithms referred to as deep neural networks (DNNs) is a subset of artificial neural networks (ANNs) with many hidden layers between the input and output. ANNs are modeled on biological neurological systems where neurons form networks such that the neurons have many to many relationships (the outputs of many neurons connect to the input of a neuron and its outputs are connected to the inputs of many more neurons). The neuron possesses an activation function that takes sum of its inputs and produces an output. The inputs are weighted such that some inputs are strongly correlated to the output and others are weakly correlated or not correlated at all.

Artificial neural networks generally organize the neurons into layers where the outputs of the neurons on one layer feed into the inputs of the neurons on the next layer. Training an ANN is the process of adjusting the weights and biases of the neuronal inputs by providing the network with training data and comparing the output to ground truth until the desired output is achieved.

A well-known result found training even a simple 2 layer, 3 node network to be NP-Complete (Blum & Rivest, 1989). This being so, how is it that the training of deep neural networks can be performed so efficiently? In the paper "On the Computational Efficiency of Training Neural Networks" the authors asked and attempted to answer this exact question finding a few factors that contribute to the efficiency of training modern neural networks: using an activation function that facilitates optimization using one of the gradient descent optimization techniques, over-specification of the networks, and regularization (Livni, Shalev-Shwartz, & Shamir, 2014).

Layers

In an ANN, a layer is a collection of neurons that all receive input from the previous layer and all provide output to the next layer. Many kinds of layers exist, some particularly well-suited to specific tasks, like convolutional layers with regards to image processing.

A layer can be configured with an activation function, allowing it to learn to operate on the data flowing through the network by a process of training, or may implement a normalization or regularization function. There are a number of different activation functions, with different characteristics including threshold, sigmoid, tanh, and linear activation functions, the rectified linear unit (ReLU), and many more. The threshold activation function does not produce results that are conducive to optimization by means of stochastic gradient descent (Livni, Shalev-Shwartz, & Shamir, 2014) and is rarely if ever used, while sigmoid and ReLU activation functions are very common in modern architectures.

A convolutional layer, for example, is a layer modeled on the visual cortex and features local receptive fields that take as input submatrices of the previous layers output in order to maintain spatial relationships. Additionally, all of the receptive fields share weights and biases so that the learned image feature is location invariant. As the name suggests, a convolutional layer essentially learns a convolutional kernel that can be applied to an image to recognize a particular type of feature.

Some layers, while still operating on the incoming data, do not possess an activation function or weights that can be adjusted and do not learn from the information that passes through them. An example of this is a max-pooling layer. A max-pooling layer implements something similar to a local receptive field (a component of convolutional layers) that takes inputs from the previous layer and produces an output equal to the maximum value its local receptive field sees. These are commonly used along with convolutional layers and reduce the dimensionality of the output.

Convolutional Neural Networks

Convolutional Neural Networks (CNNs) are a kind of ANN composed of convolutional and, commonly, pooling and fully connected layers. In general, the architecture of a CNN follows some variation of the following: an input layer feeds into some number of convolutional layers that are paired with pooling layers and which feed into a fully connected layer with a number of outputs equal to the number of classes the network is to be trained to identify. Much of the advancement in the state of the art in image analysis in recent years has been built on the back of deep CNNs and many variations on the basic architecture have been tested.

The ImageNet Large Scale Visual Recognition Challenge (ILSVRC) is a competition where research teams develop applications that classify objects and scenes in the ImageNet visual database. The first couple years of this competition the winning teams utilized a traditional approach and achieved ~25% error rate in image classification. In 2012, the winning team utilized a deep convolutional neural network (CNN) and reduced the top 5 classification error to ~16%. Since then, the competition has been dominated by CNNs and the error rates have dropped to a few percentage points – even exceeding human performance.

Currently, Google's Inception v4 network has achieved a 3.07% error rate on the ILSVRC dataset by integrating a number of recent advancements that allow very deep networks to be effectively trained. By some accounts this result is better than human classification which has been reported to be approximately 5.1% (Russakovsky et al., 2015).

BOOTSTRAPPING AN AUTOMATED IMAGE PROCESSING PIPELINE

The advancements in the techniques used to build and train deep neural networks have made it clear that, given sufficient training data, many tasks that seemed impractical only a few years ago are now clearly within the realm of possibility. Furthermore, there are a number of open source libraries and tools that have turned the development and training of deep neural networks into an approachable field, and make experimentation easy.

Possibly the most challenging part of the process is the acquisition of labeled data suitable for training these networks. Tasks that stand to benefit most from the promise of the automation promised by deep learning are often tedious or challenging to perform (this is the very reason the promise of automation is so attractive), thus acquiring sufficient data to use for training presents difficulties.

Once data is collected, a number of further steps and considerations must be taken: How should the training data be presented to the learning algorithm? What kind of network is suitable for the task? How deep does the network need to be? How many image features should be recognized at each level of abstraction (activation layer)?

Often questions about the number of image features to be identified in each layer, which will affect the design of the network, can be difficult to answer without empirical data, so network design can be an iterative process. Because of the need to train, iterate, and retrain it is important to have access to capable hardware. What might take days on a CPU (as training frequently does), can be performed in minutes or hours on a modern GPU with sufficient memory.

Deriving an Algorithm

Merriam-Webster defines an algorithm as "a step-by-step procedure for solving a problem or accomplishing some end especially by a computer". By abstracting the underlying complexities related to low-level computer vision it is possible to focus on modeling high-level conscious problem solving, and to derive an algorithm by modeling it.

Collecting and Structuring the Training Data

Because significant training data can be required to train a competent neural network, it is often impossible or difficult to automate a task using supervised learning until significant work has been performed to generate a corpus of labeled data. Early wins with deep learning owe their success to a large collection of data produced by hand, such as machine translation where U.N. documents which were produced in many languages were among the sources of training data (Halevy, Norvig, & Pereira, 2009). The core takeaway is that it is possible to begin with a manual task, and if the inputs to that task and the outputs of the task are preserved and can be correlated, it should be possible to extract the necessary information to train a suitably architected network to emulate, to some degree, the work performed by the humans, thus bootstrapping a deep learning approach to the problem.

Designing the Network

In general, deeper networks are more powerful than shallower networks, but costlier in terms of memory and computation. If a network can be executed on a GPU rather than a CPU, the run-time can be made orders of magnitude shorter, but will then be constrained by the, usually, smaller pool of memory.

When designing a deep neural network, it is important to consider the problem of exploding or vanishing gradients which can result in difficulties with training. Further, though the performance of any n layer network can theoretically be shown to be at least equal to the performance of an n-1 layer network by configuring the additional layer to simply behave as a pass-through of the preceding layer, in practice performance tends to increase to a certain network depth after which adding more layers degrades it. This is not due to overfitting, but the inability of multiple non-linear layers to learn an identity (pass-through) mapping (He, Zhang, Ren, & Sun, 2015). Residual networks are a recent advancement that have been shown to improve the trainability of deep networks by making it easier to optimize blocks consisting of a few stacked layers to the identity mapping.

Determining general aspects of an appropriate network architecture for each task can be accomplished by consideration of the inputs and the outputs. As a rule, convolutional layers work well for image analysis or other tasks with data encoding spatial relationships. Recurrent layers work well for sequential data, so while a convolutional network would be a good choice to perform analysis on the individual frames of a video, a recurrent network would be a good choice for performing analysis on the change over time occurring in the entire sequence. For low dimensional data without spatial or sequential relationships, a fully connected network might be a good choice. However, because of the density of connections (the output of every neuron in layer n-1 connects to the input of every neuron on layer n), the size of the input data should be minimized to avoid networks that require more computational resources than are available. Furthermore, a simple dense network has no effective means of representing or preserving spatial or sequential information in the input data, and so is not the ideal choice for these tasks.

Many networks will be a combination of these different types of layers along with other layers which may or may not implement an activation function (meaning that they simply perform some arithmetical operation and are not capable of learning). The output layer, for instance, of many convolutional neural networks is a fully connected layer.

Training the Network

Once the data has been organized and the network has been designed, there are a number of variables to consider when deciding how to train the network.

On a macro scale, training consists of two stages: training and validation. During training, the optimizer updates the weights and biases of the neurons in order to minimize the error between the prediction and the ground truth (as described by the labels). During validation, this error is calculated for the validation data but the weights and biases are not updated. The validation step is intended to minimize the degree to which the data is being overfitted. Because validation data is never presented to the network while it is being trained, it cannot anticipate what form it will take and if overfitting occurs with the training data, the error computed for the validation data should move inversely to the error during training.

For this reason, the stability of the selection of training versus validation data is essential. So, whatever means is used to segment the training data from the validation data it must perform the segmentation the same way each time. This should be balanced by the need to ensure that biases inherent in the ordering of the data are minimized (in the case of the application described in this chapter, the order of the extraction of the patches was driven by the numerical identity of the sensors ensuring that these relationships were encoded in the extracted images, for example). A simple solution is to select the inputs for each category using a random function, but specify an unchanging seed for the function.

Objective Function

The objective functions used during training determines the nature of the error calculation. A categorical cross-entropy objective function, for instance, measures the error in classification for outputs with mutually exclusive classes, whereas a binary cross-entropy objective function measures the classification error of non-mutually exclusive classes. If the task is to identify objects in an image that were members of a single class (either cats or dogs, for example), categorical cross-entropy would produce the desired results, whereas attempting to identify an object that might belong to several classes would suggest binary cross-entropy (cats, dogs, snakes, reptiles, or mammals, where cats and dogs are distinct classes, but also belong in the mammal class).

Optimization Function

The optimization function determines how the training will proceed. Most optimization functions implement some form of gradient descent where the gradient of the objective function's domain is estimated in order to compute the update values that are then backpropagated through the network. Backpropagation is the process by which the contribution to the total error from each neuron is estimated and then used to update those neurons (Hecht-Nielsen, 1988). Plain gradient descent has a number of problems as applied to the training of deep neural networks, so a number of variants such as Stochastic Gradient Descent (SGD), Momentum, and Adaptive Moment Estimation (Adam) have been proposed and tested.

Stochastic Gradient Descent performs a parameter update for each training sample, foregoing the need to compute the parameters for the entire training set as plain gradient descent, resulting in significant performance improvements and reduced memory requirements. Momentum is a variant of SGD where the concept of momentum is used to constrain the size of the update, addressing the tendency of SGD to oscillate around optima with steep gradients because the optimum value is overshot. Adam, similarly, attempts to adapt the learning rate to the local gradient by computing both the momentum and an exponentially decaying average of all past gradients.

Convergence

Convergence is the goal of training, where the loss is minimized and further training does not improve performance. Depending on the architecture and complexity of the network and the nature of the data it may take many epochs (or iterations through the training data) to converge or just a few. At the completion of an epoch, a loss/accuracy vs. validation data is typically performed. The average of the change trajectory of these metrics will asymptotically approach zero as the network approaches convergence.

Non-optimal convergence can be symptomatic of one of a few problems: The training data may be too noisy, represented by a form inappropriate for training, or insufficient. The network can be too shallow to represent the desired mappings. There might be issues with class imbalance, or with class definition.

Memory Consumption/Mini Batch Training

With the advent of deeper networks and large training sets it is commonly impossible to maintain the network and training data in memory and becomes essential to train the network using an approach commonly referred to as mini batch training, where a small subset of the training data is provided to the network at once.

MODELING THE HUMAN APPROACH

Human experts frequently perform image analysis tasks that we don't currently have automated solutions to. Normally these experts have developed an internal algorithm used to solve the problem, but which cannot be directly translated to computer automation because this algorithm relies on a number of non-linear and pre-conscious steps that the operator has no intuition about. If this approach to the problem can be specified such that the non-linearities are partitioned to discrete steps that function in the context of a deterministic algorithm, the inputs to and outputs from the non-linear functions can be defined. If the inputs and outputs are known, it is generally possible to collect labeled examples of such, design a network appropriate for problem, train it, and drop it into the algorithm as long as the complexity of the task is not too great to be modeled by the network.

Deriving an Algorithm

In the case of identifying and locating EEG sensors in the images, users have to overcome a number of challenges that have frustrated current approaches to automating the task.

The first stage in the location and identification of sensors is segmenting the image into foreground (sensors) and background (everything else). Once the images have been segmented into foreground elements and background elements, the sensors in the foreground are examined individually in order to determine the unique identity. There are, then, usually sensors that cannot be identified by examining the label as it might be overexposed or obscured by hair, wires, or ears. For these situations users are provided with a map of the positions of sensors they use to determine the identity of an unidentifiable sensor through an examination of nearby sensors.

From these stages, a high-level algorithm was derived and is listed here.

Computer Code

Algorithm 1: *Identify Sensors*

```
def identify_sensors (image):
  segmented_image = segment(image)
  sensor_locations = find_centroids(segmented_image)
  results = []
  unidentified_sensors = []
  For sensor_location in segmented_image:
    sensor_identity = identify_sensor(sensor_location)
```

```
   If sensor_identity != unknown:
     results.append((sensor_identity, sensor_location))
   else:
     unidentified_sensors.append(sensor_location)
 For sensor_location in unidentified_sensors:
   predictions = {}
   for known_sensor in Results.nearest_sensors(sensor_location, 6):
     prediction, response = identity_given_displacement(camera, known_sensor,
sensor_location)
     If response > response_threshold:
       try:
         predicted_identity[prediction] += response
       except KeyError:
         predicted_id[identity] = response
     predicted_identity = sorted(predicted_id, key=predicted_id.get,
reverse=True)[0]
     if predicted_id > identification_threshold:
       results.append(predicted_identity, sensor_location)
 return results
```

The bolded function calls represent steps that are complex operations and well suited to the application of deep learning, while the rest of the algorithm consists of fairly simple applications of Boolean logic or geometry. The segment(image) function performs *image segmentation*, the function identify_sensor(sensor_location) performs *visual sensor identification*, and the identity_given_displacement(camera, known_sensor, sensor_location) function performs *geometric sensor identification*.

IMPLEMENTATION OVERVIEW

The sensor identification algorithm is a high-level description of the process by which the sensors were located and identified in the input images. The important elements of this algorithm are the three steps, represented in the algorithm as function calls to undefined functions, that were implemented using neural networks. In each case, the core functionality is performed by a neural network trained on labeled data, with a little code acting as glue to return the result in the required format.

Image Segmentation

Image segmentation is accomplished by first training a convolutional neural network to recognize sensors in images, then extracting patches from a test image and asking the neural network to classify those patches as containing the depiction of a sensor or not.

In Figure 1 the overlaid segmentation of the output image is inset by 25 pixels from each edge of the image: this is because the image is broken up into a set of image patches 50 pixels by 50 pixels, and the classification of this image patch becomes the value of the segmented image at the center of the patch.

Figure 1. After the network is trained, the input image is divided into image patches and then the trained CNN classifies these patches and generates the segmented image. The segmented image is shown, here, overlaid on the original image on the right.

The actual value of a pixel in the segmentation is derived from the confidence of the prediction, so less confident predictions result in darker grays. After the segmented image is fully derived a few image processing steps are performed in order to derive a set of coordinates that are provided to the second stage (sensor identification): Gaussian blurring and morphological erosion and dilation are used to filter the results, then the centers of all remaining foreground regions are extracted.

Visual Sensor Identification

Once the coordinates of the foreground objects have been extracted, these are used to determine the locations in the original image to extract image patches from. As in the previous step, the image patches for visual sensor identification are each 50 pixels by 50 pixels. The CNN is trained to classify these patches into one of 259 classes (0 for regions not depicting a sensor and 1 through 258 for sensors).

In the resulting image of Figure 2, it can be seen that many of the sensors have been correctly identified, as indicated by the numbers drawn atop the image. If the confidence of the prediction for classes 1 through 258 exceeds a threshold, that class is used taken as the identification of the sensor. If the

Figure 2. After the network is trained and the foreground object centers are found, image patches centered at the object coordinates are extracted and the trained CNN classifies these as shown in the resulting image on the right.

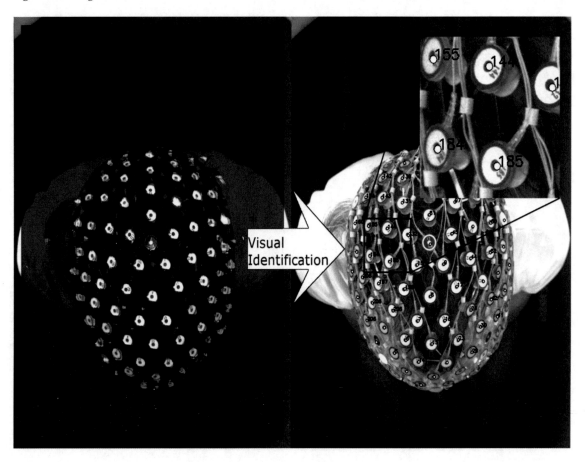

confidence of the prediction for class 0 exceeds the threshold, the sensor location is eliminated. If the confidence of the prediction does not exceed the threshold, the sensor is left unidentified.

Once the sensors that can be identified are identified, the coordinates and identities of the sensors are collected and provided to the geometric sensor identification stage.

Geometric Sensor Identification

After the sensor locations have been assigned initial identifications, the location and identity are used to update misidentified sensors and infer the identification of sensors that could not be identified by visual inspection alone. The network used for this step, once trained, is provided the location of an unidentified sensor and the relative locations and identities of the 6 nearest sensors. Based on this information, the identity of the unknown sensor is predicted and the identity assigned if the confidence of the prediction exceeds a threshold.

In Figure 3, several of the sensors not identified by the visual identification step (marked by circles with no sensor identity overlay) or misidentified (in the first image, sensor 80 is identified as sensor

Figure 3. After the network is trained and the sensor locations are assigned initial identifications, 2D geometric relationships between sensors are utilized to update sensor identifications.

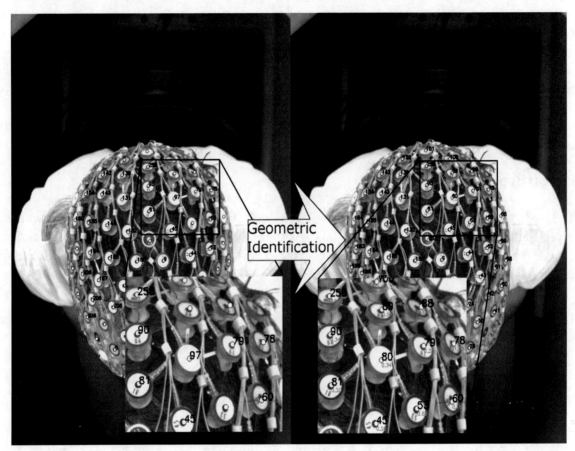

97) have been assigned new identification by the function based on the prediction of the trained neural network. Each sensor in the input (identified or not) is presented to the network along with the relative location and identities of the nearest six sensors. An optimization is performed to update and assign identities to the sensors that maximize the confidence of the predictions overall sensors. Changing the identity of one sensor will change the confidences and potentially the classifications of all sensors that rely upon it for identification, so the aggregate confidence has to be used for the metric during this step.

COLLECTING AND STRUCTURING THE TRAINING DATA

The derived algorithm defines a solution to the problem of labeling the sensors in an image but relies on three undefined functions. These functions could be fulfilled by human effort, classical computer vision, or deep learning approaches, but framing the problem in this fashion exposes the nature of the problem in a way that exposes the aspects of the problem that can benefit from recent advancements in machine learning.

In the case of the GPS, the output of the work of users is a set of files containing the 2D coordinates of the sensors within the individual camera images as verified by the users, the 3D coordinates derived using the techniques of photogrammetry, and the images captured by the cameras. Using these three pieces of information, it was possible to extract a set of data that was used to train the neural networks to recognize sensors in the images with a high degree of accuracy.

For the problem of labeling sensor positions in images, a software application has been in use for some time. While the operation of the application is imperfect and requires a not insignificant amount of user effort, its outputs can be used to train a neural network to perform the undefined functions.

An examination of the inputs to and outputs from the undefined functions reveals what information is required for each.

Image Segmentation

The input to the segmentation function is a grayscale image produced by one of the cameras, while the expected output of the function is a binary image with sensor positions highlighted. This suggests that a neural network be trained to classify each pixel in the image into one of three classes: normal or cardinal sensors or background (everything else) as shown in Figure 4.

Figure 4. Examples of the three classes of image patches extracted for the training of the segmentation CNN

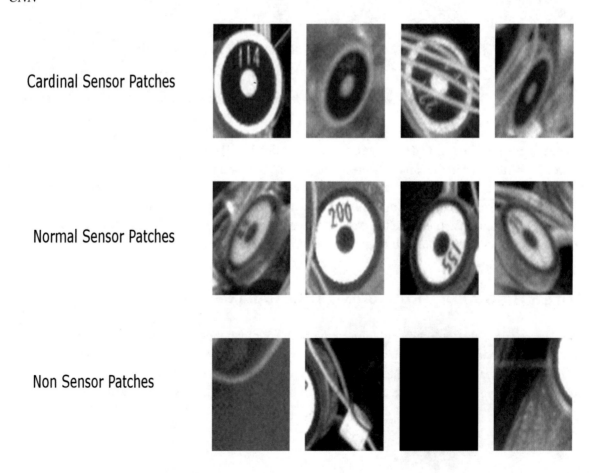

From each of the images three sets of patches, two centered around the cardinal and normal sensor locations and a third centered at random points not containing sensors, were extracted. The network was trained with these three classes to discriminate between sensors and background texture. Once the network was trained, a sliding window was moved over the test image, classifying each pixel as belonging to a sensor or to the background and yielding the segmented image as required. Figure 5 demonstrates the segmentation of two pixels from the image where one contains a sensor and one does not. Only the central white or black pixel is updated as a result of this operation.

One important consideration when collecting the data is that of class balance. If there is an imbalance in the size of the classes, the network will learn to favor the larger class, especially in indistinct cases. This can be desirable if the larger class should be favored, but should be considered. In the case of the images the segmentation network is to operate on, there is significantly more background information than foreground, so the size of the non-sensor class was expanded to approximately 4 times the size of the sensor class.

Visual Sensor Identification

The input to the visual sensor identification function is the location of sensors in the image. This information comes from the segmented image (determine the centroid of each region marked as containing a sensor by the segmentation step). The output of this function is the identity of the sensor as determined by an examination of the label.

Figure 5. The segmentation operation examines an image patch, and assigns the predicted class to the central pixel of the patch

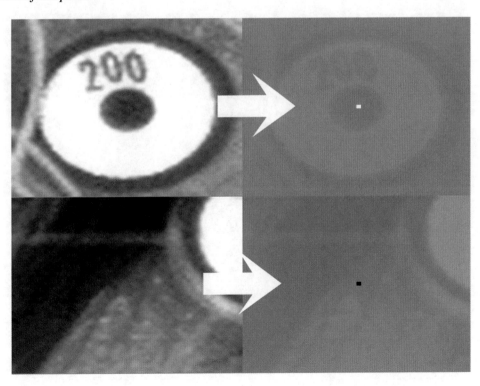

The nets used for this research work contained 258 unique electrodes (256 EEG sensors, one common, and one reference), suggesting that the network should be trained to classify the input data into one of 258 classes. To allow this network to eliminate false positives from the segmentation a 259th class was added to again represent background noise.

For each of the images in the training data, patches were again extracted, but this time they were separated into the 259 classes described. The network trained on this information was then presented with image patches extracted from an image at the points determined by the output of the segmentation function. Figure 6 shows the prediction of the visual identification CNN for one image patch containing a sensor and one containing only background noise. In the example with only background elements, the sensor response is eliminated.

Geometric Sensor Identification

The input to the geometric sensor identification function is the camera number, the 2D location of an unidentified sensor and the 2D location of a known sensor. A displacement vector between these sensors is computed, and based on the known geometric relationships between sensors, an identity is predicted for the unknown sensor, modeling the strategy applied by users to the identification of sensors with unreadable labels.

The solved files were, again the source for this information. For each recorded user verified 2D sensor location, it and the nearest 6 sensor locations were extracted. For each of the nearby sensors, the

Figure 6. The visual identification CNN examines the sensor patches identified by the segmentation and attempts to read the label. If the resultant class is not a sensor, the location is eliminated.

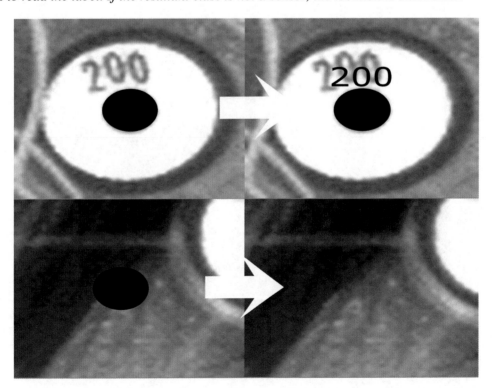

displacement vector to the sensor in question was provided as input, and the identity of the unknown sensor was provided as the label.

In Figure 7, the vector of displacements from known sensors to the central unknown sensor are represented as the black lines joining the centers of the 6 identifiable sensors to the center of the unidentifiable sensor.

DESIGNING THE NETWORK(S)

In the algorithm described, three functions were laid out that were appropriate for the application of a deep learning approach. An understanding of the input data, the operations being emulated, and the output data suggested the appropriate network architecture for each. As some networks preserve spatial relationships while others preserve sequential relationships, understanding the nature of the problem and the operation of the network is essential to attaining good results.

Figure 7. By comparing the displacements from a number of known sensors to the center of unknown sensors it is often possible to infer the identity of the unknown sensor.

Image Segmentation

As formulated, the image segmentation task is ideally suited to an emerging variation on the standard CNN: a fully convolutional network (FCN). Current results suggest that these offer a means of performing dense semantic segmentation of arbitrarily sized images (Shelhamer, Long, & Darrell, 2016).

Instead, the authors implemented this step using a normal convolutional network (as shown in Table 1) and reformulated segmentation as a classification task. The entire image is covered by a sliding window that computes, for every pixel it examines, a class: either the pixel is part of a sensor or not. Table 2 shows how the trained network is used for segmenting input images.

This is a computationally expensive task, requiring about 6 hours per image on a CPU or about 5 minutes on a high-end GPU. Improvements to the computational run time could be obtained by downsampling the input image, but selecting an ideal network architecture (in this case, an FCN) could potentially achieve similar results to the current implementation at full resolution with a sub-second runtime.

Visual Sensor Identification

The input to the visual sensor identification function is an image patch extracted from the original image at the location determined by an examination of the segmented image. The output of this function is a classification of the patch into one of 259 categories. Current results with image classification demonstrate the suitability of CNNs to this class of problems.

The traditional architecture of CNNs consists of an input layer followed by a number of convolutional layers paired with max pooling layers, feeding into one or more fully connected layers with output dimensionality equivalent to the number of classes the network is to be trained for (Krizhevsky, Sutskever, & Hinton, 2012).

Table 3 lists the definition of the visual sensor identification CNN. This is virtually identical with the exception that it contains 259 output classes.

Geometric Sensor Identification

The input to this function is a one dimensional vector of length 272. A one-hot encoding is used to specify the camera number (the first 11 elements), the known sensor id (the next 259 elements), and the displacement vector (the final 2 elements). For any unidentified sensor, the six nearest sensors are selected, and for each a displacement vector from it to the unidentified sensor is computed. This is fed to the network, which will make a prediction. The predictions from these tests are, then, counted and the winning identification is assigned to the unknown sensor.

As there are no spatial relationships within the data to be preserved and as the input vectors are relatively small, a fully connected network was deemed appropriate. A listing of the definition of the geometric sensor identification network is included in Table 4.

Table 1. Segmentation CNN definition

```
# TFLearn is a deep learning library, like Keras, built on top of TensorFlow
import tflearn
from tflearn.layers.core import input_data, dropout, fully_connected
from tflearn.layers.conv import highway_conv_2d, conv_2d, max_pool_2d
from tflearn.layers.normalization import local_response_normalization, batch_normalization
from tflearn.layers.estimator import regression
from tflearn.data_preprocessing import ImagePreprocessing
from tflearn.data_augmentation import ImageAugmentation

# gpspatch is a python module that reads the training patches and returns them
# as the tuple (training_patches, training_labels, validation_patches, validation_labels)
# where the patches are the actual image patches and the labels specify the class that
# each patch is a member of. Training data is used for training the network and validation
# data is held back during training and used to validate the performance of the neural
# network against data it has not seen before.
import gpspatch

SIZE = 50

# define the shape up the input data. In the case of image segmentation
# network, the image patches are 50 by 50 pixel single channel images.
# the first element of the shape list, here, specifies the number of images:
# leaving it set to None means it will be computed (at most, a single
# dimension can be left unspecified).
network = input_data(shape=[None, SIZE, SIZE, 1], name='input_main')

# Define the layers of the network, here we start with a plain 2d convolutional
# layer with 64 filters that feeds into a series residual bottleneck layers.
# See http://tflearn.org/layers/conv/ for a full breakdown of the input
# parameters.
network = tflearn.conv_2d(network, 64, 3, activation='relu', bias=False)
network = tflearn.residual_bottleneck(network, 4, 32, 64)
network = tflearn.residual_bottleneck(network, 1, 64, 256, downsample=True)
network = tflearn.residual_bottleneck(network, 2, 64, 256)
network = tflearn.residual_bottleneck(network, 1, 128, 512, downsample=True)
network = tflearn.residual_bottleneck(network, 2, 128, 512)
network = tflearn.residual_bottleneck(network, 1, 256, 1024, downsample=True)
network = tflearn.residual_bottleneck(network, 2, 256, 1024)

# Add batch normalization, ReLU activation and a global average pooling layer
network = tflearn.batch_normalization(network)
network = tflearn.activation(network, 'relu')
network = tflearn.global_avg_pool(network)

# Regression: Defines 3 output classes and the use of softmax activation to ensure the sum of
# the confidence for the prediction of each of the 3 classes is 1.0
network = tflearn.fully_connected(network, 3, activation='softmax')
network = tflearn.regression(network, optimizer='adam',
loss='categorical_crossentropy',
learning_rate=0.0005,
name="target_main")

model = tflearn.DNN(network, tensorboard_verbose=0)
```

Table 2. Image segmentation using the trained CNN. The output of this stage is saved to an 8-bit image where white regions represent foreground objects.

```
# gps_image_patch_source is a python module that takes an image to be segmented and breaks
# it into patches suitable for the segmentation CNN.
from gps_image_patch_source import get_patches_in_region as patches

# OpenCV python module
import cv2
import numpy as np
from scipy.misc import imread

def dense_sensor_search(path, camera_number):
"""Given the path to an image, the number of the camera the image was captured by
and a trained CNN, this function will generate a segmented image where sensors
are represented by white regions and everything else is black."""

# Determine the shape (width, height) of the input image
image_shape_full = imread(path).shape
image_shape = (image_shape_full[0], image_shape_full[1])

# Generate the output image and fill it with zeroes.
full_image = np.zeros((image_shape[0]-SIZE,image_shape[1]-SIZE))

# Depending on the memory available to the CNN, a different
# number of image patches can be classified concurrently.
# An NVidia GTX 1080ti has 11GB of memory and can handle approx.
# 360 50x50 patches at a time, given the complexity of the CNN.
width_ = 60
height_ = 60

# For each width_ by height_ patch in the output image, extract a width_ by height_
# block of image patches from the input image (respecting image boundaries) to be
# classified by the CNN. Each prediction will become one pixel in the output image.
for y_ in range(1,int(image_shape[0]/height_)-1):
for x_ in range(1,int(image_shape[1]/width_)-1):
# Determine the actual dimensions of the input vector of patches so as not to
# attempt to extract patches from outside the boundaries of the input image.
width = width_
height = height_
x = x_*width
y = y_*height
if x + width + SIZE >= image_shape[1]:
width = image_shape[1]-(x+SIZE)
if y + height + SIZE >= image_shape[0]:
height = image_shape[0]-(y+SIZE)

# if the dimensions are valid:
if height > 0 and width > 0:

# Extract the patches from the input image. The patches method returns
# a tuple consisting of a list of image patches and a list of the coordinates
# of the patches: (image_patches, patch_coordinates).
tp,tc = patches(path, SIZE, (int(x), int(y), int(width), int(height)), step=1)

# Predict the classes of each patch
prediction = model.predict(tp)

# Generate the prediction image for the width_ by height_ patch of the output.
# Because we don't care whether the sensor is a normal sensor or a cardinal
# sensor, and because the sum of the confidence in the three classes will sum
# to 1.0, we combine the confidences for normal and cardinal sensor (p[0],
# p[1]), and subtract the confidence that the patch is not a sensor, clamped to
# the range [0, 1].
prediction_image = [max(p[0]+p[1]-p[2],0) for p in prediction]
prediction_image = np.array(prediction_image).reshape(int(height),int(width))

# Add this patch to the full output image.
full_image[int(y):int(y+height),int(x):int(x+width)] = prediction_image

# Scale by 255 to generate an 8bit grayscale image
full_image = 255 * full_image
full_image = full_image.astype(np.int)

# Save the results
cv2.imwrite("./prediction_image_{}.tiff".format(camera_number), full_image)
```

Table 3. Visual identification CNN.

```
# TFLearn is a deep learning library, like Keras, built on top of TensorFlow
import tflearn
from tflearn.layers.core import input_data, dropout, fully_connected
from tflearn.layers.conv import highway_conv_2d, conv_2d, max_pool_2d
from tflearn.layers.normalization import local_response_normalization, batch_normalization
from tflearn.layers.estimator import regression
from tflearn.data_preprocessing import ImagePreprocessing
from tflearn.data_augmentation import ImageAugmentation

# gpspatch is a python module that reads the training patches and returns them
# as the tuple (training_patches, training_labels, validation_patches, validation_labels)
# where the patches are the actual image patches and the labels specify the class that
# each patch is a member of. Training data is used for training the network and validation
# data is held back during training and used to validate the performance of the neural
# network against data it has not seen before.
import gpspatch

SIZE = 50

# define the shape up the input data. In the case of the visual sensor identification
# network, the image patches are 50 by 50 pixel single channel images.
# the first element of the shape list, here, specifies the number of images:
# leaving it set to None means it will be computed (at most, a single
# dimension can be left unspecified).
network = input_data(shape=[None, SIZE, SIZE, 1], name='input_main')

# Define the layers of the network, here we start with a plain 2d convolutional
# layer with 64 filters that feeds into a series residual bottleneck layers.
# See http://tflearn.org/layers/conv/ for a full breakdown of the input
# parameters.
network = tflearn.conv_2d(network, 64, 3, activation='relu', bias=False)
network = tflearn.residual_bottleneck(network, 4, 32, 64)
network = tflearn.residual_bottleneck(network, 1, 64, 256, downsample=True)
network = tflearn.residual_bottleneck(network, 2, 64, 256)
network = tflearn.residual_bottleneck(network, 1, 128, 512, downsample=True)
network = tflearn.residual_bottleneck(network, 2, 128, 512)
network = tflearn.residual_bottleneck(network, 1, 256, 1024, downsample=True)
network = tflearn.residual_bottleneck(network, 2, 256, 1024)

# Add batch normalization, ReLU activation and a global average pooling layer
network = tflearn.batch_normalization(network)
network = tflearn.activation(network, 'relu')
network = tflearn.global_avg_pool(network)

# Regression: Defines 3 output classes and the use of softmax activation to ensure the sum of
# the confidence for the prediction of each of the 259 classes is 1.0
network = tflearn.fully_connected(network, 259, activation='softmax')
network = tflearn.regression(network, optimizer='adam',
loss='categorical_crossentropy',
learning_rate=0.0005,
name="target_main")

model = tflearn.DNN(network, tensorboard_verbose=0)
```

Table 4. Geometric sensor identification network

```
# TFLearn is a deep learning library, like Keras, built on top of TensorFlow
import tflearn
from tflearn.layers.core import input_data, dropout, fully_connected
from tflearn.layers.conv import highway_conv_2d, conv_2d, max_pool_2d
from tflearn.layers.normalization import local_response_normalization, batch_normalization
from tflearn.layers.estimator import regression
from tflearn.data_preprocessing import ImagePreprocessing
from tflearn.data_augmentation import ImageAugmentation

# The sensor vector is a 1-hot encoding of the following information:
# 1. Camera Number: 1-11
# 2. Sensor Number: 1-258 (+1 for non-sensors)
# 3. 2D displacement vector: For each sensor in an image, a sensor vector is produced for
# 6 neighboring sensors such that the displacement vector is a representation of the
# (x, y) displacement from the neighbor sensor to the sensor in question.
SENSOR_VECTOR_LENGTH = 11+259+2

# Define the shape up the input data. In the case of the geometric sensor identification
# network, the input consists of vector of 1D vectors, which are the 1-hot encoding for sensors
# where the vector represents the displacement between the known sensor and an unknown sensor
# in a particular image. The network will be trained to predict the identity of the unknown
# sensor given this vector.
id_network = input_data(shape=[None, SENSOR_VECTOR_LENGTH], name='input_id')

# As this network should not encode spatial relationships between different columns,
# most layers are fully connected, and none are convolutional.
id_network = highway(id_network, SENSOR_VECTOR_LENGTH, activation='elu')

# Batch normalization
id_network = batch_normalization(id_network)
id_network = fully_connected(id_network, 512, activation='tanh')
id_network = fully_connected(id_network, 256, activation='prelu')
id_network = fully_connected(id_network, 128, activation='tanh')

# The output is one of 259 classes (258 Sensors + 1 non-sensor class). Softmax activation ensures
# that the sum of the confidences across all classes is 1.0.
id_network = fully_connected(id_network, 259, activation='softmax')
id_network = regression(id_network, optimizer='momentum', learning_rate=0.01,
loss='categorical_crossentropy', name='target_id')
id_model = tflearn.DNN(id_network, tensorboard_verbose=0)
```

REFINING THE NETWORKS

A careful analysis of the results of training the networks suggested improvements to the network architecture and helped to identify problems with the training data.

Image Segmentation

As the patches used for the training of the segmentation network are small grayscale images, the architecture of the network model used to perform the operation was based on a simple convolutional network that was capable of achieving good results on the (similarly small grayscale) MNIST database of handwritten digits. This produced reasonable results, but because the goal of the segmentation stage was to segment photographic images it became clear that a deeper network capable of capturing more abstract image features was necessary. Due to the challenges with training deeper networks, this also meant changing the type of layers. Very good results have been achieved with the training of deep Residual Networks (He et al., 2015), so the network for segmentation was redesigned with many more layers, most of which were residual layers.

In looking at the data, it also became clear that there were more than two classes. The non-sensor patches could be divided into a discrete set of expected elements (wires, bird bands - used in wire management, net structural elements, as well as features of the subject: eyes, ears, nose, etc.) and the sensors were best described as belonging to one of two classes normal sensors and cardinal sensors. This distinction is important because the white/black of the labels of the two classes is inverted.

No data existed capable of automatically classifying background (non-sensor) features, but it was possible to separate normal and cardinal sensors. After extracting data with the sensors separated into these two classes the network was adjusted to output three classes (normal sensor, cardinal sensor, non-sensor) instead of two and retrained.

Visual Sensor Identification

Because the development of the sensor visual identification stage benefitted from the development of the segmentation stage very few iterative changes were made to the design of the network and the it is essentially identical to the design of the segmentation stage network with the exception that the number of classes in the output is 259 rather than three.

Refining the Training Data

The extracted patches consisted of about 21,000 images of sensors and nearly 80,000 examples of background texture, too much data to individually verify. Interestingly, after training the network, it was possible to use it to predict the classes of the training data patches. This result allowed the authors to search the images and remove the misclassified patches. If an image was part of the training data for the sensor class, but the network failed to predict this class it was highlighted. The same operation was performed over the non-sensor patches and a user went through the ~2000 patches this operation returned to verify the performance of the search. The accuracy of this search was about 50% in that nearly 1000 of the selected patches were returned to their category of origin after examination. This high error was intentional and due to the decision to relax the classification thresholds for this operation: the authors were not concerned that some positive results might be returned, but wanted to avoid negative results (where a misclassified patch was left in the wrong class).

With the networks trained, the algorithm was run against separate input that had not been a part of the training or validation data. This highlighted a few deficiencies in the networks. Specifically, during segmentation, cardinal sensors, overexposed sensors, sensors partially obscured by ears, and sensors behind wires were regularly missed. The reason for these issues was related to class imbalance: though there were a number of examples of these types in the training data there were few enough of them that the penalty for ignoring them entirely was small enough that the network was unable to learn them. To address this issue, a number of examples of each of these types were duplicated, with no change made to them at all, and the network was retrained.

During the sensor identification stage, it became clear that a number of individual sensors, though clearly in the segmented data were identified as background noise. An examination of the training data for these sensors made it obvious that this was again an issue of class imbalance. In some cases, there were as few as 32 example patches for a given sensor (while most were in the range of 150 - 200). These were duplicated to improve the balance and ensure that the network could learn to identify them.

The results obtained during the geometric identification stage illustrated the importance of understanding the data and considering how the network will interpret it. It was initially impossible to train this network to achieve anything approaching useful results with a validation accuracy never exceeding 50%. This was due to the decision to represent the displacement vector as distance and angle causing vectors in the first 90 degrees and the last 90 degrees to look very different to the network despite being very similar in many cases. After changing this vector representation to use normalized x and y displacement values the network was quite easily trained to a validation accuracy of about 98%.

Issues, Controversies, Problems

The implementation of the technique described in this chapter was capable of very nearly completely automating the process of locating the sensors within the images as of writing given a short development time and limited data. This being the case, there were still a number of deficiencies the authors identified.

The sliding window approach taken to compute the segmentation of the image is an inefficient means of performing this operation. Newer techniques like fully convolutional networks can perform the same task far more efficiently and are a better fit for the work that is being done.

Due to the sequential nature of the solution, where the output of one step acts as the input of the next, any errors that occur early on in the process are exaggerated by later stages. By design, sensors not properly located during the segmentation stage could not be added by the identification stage, and these "holes" in the sensor net could lead to the geometric sensor identification to misidentify sensors. Because this step is performed recursively in order to build out a full solution, a single misidentification could lead to several more.

The nets contain several items that resemble sensors when overexposed; in particular, the net is populated with a few unlabeled pedestals similar in shape and size to the sensors. In the images these appear as nearly pure white. Additionally, there are a number of "bird bands" used for wire management the, depending on their angle versus the camera can appear as small white circles, similar to the central dots of the cardinal sensors.

Another challenge resulted from class imbalances as described previously. Within the nets there are a few "cardinal" sensors whose labels are inverted (a black donut around a white circle, rather than a white donut around a black circle), and even when clearly visible, they were identified at a much lower rate than the non-cardinal sensors.

SOLUTIONS AND RECOMMENDATIONS

Tools like TensorFlow and Keras (or TFLearn) have made deep learning very approachable, but there are still a number of complexities and details that need to be taken into account.

Plan Ahead

One of the roadblocks to the adoption of deep learning systems is the lack of appropriate training data. Oftentimes, the output of non-automated or semi-automated approaches can be used as training data, but considering the formulation of a deep learning approach to a problem ahead of time can improve the quality or quantity of data or reduce the pre-processing work required. In the case of the work described herein, the semi-automated solution used to generate the training data did not initially capture the user specified 2D sensor positions. Though the 2D positions of the sensors could be recovered by projecting the 3D coordinates into the images, the nature of the problem meant that the 2D projection of a sensor in an image might be obscured by hair or other objects that would then be labeled as belonging to the sensor class.

Mind the Learning Rate

The learning rate is an important hyperparameter. A learning rate set too low can negatively impact the time required to train the network, but too high, and the optimizer will overshoot the optima, preventing the network from being trained to the accuracy it would otherwise be capable of demonstrating. If a network converges prematurely this can be a sign that the learning rate is set too high. Before re-examining the training data or the network architecture, lowering the learning rate can be an effective means of achieving (usually) small increases in accuracy performance.

Overfitting

If training accuracy is very high and stable and validation stability is similarly stable, but lower, this can indicate that the network has been trained to recognize the features of the training set, but that there is variance in the data for which it has not been trained. The network is overfitting the training data in this case. If this is the case, a solution would be to provide more training data better capable of capturing this variance.

Understand the Data to Formulate an Efficient and Effective Solution

It is important to consider the characteristics of the data and the desired operation in order to determine the appropriate network architecture. The segmentation step was formulated as a classification task for a small image patch, outputting a class that was assigned to the pixel at the center of the image. This approach was capable of segmenting the images with a high degree of accuracy, but was slow and memory intensive. Where it slid a window over the whole image with a stride of one pixel, an FCN could examine the entire image by dividing it into just a few overlapping patches.

On the other hand, by selecting only a few points in the images, and utilizing a network well-suited to the problem (A deep Residual CNN), the visual identification step performs its work efficiently.

Class Balance

Classes should be balanced in such a way as to produce the desired results. Underrepresented classes will often not impose enough of a penalty during training to ensure that the network learns to identify them. In the absence of more training data to fill out these classes, it can be effective to duplicate the data on hand to ensure that the class will be learned. Doing this will, of course, suffer from a lack of robustness to all of the possible variations that exist.

Consider the cost of a false positive result versus a false negative. If one is preferable to the other, class balances could be adjusted to prefer one over the other.

Understand the Data and How the Network Interprets It

When poor results are obtained, the training data is commonly the problem and understanding the form of the data and how it will be interpreted during training is absolutely essential. As previously described, when training the geometric sensor identification network, the displacement vector was originally specified as an unnormalized vector of angle and distance. By reworking the vector representation to use a simple normalized x-displacement, y-displacement format, the results were quickly improved and the training converged at about 95% versus the validation data set.

FUTURE RESEARCH DIRECTIONS

The authors have attempted to identify as many deficiencies as possible, and have devised approaches to address these deficiencies.

The computational cost of the segmentation step is unnecessarily significant. Fully convolutional networks are defined for exactly this task and have much better computational performance characteristics (Shelhamer, Long, & Darrell, 2016). Given the 11 images and approximately five-minute computational time (on a GPU) for the current approach, deriving a solution with would take nearly an hour, almost all of it for computing the dense segmentation.

The current approach is, in part, a workaround for a lack of sufficient training data. If sufficient data existed, the segmentation step could be extended to segment directly to sensor identity instead of breaking this into two steps. When considering all sensors as being of the same class, the solved files contained at least 21,000 examples, but for each sensor, there were generally less than 200 examples. Once the computational performance of the segmentation step has been improved, the current implementation can be used to drastically reduce the amount of time required to compute solutions to these files; at this time, more data can be acquired allowing a segmentation step to be trained that can segment the images by sensor identity, reducing the depth of the pipeline and the potential for an early failure to be exaggerated by a later step. In regard to the sensor misidentification problem in the output of the geometric identification stage in particular, utilizing the predicted sensor identities as suggestions and then testing the suggestion by taking the RMS error of the triangulation of the sensor given the suggestion might significantly improve these results by highlighting results that exhibit high triangulation error (which typically indicates a sensor misidentification).

The issue with overexposed sensors being mistaken for extraneous net structures is thorny. It is almost always possible to determine by careful examination of the image which is which, so training a network to differentiate the two should not be impossible. Furthermore, there is a lot of geometric information and inter-image correlations that can also be used to differentiate the two. If the segmentation step selects these as likely sensors, a filtering step could be implemented that would utilize this information to sort them into the appropriate class.

CONCLUSION

Many problems arising in the domain of image analysis can be formulated in terms of algorithms that rely on neural networks trained on human effort to solve non-linearities in the input data. This is, the authors argue, the way humans solve these problems; relying upon pre-conscious regions of the brain to perform operations that feel effortless while devising a set of higher order strategies to operate on the outputs of the pre-conscious work the brain has performed.

While translating directly from input to desired output is ideal, the approach described here is capable of reducing the problem scope to the level of human intuition and facilitates designing approaches to these problems that draw upon the insights of subject-matter experts.

The paucity of training data in some more specialized domains is the biggest gating factor to the implementation of deep learning approaches to these problems. The authors have attempted to show, here, how scarce training data can be used to train a neural network to simulate the human visual cortex in the performance of a task, while using the conscious intuitive approach employed by human volunteers to define a deterministic algorithm capable of performing tasks that were previously difficult to automate. In so doing, the amount of available training data can be rapidly increased (as much of the work of producing this data can then be automated) opening the door to the possibility of designing and training a network capable of moving from input to desired output without the requirement of additional machinery.

REFERENCES

Blum, A., & Rivest, R. L. (1989). Training a 3-node neural network is NP-complete. In Advances in neural information processing systems (pp. 494-501). Academic Press.

Fernández-Corazza, M., Turovets, S., Luu, P., Anderson, E., & Tucker, D. (2016). Transcranial electrical neuromodulation based on the reciprocity principle. *Frontiers in Psychiatry*, 7. PMID:27303311

Groppa, S., Bergmann, T. O., Siems, C., Mölle, M., Marshall, L., & Siebner, H. R. (2010). Slow-oscillatory transcranial direct current stimulation can induce bidirectional shifts in motor cortical excitability in awake humans. *Neuroscience*, *166*(4), 1219–1225. doi:10.1016/j.neuroscience.2010.01.019 PMID:20083166

Halevy, A., Norvig, P., & Pereira, F. (2009). The unreasonable effectiveness of data. *IEEE Intelligent Systems*, *24*(2), 8–12. doi:10.1109/MIS.2009.36

He, K., Zhang, X., Ren, S., & Sun, J. (2015). *Deep Residual Learning for Image Recognition*. Retrieved from https://arxiv.org/abs/1512.03385

Hecht-Nielsen, R. (1988). Theory of the backpropagation neural network. *Neural Networks, 1*(Supplement-1), 445–448. doi:10.1016/0893-6080(88)90469-8

Krizhevsky, A., Sutskever, I., & Hinton, G. E. (2012). Imagenet classification with deep convolutional neural networks. In Advances in neural information processing systems (pp. 1097-1105). Academic Press.

Livni, R., Shalev-Shwartz, S., & Shamir, O. (2014). *On the Computational Efficiency of Training Neural Networks.* Retrieved from https://arxiv.org/abs/1410.1141

Russakovsky, O., Deng, J., Su, H., Krause, J., Satheesh, S., Ma, S., . . . Fei-Fei, L. (2015). *ImageNet Large Scale Visual Recognition Challenge.* Retrieved from https://arxiv.org/abs/1409.0575

Shelhamer, E., Long, J., & Darrell, T. (2016). *Fully Convolutional Networks for Semantic Segmentation.* Retrieved from https://arxiv.org/abs/1605.06211

Swartz, B. E. (1998). The advantages of digital over analog recording techniques. *Electroencephalography and Clinical Neurophysiology, 106*(2), 113–117. doi:10.1016/S0013-4694(97)00113-2 PMID:9741771

ADDITIONAL READING

He, K., Zhang, X., Ren, S., & Sun, J. (2016, October). Identity mappings in deep residual networks. In *European Conference on Computer Vision* (pp. 630-645). Springer International Publishing.

Ioffe, S., & Szegedy, C. (2015, June). Batch normalization: Accelerating deep network training by reducing internal covariate shift. In *International Conference on Machine Learning* (pp. 448-456).

Jegou, S., Drozdzal, M., Vazquez, D., Romero, A., & Bengio, Y. (2016). *The One Hundred Layers Tiramisu: Fully Convolutional DenseNets for Semantic Segmentation.* Retrieved from https://arxiv.org/abs/1611.09326

Ren, S., He, K., Girshick, R., & Sun, J. (2015). Faster R-CNN: Towards real-time object detection with region proposal networks. In Advances in neural information processing systems (pp. 91-99).

Simoyan, K., & Zisserman, A. (2014). *Very Deep Convolutional Networks for Large-Scale Image Recognition.* Retrieved from https://arxiv.org/abs/1409.1556

Szegedy, C., Ioffe, S., Vanhoucke, V., & Alemi, A. A. (2017). *Inception-v4, Inception-ResNet and the Impact of Residual Connections on Learning* (pp. 4278–4284). AAAI.

Szegedy, C., Liu, W., Jia, Y., Sermanet, P., Reed, S., Anguelov, D., ... Rabinovich, A. (2015). Going deeper with convolutions. In *Proceedings of the IEEE conference on computer vision and pattern recognition* (pp. 1-9).

Xie, S., Girshick, R., Dollar, P., Tu, Z., & He, K. (2016). *Aggregated Residual Transformations for Deep Neural Networks.* Retrieved from https://arxiv.org/abs/1611.05431

KEY TERMS AND DEFINITIONS

Backpropagation: Process of estimating the part of the total error attributable to each neuron and updating the weights to minimize that error.

Convolutional Neural Network: A neural network consisting (at least partially) of convolutional layers, which learn convolutions that can be applied to an image (or other data) to extract features.

EEG (Electroencephalogram): Non-invasive EEG recovers electrical signals generated by the firing of neurons using sensors at the surface of the scalp, while with intracranial EEG recovers those signals using sensors emplaced in actual brain tissue.

GPS (Geodesic Photogrammetry System): A system consisting of 11 cameras arranged in a geodesic dome structure mounted on a gantry which is used to derive the 3D positions of EEG sensor nets as applied to the heads of subjects.

Photogrammetry: The technique of recovering 3D information through the process of 2D image analysis.

ReLU: Rectified linear unit. The non-linear activation function $A(x) = max(0, x)$ with a range of $[0, \infty)$.

Threshold Activation Function: The function $A(x) = 1$ if $x > 0$ else 0 with outputs of either 1 or 0. Because of this, the derivative of the threshold activation function is zero almost everywhere and is not a good candidate for optimization using stochastic gradient descent.

Chapter 6
Object–Oriented Logic Programming of Intelligent Visual Surveillance for Human Anomalous Behavior Detection

Alexei Alexandrovich Morozov
Kotel'nikov Institute of Radio Engineering and Electronics of RAS, Russia

Olga Sergeevna Sushkova
Kotel'nikov Institute of Radio Engineering and Electronics of RAS, Russia

Alexander Fedorovich Polupanov
Kotel'nikov Institute of Radio Engineering and Electronics of RAS, Russia

ABSTRACT

The idea of the logic programming-based approach to the intelligent visual surveillance is in usage of logical rules for description and analysis of people behavior. New prospects in logic programming of the intelligent visual surveillance are connected with the usage of 3D machine vision methods and adaptation of the multi-agent approach to the intelligent visual surveillance. The main advantage of usage of 3D vision instead of the conventional 2D vision is that the first one can provide essentially more complete information about the video scene. The availability of exact information about the coordinates of the parts of the body and scene geometry provided by means of 3D vision is a key to the automation of behavior analysis, recognition, and understanding. This chapter supplies the first systematic and complete description of the method of object-oriented logic programming of the intelligent visual surveillance, special software implementing this method, and new trends in the research area linked with the usage of novel 3D data acquisition equipment.

INTRODUCTION

A research area of rapidly growing is the human behavior recognition on the base of the intelligent visual surveillance. This research area is very important for applications such as security and anti-terrorism issues. A promising approach for dynamic visual scenes analysis is the logic programming. The idea of

DOI: 10.4018/978-1-5225-5751-7.ch006

the logic programming approach to the intelligent visual surveillance is in the usage of logical rules to formulate the description of context information and the analysis of people behavior. The behavior and activity notions differ in that the behavior of an object is the activity of the object related to the context information about the place, time, object attributes, etc. The information about the context allows deciding, for instance, whether the behavior of the object is abnormal or dangerous. Thus, the analysis of the behavior is a more complicated problem than the analysis of the activity. It is necessary to describe and analyze the information about the context of the activity and the mathematical logic is perhaps the best instrument that can be used for this purpose. The logic-based approach to the people behavior analysis and the intelligent visual surveillance has the following advantages:

1. One can easily incorporate domain knowledge into the recognition process in the form of logical formulae.
2. Reasoning about the behavior of people can be very difficult, but for the developer of the system of the intelligent visual surveillance, it always remains fundamentally understandable.
3. One can easily define a complex activity in terms of simpler activities.

In this chapter, research software for logic programming of the intelligent visual surveillance systems is discussed and examples of logic programs for video analysis are represented. The software is based on the Actor Prolog concurrent object-oriented logic language and a state-of-the-art Prolog-to-Java translator. Built-in classes of the Actor Prolog logic programming system are implemented in Java. The low-level stage of the video processing is implemented in special built-in classes of Actor Prolog. The purpose of the software developed is to facilitate research in the area of intelligent video monitoring especially for the evaluation of anomalous people activity and the study of logical description and analysis of people behavior. The open source library of the built-in classes is published in GitHub (Morozov, 2018).

A promising approach to implement the logic programming in the video analysis is the translation from the concurrent object-oriented logic language to Java. The means of the object-oriented logic programming enable fast and effective processing big arrays of video data because the arrays of data can be encapsulated in the instances of specialized built-in classes in the object-oriented logic language. The Actor Prolog logic programming system is suitable for this purpose since it is fast enough for the real-time video processing and ensures essential separation of the recognition process into concurrent sub-processes implementing different stages of the high-level video analysis. Generation of Java intermediate code ensures platform independence of the application software and guarantees absence of errors, such as the difficult-to-locate bugs (for instance, caused by memory leaks) and the out-of-range array operations.

The use of an industrial Java virtual machine for the logic programming system enhances its flexibility and simplifies adaptation to new operating systems and processor architectures. New prospects in the logic programming of the intelligent visual surveillance are connected with the usage of 3D machine vision methods and adaptation of the multi-agent approach to the intelligent visual surveillance. The usage of 3D data acquisition equipment like time-of-flight (ToF) cameras, structured light sensors, stereo cameras, laser scanners (LiDARs), and Flash Ladars open new prospects in the area of the intelligent visual surveillance. 3D data make possible to solve complicated problems of computer vision caused by the presence of shadows, variations in illumination, viewpoint changes, cluttered background, and occlusions; however, there is no universal 3D data acquisition equipment for solving all computer vision problems. The main advantage of 3D vision is in that it can supply essentially more complete information about the video scene. The availability of exact information about the coordinates of the parts of the body and scene geometry provided by means of 3D vision is a key to the automation of behavior analysis, recognition, and understanding. A disadvantage of 3D vision is in that the size of 3D video data is usually huge and requires high computational capability. A method of object-oriented logic programming for processing the data obtained from the equipment for 3D data acquisition in the intelligent visual surveillance systems was developed on the base of the Actor Prolog language. For this purpose, the Actor Prolog language is linked with the Kinect 2 device using the J4K library (Barmpoutis, 2013) and preprocessing procedures implemented in Java.

The conventional approach to human behavior recognition includes two stages of video processing. In the case of conventional 2D intelligent visual surveillance, the low-level stage can include background subtraction, detection of people, detection of vehicles, construction of the trajectories of moving objects, estimation of the velocities of the objects, etc. The high-level stage can include logical analysis of graphs of the trajectories that are supplied by the low-level analysis algorithms and displaying the results on the screen. In the case of 3D intelligent visual surveillance, the goals of the low-level video analysis are essentially different. For instance, the standard software of the Kinect 2 time-of-flight camera can provide ready-for-use foreground blobs, as well as skeletons of people to be monitored. Thus, the minimal preprocessing of the data can include just a conversion of the images of the skeletons into the terms of the logic language. The high-level stage of the video data processing can be implemented in accordance with the former principles, but the information about the video scene is essentially more complete, that gives an opportunity to infer much more of logical conclusions on the semantics of the video scene.

In the recent decade, the declarative approach to the development of multi-agent systems is recognized as a basic idea in this research area; a set of excellent declarative multi-agent platforms and languages are developed and implemented. Unfortunately, in the framework of the intelligent visual surveillance systems, the agents are to perform very specific operations on big arrays of binary data that are out of the framework of the conventional symbolic processing operations typical for declarative languages. Thus, there is a reason for the development of new means of the multi-agent-programming to experiment with intelligent visual surveillance systems. In this chapter, the multi-agent approach to the visual surveillance is considered on the base of the distributed version of the Actor Prolog language. This extension of the Actor Prolog language implements a combined type system, which provides a solution to the problem of the strong typing in the multi-agent systems. This type system ensures the advantages of the static type-checking that are necessary for generation of fast executable code and the flexibility of the dynamic type-checking that is necessary for multi-agent systems design.

This chapter contains the first systematic and complete description of the method of object-oriented logic programming of the intelligent visual surveillance systems, special software implementing the method, and new trends in this research area linked with the usage of novel 3D data acquisition equipment. In the Background section, the history of the research area is considered and main bibliographic references are supplied. In the Conventional 2D Intelligence Visual Surveillance Logic Programming section, the basic principles of the logic programming of the intelligent visual surveillance systems and statistical methods for reliable and quick discrimination of running pedestrians, bicycles, and vehicles during the logical analysis are considered. In the Implementation Issues of the Object-Oriented Logic Programming of the Intelligent Visual Surveillance section, necessary technical details of the implementation of the method including translation of the Actor Prolog logic language to Java, agent logic programming of video processing, and low-level video processing are discussed. In the 3D Intelligent Visual Surveillance Logic Programming section, new trends in the research area linked to the usage of

3D computer vision equipment are considered. Several Actor Prolog coding examples in the intelligent visual surveillance, agent programming, and low-level video processing are introduced in the chapter; a very simple logic programming example of applying Actor Prolog in 3D visual surveillance is presented, which introduce some practical insights to the readers.

BACKGROUND

Mathematical logic and logic programming were recognized as convenient means for video scene context description and analysis. The base of the logic programming approach to the people behavior analysis is in the usage of logical rules for description and analysis of people activities. Knowledge about the human body structure, object properties, time constraints, and scene geometry is encoded in the form of the rules in a logic language and is applied to the output data of some low-level procedures for detection of people and objects. This idea was developed in the context of the conventional 2D intelligent visual surveillance, and there are many studies based on it. The usage of temporal logic programs in the W^4 real-time system for representation of actions and analysis of people movements was reported in Haritaoglu, Harwood, and Davis, (1998). The VidMAP video surveillance system that combines computer vision algorithms with Prolog-based logic programming had been announced in Shet, Harwood, and Davis (2005). Later this research group has proposed an extension of the predicate logic with the bilattice formalism that takes into consideration the uncertainty in the reasoning about the video scenes (Shet, Singh, Bahlmann, Ramesh, Neumann, & Davis, 2011). The VERSA general-purpose framework for defining and recognizing events in live or recorded surveillance video streams was described in O'Hara (2008). The VERSA system is based on SWI-Prolog (Wielemaker, Schrijvers, Triska, & Lager, 2012). A real-time complex audio-video event detection based on the Answer Set Programming approach (Baral, Gelfond, Son, & Pontelli, 2010) was proposed in Machot, Kyamakya, Dieber, ans Rinner (2011). Heuristic rules based on a temporal logic are used for describing the semantics of video scenes in Lao, Han, and de With (2010). The LTAR system (Artikis, Sergot, & Paliouras, 2010) was designed for recognition of so-called long-term activities (such as "fight" or "meeting") consisting of sequences of short-term activities (such as "walking", "running", "active movement on the spot", "standstill", and "sharp movements") using a logic-programming implementation of the Event Calculus (Kowalski & Sergot, 1986). Later this approach was extended to handle the uncertainty that occurs in human activity recognition (Skarlatidis, Artikis, Filippou, & Paliouras, 2014). The ProbLog probabilistic logic language (Kimmig, Demoen, Raedt, Costa, & Rocha, 2011) was used for this purpose. A Prolog-based video surveillance framework for robust detecting abandoned objects was described in Ferryman et al. (2013). It is believed that the system automatically identifies interrelations between the observed people in order to draw a conclusion about who exactly owns the observed objects, and thus reduce the probability of a false alarm.

There are not so many papers that utilize logic programming for 3D video data processing. The method of recognition of long-term activities (Artikis, Sergot, & Paliouras, 2010) was adapted in Johanna (2013) for processing the data collected with the help of Kinect. The monitoring system is intended for the analysis of the behavior of the person in the support room. As reported in Johanna (2013), the skeleton data were not used for the analysis of long-term activities though the author calculates so-called Activity Level coefficient on the base of the average displacement of all tracked and inferred joints of the skeleton. The probabilistic approach to the long-term activity recognition (Skarlatidis, Artikis, Filippou, & Paliouras, 2014) was adapted to the distributed monitoring system (Katzouris, Artikis, & Paliouras,

2014) that provides health-care assistance in a smart-home setting. A reasoning engine based on the Description Logics (DL) formalism (Baader, Calvanese, McGuinness, Nardi, & Patel-Schneider, 2002) was used for Kinect-based posture and gesture recognition in Ruta, Scioscia, Summa, Ieva, Sciascio, and Sacco (2014). Prolog rules were used in the RoboSherlock system (Worch, Bálint-Benczédi, & Beetz, 2016) to distinguish between actions such as "take", "put", "fill", and "push". The logical inference takes into account context information about the properties of the objects. For example, to recognize the "fill" action, it is necessary to know that the object manipulated by a person is a container, food, or an ingredient (Worch, Bálint-Benczédi, & Beetz, 2016). The 3D video surveillance term has different meanings in research papers:

1. Initially, this term was associated with advanced video surveillance systems supporting real-time 3D visualization of indoor environment, buildings, and people to be monitored, as well as application of virtual and augmented reality technologies in the video surveillance (Sebe, Hu, You, & Neumann, 2003; Ott, Gutiérrez, Thalmann, & Vexo, 2006; Barnum, Sheikh, Datta, & Kanade, 2009; DeCamp, Shaw, Kubat, & Roy, 2010).
2. This term could be also associated with so-called 3D approaches to motion recovery from 2D images (Gavrila, 1999).
3. Nowadays, this term could be associated with using 3D/2.5D data acquisition equipment (Chen, Wei, & Ferryman, 2013) like time-of-flight (ToF) cameras, structured light sensors, stereo cameras, laser scanners (LiDARs) (Benedek, 2014), and Flash Ladars (Stettner, Bailey, & Silverman, 2008) in video surveillance.

Availability of 3D data acquisition equipment leads to considerable progress in video-surveillance-related research areas such as background subtraction and tracking people (Hansen, Hansen, Kirschmeyer, Larsen, & Silvestre, 2008), person identification (Satta, Pala, Fumera, & Roli, 2013; Sinha, Chakravarty, & Bhowmick, 2013), face recognition (Krishnan & Naveen, 2015), posture recognition (Diraco, Leone, & Siciliano, 2013; Ibañez, Soria, Teyseyre, & Campo, 2014), gesture recognition (Raheja, Minhas, Prashanth, Shah, & Chaudhary, 2015), people counting (Hsieh, Wang, Wu, Chang, & Kuo, 2012), human activity recognition (Aggarwal & Xia, 2014), etc. because the use of 3D/2.5D data makes possible to solve many complicated problems of computer vision caused by presence of shadows, variations in illumination, view-point changes, cluttered background, and occlusions. And a more important point is in that this technological advance enables progress in solving complex intelligent visual surveillance problems (Lun & Zhao, 2015) like human-human interaction recognition and prediction (Alazrai, Mowafi, & Lee, 2015), abnormal event detection for healthcare systems (Rougier, Auvinet, Rousseau, Mignotte, & Meunier, 2011; Lee & Chung, 2012; Mastorakis & Makris, 2014), unusual events detection (Wang & Liu, 2013), the behavior of customers (Popa, Koc, Rothkrantz, Shan, & Wiggers, 2011), monitoring of people with disabilities (Johanna, 2013; Lau, Ong, & Putra, 2014), aggressive behavior and anger detection (Patwardhan & Knapp, 2016), gait recognition (Preis, Kessel, Werner, & Linnhoff-Popien, 2012), biometric surveillance (Savage, Clarke, & Li, 2013), and analyzing the human motion to aid clinical decision making (Chaaraoui, Padilla-López, & Flórez-Revuelta, 2015; Leightley, Yap, Hewitt, & McPhee, 2016).

In contrast to the conventional 2D intelligent visual surveillance, the methods of 3D vision provide reliable recognition of the parts of the human body that makes possible a new statement of the problem and efficient practical application of methods of people behavior analysis in the visual surveillance systems. According to Borges, Conci, and Cavallaro (2013), people behavior is a kind of activity, which is a response to some stimuli (internal, external, conscious, or unconscious). For example, such interactions of people as a friendly handshake, a fight, the stealing of a suitcase from another person, etc. could be forms of behavior. Analysis of the behavior requires information about the context and other factors influencing it (Borges, Conci, & Cavallaro, 2013). This information can include video scene geometry, place and presence of objects, time constraints, etc. Thus, the availability of exact information about the coordinates of the parts of the body and scene geometry provided by means of 3D vision is a key to the automation of people behavior analysis, recognition, and understanding.

The method of the object-oriented logic programming of the intelligent visual surveillance systems was developed in Morozov et al., (2014); Morozov and Polupanov (2014); Morozov et al. (2015); Morozov and Polupanov (2015); Morozov, Sushkova, and Polupanov (2015a); Morozov (2015); Morozov and Sushkova (2016); and Morozov and Sushkova (2018). The distinctive features of the method are usage of the Actor Prolog object-oriented logic language (Morozov, 1994,1999,2007) and translation of the logic programs for intelligent visual surveillance to the Java language (Morozov & Polupanov, 2014; Morozov, Sushkova, & Polupanov, 2015b).

The Actor Prolog language is an object-oriented logic language, that is, it combines expressiveness of the logical and object-oriented approaches to the programming (Morozov, 1994,1999, 2002, 2003). This combination has increased the area of application of the logic programming. In particular, the object-oriented features enable to effort an opportunity to solve the problem of storing and processing big arrays of binary data (such as audio/video data) in the logic languages. The problem is in that plain logic languages do not implement data arrays directly, but use lists and structures for storing data because these data structures correspond to the Skolem functions in the first-order Predicate Calculus (Chang & Lee, 1973). In the object-oriented logic languages, the arrays of data can be encapsulated in the instances of specialized built-in classes. This enables fast and effective processing of the big data arrays in the logic languages.

In Actor Prolog, the text of the logic program consists of separate classes (Morozov, 1999). Parallel processes represent a kind of instances of classes (Morozov, 2003). The object-oriented means of the logic language allow one to split the program to interacting parallel processes that implement various stages of image processing and scene analysis while translation to the Java language provides a high performance sufficient for analyzing real-time video, as well as reliability and stability of the work of the intelligent visual surveillance software.

Recently, a method of the object-oriented logic programming for processing the data obtained from the equipment for 3D data acquisition in the intelligent visual surveillance systems was developed (Morozov, Sushkova, & Polupanov, 2017b, 2017c). Preliminary results of the research demonstrate that the application of methods of the object-oriented logic programming to 3D data processing gives exciting prospects in the area of intelligent visual surveillance. Progress in this research area can be achieved by the development of efficient low-level 3D data processing algorithms, logic programming languages and systems for general purpose, and advanced logical methods of 3D data processing.

CONVENTIONAL 2D INTELLIGENT VISUAL SURVEILLANCE LOGIC PROGRAMMING

Basic Principles of Intelligent Visual Surveillance Logic Programming

Let us consider an example of logical inference on video. The input of a logic program written in Actor Prolog is a standard sample provided by the BEHAVE team (Blunsden & Fisher, 2010; Fisher, 2007). The program only uses the coordinates of reference points in the ground plane (the points are provided by BEHAVE) that are necessary for estimation of physical distances in the video scene; no additional meta-information about the content of the video scene is used.

The video (see Figure 1) demonstrates a case of a street offense – a probable conflict between two groups of persons. These people meet in the scope of the video camera, then one group attacks another one, they fight, then people run away. This scene could be quickly recognized by a human; however, it is an interesting problem in the area of pattern recognition and video analysis.

First of all, note that probably the main evidence of an anomalous human activity in this video is the so-called abrupt motion of the persons. Abrupt motions can be easily recognized by the human as

Figure 1. An example of BEHAVE (Blunsden & Fisher, 2010; Fisher, 2007) video with a case of a street offense: one group attacks another

motions of a person's body and/or arms and legs with abnormally high velocity/acceleration. Thus, a logic programmer has a temptation to describe an anomalous human activity in terms of abrupt motions, somehow like this: "Several persons have met in the video scene. After that, they perform abrupt motions." It is not a problem to implement this definition using a set of logical rules. However, experiments with video samples demonstrate that this naive approach does not work. The problem is in that computer procedures recognize abrupt motions much worse than the human does and there are several reasons for this:

1. In the general case, to determine abrupt motions in a reliable and accurate way the program should recognize separate parts of person's body.
2. It is difficult to determine the exact coordinates of the person in the video scene. A common approach to the problem is in the usage of so-called ground plane assumption, that is, the computer program determines coordinates of body parts that are situated inside a predefined plane and this predefined plane usually is the ground. Thus, one can estimate properly the coordinates of person's feet, but a complex surface of the ground make the problem much more complex.
3. Computing the instantaneous velocity (that is the first derivative of moving person's coordinates) is a separate problem because the silhouette of the person changes unpredictably in different lighting conditions and can be partially overlapped by other objects. Therefore, the trajectory of the person contains a big amount of false coordinates that makes numerical differentiation senseless.

All these issues illustrate a close connection between the principles to be used for the logical description of the anomalous human activity and the output of low-level video processing procedures. One should take into account this connection when addresses the problem of the high-level (semantic) analysis of people activities.

In the example under consideration, the problem of anomalous human activity recognition will be solved using a logic program that describes a scenario of complex people behavior. The input data for the logic program will be supplied by low-level algorithms that trace objects in the video scene and estimate average velocity in different segments of the trajectories (Morozov et al., 2015). This low-level processing is implemented in Java and includes extraction of foreground blobs, tracking the blobs over time, detection of interactions between the blobs, the creation of connected graphs of linked tracks of the blobs, and estimation of the average velocity of the blobs in separate segments of the tracks (see Figure 2). The input data include a special set of blob motion statistics to discriminate running pedestrians, bicycles, and vehicles during the logical inference.

Introduction of Blob Motion Statistics

In contrast to the previously developed motion analysis methods of pedestrian detection that consider the periodicity and cyclic motion in the way humans move (Borges, 2013), a set of blob motion metrics based on the windowed coefficient of determination of the temporal changes of the length of the contour of the blob is developed in the Actor Prolog language. These metrics are more reliable when images of moving objects are noised and/or fuzzy, that is necessary for real applications.

The coefficient of determination R^2 indicates the proportionate amount of variation in the given response variable Y explained by the independent variable X in a linear regression model. The larger R^2 is, the more variability is explained by the linear regression model. Thus, the R^2 metric is supposed to be

Figure 2. The low-level processing of the video. Blobs are depicted by light (cyan) rectangles; multicolored lines denote the tracks of the blobs. The program estimates the velocity of the blobs and depicts it by different colors. Direct dark (blue) lines depict possible links between the blobs.

useful for discrimination of vehicles and running pedestrians when variable X is the time and variable Y is either the area or the length of the blob contour. An example is considered below to demonstrate that in the general case vehicles will be characterized by larger values of R^2 than running persons (the X variable is the time and the Y variable is the length of the contour of the blob) because the running person contour changes permanently in the course of motion when he waves his arms and moves his legs.

The standard definition of the (unadjusted) coefficient of determination is the following one:

$$R^2 = 1 - SSE/SST \tag{1}$$

where *SSE* is the sum of squared error and *SST* is the total sum of squares (Draper & Smith, 1998). A windowed modification of the R^2 metric is used in the Actor Prolog language, that is, the trajectory of the moving blob is characterized by a set of instantaneous values of the R^2 metric computed in each point of the trajectory. Suppose t_B is the beginning time point of the trajectory and t_E is the termination time point. Thus, the windowed R^2 metric is a set:

$$wR^2 = \left\{ R^2_{t,w} \right\} \tag{2}$$

where t is the time $\left(t \in \left\{ t_B + w/2, \ldots\ldots, t_E - w/2 \right\} \right)$ $t \in \left\{ t_B + w/2, \ldots\ldots, t_E - w/2 \right\}$ and w is the width of the window (the neighborhood of t) to be used for computation of R^2. In Actor Prolog, two statistical metrics that characterize the motion of the blob are supported, namely, the mean of the wR^2 distribution:

$$mean\left(wR^2 \right) = \frac{\sum_{i=1}^{n} wR^2_i}{n} \tag{3}$$

and the bias-corrected skewness of the wR^2 distribution:

$$skewness\left(wR^2 \right) = \frac{\sqrt{n\left(n-1 \right)}}{n-2} s\left(wR^2 \right) \tag{4}$$

$$s\left(wR^2 \right) = \frac{\frac{1}{n} \sum_{i=1}^{n} \left(wR^2_i - \overline{wR^2} \right)^3}{\left(\sqrt{\frac{1}{n} \sum_{i=1}^{n} \left(wR^2_i - \overline{wR^2} \right)^2} \right)^3} \tag{5}$$

where n is the number of elements in the wR^2 set. Skewness is a measure of the asymmetry of the data around the mean of the sample. If skewness is negative, the data are spread out more to the left of the mean than to the right and vice versa. Thus, in the framework of the moving blobs discrimination problem, one can expect that the vehicles will be characterized by larger values of the $mean\left(wR^2 \right)$ metric and smaller values of the $skewness\left(wR^2 \right)$ metric than running persons.

The properties of the $mean\left(wR^2 \right)$ and the $skewness\left(wR^2 \right)$ metrics can be illustrated by the example of the BEHAVE dataset (Fisher, 2013). For that, the tracks of moving blobs in this dataset were generated by the blob extraction methods implemented in Actor Prolog (Morozov et al., 2015; Morozov & Polupanov, 2014). Then 193 samples of tracks including 58 groups of walkers, 85 alone walkers, 15 groups of running persons, 20 alone running persons, 9 cars, and 6 bicycles were selected. Each blob has the following attributes: the trajectory of the central point of the rectangle blob; the area and the length of the contour of the blob in each point of time; the lower boundary estimation of the velocity of the blob (Morozov et al., 2015) in each point of time; and others. The values of the $mean\left(wR^2 \right)$ and the $skewness\left(wR^2 \right)$ metrics of these blobs are computed on the base of blob contour length values (Figure 3) with full window width $w = 0.4$ second (10 points) and frame rate 25 Hz. The following fast method of blob contour extraction is used: eight surrounding pixels are considered for every pixel of the

Figure 3. The wR² metric is computed on the base of changes of the blob contour length

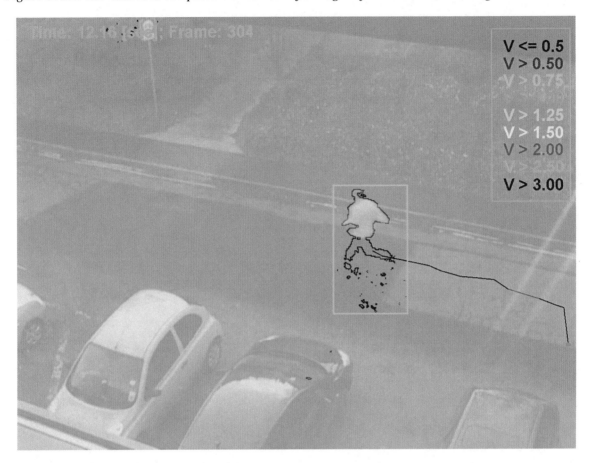

blob; all pixels of the blob that are enclosed by surrounding eight pixels of the blob are considered as internal points of the blob; all other pixels of the blob are considered as the contour of the blob.

Note that in the general case the frame rate and the window width are to be selected in accordance with the required response rate of the intelligent visual surveillance system. Theoretically speaking, a bigger width of the window supplies a better reliability of the recognition. On the other hand, the $mean\left(wR^2\right)$ and the $skewness\left(wR^2\right)$ metrics have sense only when the sample is big enough, because a small cardinality of the wR^2 set (the number of the values in the set) decreases the reliability of the recognition. The bigger width of the window decreases the cardinality of the wR^2 set in the segment of the trajectory (see equation 2) and, therefore, can decrease the reliability of the recognition in short segments of the trajectory. At the same time, the analysis of the short segments of the trajectory is necessary to supply a high response rate of the video surveillance system. A smaller width of the window increases the cardinality of the wR^2 set in short segments and, therefore, can supply a better response rate. A bigger frame rate of the video increases the cardinality of the wR^2 set and, therefore, can supply a better response rate also. At the same time, the bigger frame rate of the video increases the computational complexity of the recognition that can decrease the response rate of the visual surveillance system. Thus, a proper selection of these attributes implies a complex compromise between the reliability, the response rate, and the computational complexity of the recognition. Besides, one should use the cardinality as an additional argument in the fuzzy logical inference.

The "$velocity - mean\left(wR^2\right)$" and the "$velocity - skewness\left(wR^2\right)$" diagrams (Figures 4, 5) show clearly that these metrics allow the discrimination of fast-moving persons and vehicles, but are rather useless for the discrimination of slow or motionless objects. In Figure 4, the x coordinate is the velocity of the object and the y coordinate is the "$velocity - mean\left(wR^2\right)$" metric value. Pedestrians are depicted by circles: small light circles denote single walking persons; big light circles denote groups of walking persons; small dark circles denote single running persons; big dark circles denote groups of running persons. Vehicles are depicted by diamonds: small dark diamonds denote bicycles and big light diamonds denote cars. In Figure 5, the x coordinate is the velocity of the object and the y coordinate is the "$velocity - skewness\left(wR^2\right)$" metric value. Pedestrians and vehicles are depicted in the same manner as in Figure 4.

The size of blobs can be used as an additional statistical metric for the discrimination of running people and vehicles. It is difficult to estimate the real physical size of objects in the video scene, but one can use a standardized size of the blobs that is a ratio:

Figure 4. The values of the $mean\left(wR^2\right)$ metric of the blobs

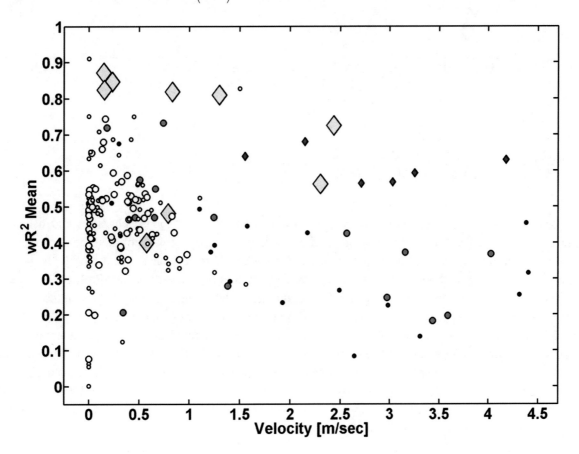

Figure 5. The values of the $kewness\left(wR^2\right)$ *metric of the blobs*

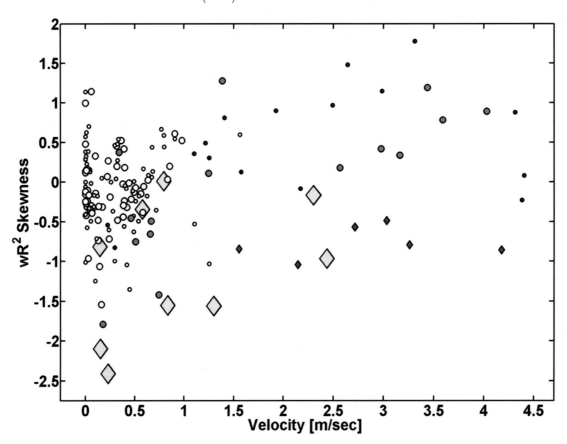

$$StdArea\left(t\right) = \frac{BlobArea\left(t\right)}{CharactLength\left(x\left(t\right),y\left(t\right)\right)^2} \qquad (6)$$

where *BlobArea(t)* is the area of the blob (in pixels) and *CharactLength(x, y)* is a characteristic distance. The characteristic distance is a function of *x* and *y*, where *(x, y)* are coordinates of the blob anchor point that is a function of time. It is convenient to define the anchor point *(x, y)* as the centre of the rectangle blob. The following method of characteristic distance computation is used in Actor Prolog. Let *C* be a 1-meter diameter circle created in the ground plane and the centre of the circle is *(x, y)*. Then the characteristic distance in the *(x, y)* point is the maximal diameter (in pixels) of *C* in the pixel space (see Figure 6). The transfer between the real space and the pixel space is implemented using the projective transform matrix. In Figure 6, the (green) ellipse corresponds to the one-meter diameter circle created around the position in the physical space. The (red) diameter line corresponds to the characteristic distance, i.e., maximal diameter of the ellipse in the pixel space. The (blue) stick demonstrates the direction of the minimal radius of the ellipse in pixel space; it is an estimation of the vertical direction in the physical space. The data are coming from (Fisher, 2007).

Figure 6. An example of computation of the characteristic distances in given positions of the ground plane

In the example below, the average value of the standardized area will be used for the discrimination of vehicles and other objects:

$$mean\left(StdArea\right) = \frac{\sum_{i=1}^{n} StdArea\left(i\right)}{n} \tag{7}$$

The diagram in Figure 7 demonstrates that the metric really does work. It provides a good discrimination of the vehicles in the video scene. In Figure 7, the *x* coordinate is the velocity of the object and the *y* coordinate is the value of the metric. Pedestrians and vehicles are depicted in the same manner as in Figure 4 notes that the $mean\left(StdArea\right)$ metric is useful only for detection of the vehicles and other objects of a big size, whereas the pedestrians and running people are not distinguished. Thus, the $mean\left(StdArea\right)$ metric is to be used jointly with the $mean\left(wR^2\right)$ and the $skewness\left(wR^2\right)$ metrics for the running people identification.

The $mean\left(wR^2\right)$, the $skewness\left(wR^2\right)$, and the $mean\left(StdArea\right)$ metrics are implemented in the Vision standard package of the Actor Prolog language (Morozov & Sushkova, 2018).

Figure 7. The values of the $mean(StdArea)$ *metric of the blobs*

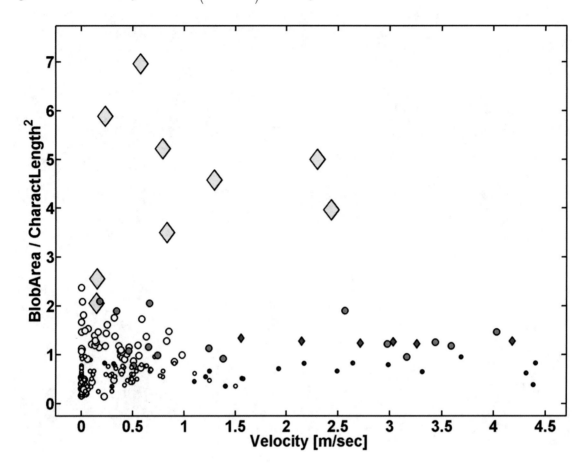

An Example of the Logical Inference

All results of the low-level analysis are received by the logic program in a form of Prolog terms describing a list of connected graphs of the tracks of blob movements. The following data structures will be used for describing the connected graphs of tracks. Note that the DOMAINS, the PREDICATES, and the CLAUSES program sections in Actor Prolog have traditional semantics developed in the Turbo/PDC Prolog system (Turbo Prolog Owner's Handbook, 1986).

Computer Code

```
DOMAINS:
BlobCoordinates = {
 frame: INTEGER,
 x: INTEGER, y: INTEGER,
 width: INTEGER, height: INTEGER,
 foreground_area: INTEGER,
```

```
     characteristic_length: REAL,
     contour_length: INTEGER,
     r2: REAL,
     velocity: REAL
     }.
TrackOfBlob = BlobCoordinates*.
EdgeNumber = INTEGER.
EdgeNumbers = EdgeNumber*.
ConnectedGraphEdge = {
     identifier: INTEGER,
     coordinates: TrackOfBlob,
     mean_velocity: REAL,
     wr2_cardinality: INTEGER,
     wr2_mean: REAL,
     wr2_skewness: REAL,
     mean_standardized_area: REAL,
     frame1: INTEGER,
     x1: INTEGER, y1: INTEGER,
     frame2: INTEGER,
     x2: INTEGER, y2: INTEGER,
     outputs: EdgeNumbers,
     inputs: EdgeNumbers
     }.
ConnectedGraph = ConnectedGraphEdge*.
```

That is, the connected graph of tracks is a list of underdetermined sets (Morozov, 1999) denoting separate edges of the graph. The nodes of the graph correspond to the points where tracks cross, and the edges are pieces of tracks between such points. Every edge is directed and has the following attributes: the identifier of corresponding blob (an integer *identifier*); the list of sets describing the coordinates and the velocity of the blob in different moments of time (*coordinates*); the average velocity of the blob in this edge of the graph (*mean_velocity*); the cardinality of the wR^2 metric (*wr2_cardinality*); the $mean\left(wR^2\right)$ metric value (*wr2_mean*); the skewness $\left(wR^2\right)$ metric value (*wr2_skewness*); the $mean\left(StdArea\right)$ metric value (*mean_standardized_area*); the numbers of the first and the last frames (*frame1, frame2*); the coordinates of the first and the last points (*x1, y1, x2,* and *y2*); the list of the numbers of the edges that are direct followers of the edge (*outputs*); and the list of the numbers of the edges that are direct predecessors of the edge (*inputs*).

Let us define a logic program that checks the graph of tracks and looks for the following pattern of interaction among several persons: "If two or more persons meet somewhere in the scene and one of them runs after the end of the meeting, the program should consider this scenario as a kind of a running away and a probable case of a sudden attack or a theft." Thus, the program has to alarm when this kind of sub-graph is detected in the total connected graph of tracks. In this case, the program marks all persons in the inspected graph by yellow rectangles and outputs the "Attention!" warning in the middle of the screen (see Figure 8).

Figure 8. The logical inference has found a possible case of a street offense in the graph of blob trajectories. All probable participants of the conflict are marked by yellow rectangles. The tracks are depicted by lines.

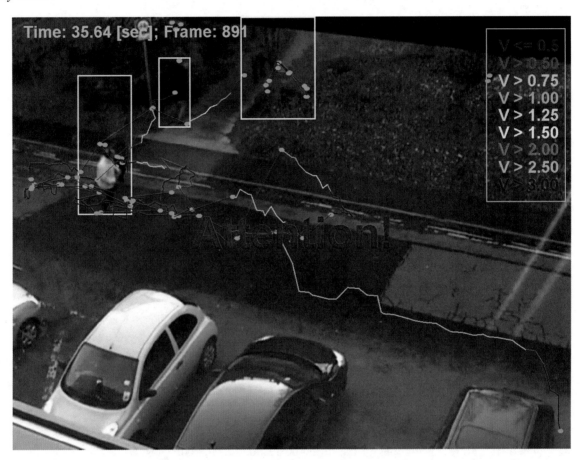

One can describe the concept of the running away formally using the defined data types:

Computer Code

```
PREDICATES:
is_a_kind_of_a_running_away(
    ConnectedGraph,
    ConnectedGraph,
    ConnectedGraphEdge,
    ConnectedGraphEdge,
    ConnectedGraphEdge) - (i,i,o,o,o);
```

The *is_a_kind_of_a_running_away(G,G,P1,E,P2)* predicate is defined with the following arguments: *G* is a graph to be analyzed; *E* is an edge of the graph corresponding to a probable incident; *P1* is an edge of the graph that is a predecessor of *E*; *P2* is an edge that is a follower of *E*. Here is an Actor Prolog program code with brief explanations. Note that in the Actor Prolog language, the "==" operator corresponds to the "=" ordinary unification of the standard Prolog.

Computer Code

```
CLAUSES:
get_edge(1,[Edge|_],Edge):-!.
get_edge(N,[_|Rest],Edge):-
    N > 0,
    get_edge(N-1,Rest,Edge).
contains_a_running_person([N|_],G,P):-
    get_edge(N,G,E),
    is_a_running_person(E,G,P),!.
contains_a_running_person([_|R],G,P):-
    contains_a_running_person(R,G,P).
is_a_meeting(O,_,E,E):-
    O == [_,_|_],!.
is_a_meeting([N1|_],G,_,E2):-
    get_edge(N1,G,E1),
    E1 == {inputs:O|_},
    is_a_meeting(O,G,E1,E2).
is_a_kind_of_a_running_away([E2|_],G,E1,E2,E3):-
    E2 == {inputs:O,outputs:B|_},
    B == [_,_|_],
    contains_a_running_person(B,G,E3),
    is_a_meeting(O,G,E2,E1),!.
is_a_kind_of_a_running_away([_|R],G,E1,E2,E3):-
    is_a_kind_of_a_running_away(R,G,E1,E2,E3).
```

In other words, the graph contains a case of a running away when there is an edge *E2* in the graph that has a follower *E3* corresponding to a running person and a predecessor *E1* that corresponds to a meeting of two or more persons. It is requested also that *E2* has two or more direct followers, that is, it is a case of branching in the graph.

A fuzzy definition of the running person concept is as follows:

Computer Code

```
is_a_running_person(E,_,E):-
    E == {     frame1:T1,
               frame2:T2,
               mean_velocity:V,
               mean_standardized_area:A,
               wr2_mean:M,
```

```
                    wr2_skewness:S,
                    wr2_cardinality:C|_},
        is_a_fast_object(T1,T2,V),
        fast_object_is_a_runner(A,M,S,C),!.
is_a_running_person(E,G,P):-
        E == {outputs:B|_},
        contains_a_running_person(B,G,P).
```

The graph edge corresponds to the running person if and only if two conditions hold:

1. The values of the $mean\left(wR^2\right)$, the skewness $\left(wR^2\right)$, the $mean\left(StdArea\right)$ metrics, and the cardinality of the wR^2 set satisfy the fuzzy definition of the running pedestrian.
2. This edge is recognized as a fast object, that is, the velocity and the length of the graph edge satisfy the fuzzy definition of the fast object.

This is the definition of the predicates that recognize fast objects and running persons:

Computer Code

```
fast_object_is_a_runner(A,M,S,C):-
        MC== ?fuzzy_metric(C,7,2),
        MA== 1 - ?fuzzy_metric(A,2.75,0.75),
        MM== 1 - ?fuzzy_metric(M,0.49,0.10),
        MS== ?fuzzy_metric(S,0.25,1.00),
        MA * MM * MS * MC >= 0.25.
is_a_fast_object(T1,T2,V):-
        M1== ?fuzzy_metric(V,1.7,0.7),
        D== (T2 - T1) / 25,
        M2== ?fuzzy_metric(D,0.5,0.25),
        M1 * M2 >= 0.5.
```

The values of fuzzy thresholds used in the rules were computed based on the blob tracks samples discussed above. It is interesting, that these fuzzy rules discriminate correctly all humans and vehicles that are fast objects in accordance with the fuzzy definition. More precisely, only 22 blobs from 193 are recognized as fast objects; the rules properly recognize 7 fast moving vehicles and 15 running pedestrians (see Table 1).

Note that Actor Prolog implements a non-standard functional notation, namely, the "?" prefix informs the compiler that the *fuzzy_metric* term is a call of a function, but not a data structure. An auxiliary function that calculates values of the fuzzy metrics is represented below. The first argument of the function is a value to be checked, the second argument is the value of a fuzzy threshold, and the third one is the width of the threshold ambiguity area. The "=" delimiter defines an extra output argument that is a result to be returned by the function:

Table 1. The quality of the running person detection

Characteristic	Value
Number of true positive	15
Number of false negative	0
Number of true negative	7
Number of false positive	0
True positive rate (sensitivity)	1
True negative rate (specificity)	1

Computer Code

```
fuzzy_metric(X,T,H) = 0.0:-
    X <= T - H,!.
fuzzy_metric(X,T,H) = 1.0:-
    X >= T + H,!.
fuzzy_metric(X,T,H) = V:-
    V== (X-T+H) * (1 / (2*H)).
```

Studies of the BEHAVE samples of anomalous people behavior have demonstrated that considered logic program recognizes several cases of the abnormal behavior. This result indicates that given behavior pattern really presents in the videos of anomalous people behavior. One can develop other sets of logical rules describing different scenarios of aggressive and/or abnormal people behavior. Note that demonstrated above 52 lines of the Actor Prolog code correspond to 351 lines of the optimized Java source code that implements the graph search operations.

This example illustrates the basic principles of the logical description of anomalous human activities and the logical inference on the video data. Surely, even the simplest scheme of a visual surveillance logic program has to contain many additional elements, including video information gathering, the low-level image analysis, a control of the logical inference, and reporting the results of intelligent visual surveillance. Note that all these stages of the video data processing can be implemented using the Actor Prolog language.

IMPLEMENTATION ISSUES OF THE OBJECT-ORIENTED LOGIC PROGRAMMING OF THE INTELLIGENT VISUAL SURVEILLANCE

Actor Prolog to Java Translation

In the intelligent visual surveillance applications, the authors use a compilation of the Actor Prolog language to Java because processing huge video data arrays requires a reliable implementation of the logic language with a clear memory management. Logic programs operating with big amounts of data should work stably for long periods. The industrial Java virtual machine is used as a basis for the logic programming system because from the authors' point of view modern processors are fast enough to neglect the speed of the executable code for the sake of robustness, readability, and openness of the logic programs. At the same time, a fast executable code that is appropriate for real-time data processing is necessary and the compilation schema of Actor Prolog ensures high performance of the executable code. It is important also that using an industrial Java virtual machine as a basis for the logic programming system ensures its flexibility and fast adaptation to the new operating systems and processor architectures.

The Actor Prolog language is significantly different from the conventional Clocksin&Mellish Prolog (Clocksin & Mellish, 2003). Turbo-Prolog style domain and predicate declarations of Actor Prolog are of importance for the industrial application programming and are helpful for the executable code optimization, but, on the other hand, object-oriented features and supporting concurrent programming make the translation to be a non-trivial problem.

The state-of-the-art compilation schema of the Actor Prolog system includes the following elements (Morozov & Polupanov, 2014; Morozov, Sushkova, & Polupanov, 2015b):

1. **Source Text Scanning and Parsing:** Methods of "thinking translation" preventing unnecessary processing of already translated source files are implemented. That is, after the update of source codes, the compiler tries to use information collected/computed during its previous run.
2. **A Global Flow Analysis:** The compiler tracks flow patterns of all predicates in all classes of the program.
3. **Determinism Check:** The translator checks whether predicates are deterministic or non-deterministic. A special kind of so-called imperative predicates is supported, that is, the compiler can check whether a predicate is deterministic and never fails.
4. **Type Check:** The translator checks data types of all predicate arguments and arguments of all class instance constructors.
5. **Inter-Class Links Analysis:** On this stage of global analysis, the translator collects information about the usage of separate classes in the program, including data types of arguments of all class instance constructors. This information is necessary for the global flow analysis and the global optimization of the program. In particular, this information is used to eliminate all unused predicates from the executable code.
6. **Generation of an Intermediate Java Code.**
7. **Translation of this Java Code by a Standard Java Compiler.**

There are three keywords in the languages for the declaration of the determinacy of the methods (predicates): *determ*, *nondeterm*, and *imperative*. The *nondeterm* keyword informs the compiler that there are no restrictions on the behavior of methods and they can produce several answers in the case of backtracking and/or terminate with failure. The *determ* keyword means that methods can produce just one answer or terminate with failure. The *imperative* keyword imposes the hardest restrictions on the methods: the predicates must succeed and produce one answer; this means that the predicate operates indeed as a usual procedure in an imperative language. All these restrictions are checked by the compiler during the translation of the program. The determinism check ensures a possibility to use different optimization methods for different kinds of predicates:

1. The non-deterministic predicates are implemented using a standard method of continuation passing. Clauses of one predicate correspond to one or several automatically generated Java classes.

2. The deterministic predicates are translated to Java procedures. All clauses of one deterministic predicate correspond to a single Java procedure. Backtracking is implemented using a special kind of lightweight Java exceptions.

3. The imperative predicates check is the most complex stage in the translation schema because it requires a check of all separate clauses as well as a mutual influence of the clauses/predicates. This check is of the critical importance, since the imperative predicates, as a rule, constitute the main part of the program and the check ensures a deep optimization of these predicates. Clauses of the imperative predicates are translated to Java procedures.

The tail recursion optimization is performed for recursive predicates. The recursive predicates are implemented using the *while* Java command. Note that Actor Prolog supports the explicit definition of ground/non-ground domains; the *reference*, *ground*, and *mixed* keywords are used for this purpose. The translator uses this information for a deep optimization of Java code.

The described compilation schema ensures a high performance of the executable code. The translator creates Java classes corresponding to the classes of an object-oriented Actor Prolog program. Given external Java classes can be declared as ancestors of these automatically created classes and this is the basic principle of the implementation of built-in classes (Morozov, 2018) and integration of Actor Prolog programs with external libraries. The possibility of easy extension of the Actor Prolog programming system by new built-in classes is a benefit of the selected implementation strategy. For instance, the Java2D, the Java3D, and the FFmpeg open source libraries are connected with the Actor Prolog system in this way.

The described approach to Prolog and Java merging has the following advantages in comparison with an approach where a logic program and a Java program communicate through an interface as two separate black boxes, e.g., a Prolog program and a Java program exchange data through a Prolog-Java interface such as in SWI Prolog (Wielemaker, Schrijvers, Triska, & Lager, 2012):

1. **Portability of the Logic Programming System:** The use of the industrial virtual machine is a basis for quick adaptation to new operating systems and processor architectures.
2. **Portability of the Programs:** The translator generates Java programs/applets that can operate on any computer without preliminary installation of Actor Prolog; only Java is necessary.
3. **Readability of the Intermediate Code:** The intermediate code can be easily inspected by a human, if necessary.
4. **Safety of the Programs:** All Java features ensuring the safety of the programs are available.
5. **Reliability and Stability of the Programs:** The single-language-approach always ensures better reliability and stability of the application programs.

The main disadvantages of the developed approach are the following:

1. The logic programming system depends on Java virtual machine.
2. Only static optimization of the code is possible because Java implements no advanced run-time optimization methods developed in the logic programming area.
3. The executable code is slower in comparison to the translation-to-the-C-language approach.

Note that previously a compilation schema based on C/C++ intermediate code generation was recognized as an appropriate way to obtain maximal speed of the executable code; the following projects are examples of using this compilation schema: Mercury (Henderson & Somogyi, 2002), KLIC (Fujise, Chikayama, Rokusava, & Nakase, 1994), wamcc (Codognet & Diaz, 1995). On the other hand, generation of Java intermediate code ensures platform independence of the application software and guarantees absence of difficult-to-locate errors caused by memory leaks and out-of-range array operations; the examples are Actor Prolog, PrologCafe (Banbara, Tamura, & Inoue, 2006), KLIJava (Kuramochi, 1999), SAE-Prolog (Eichberg, 2011), and jProlog (Demoen & Tarau, 1997). In the Actor Prolog, the second compilation schema is used to ensure robustness, readability, and openness of the executable code.

In contrast to the PrologCafe and wamcc projects, the Actor Prolog compiler does not implement Warren Abstract Machine (WAM) (Warren, 1983). In contrast to jProlog and BinProlog of Tarau (2012), it does not apply binarization of the logic programs. The Actor Prolog compiler generates a kind of idiomatic source code, but in contrast to the SAE-Prolog project (Eichberg, 2011) it uses domain and predicate declarations to process non-deterministic, deterministic, and imperative predicates in different ways. In contrast to the P# project (Cook, 2004); Actor Prolog applies non-idiomatic predicate calls from idiomatic predicates and vice versa.

Distributed Logic Programming and Agent Logic Programming

The concept of multi-agent programming came to the field of the intelligent visual surveillance from Artificial Intelligence (Russell & Norvig, 1995; Shen, Hao, Yoon, & Norrie, 2006; Baldoni, Baroglio, Mascardi, Omicini, & Torroni, 2010; Gascueña, Fernández-Caballero, 2011; Badica, Braubach, & Paschke, 2011; Kravari & Bassiliades, 2015). The idea of the multi-agent approach to the visual surveillance is in that the intelligent visual surveillance system consists of communicating programs (agents) that have the following properties: autonomy (they operate without direct control from users and other agents), social ability (they can cooperate to solve the problem), reactivity (they perceive the environment and respond to external events), and pro-activity (they demonstrate a goal-directed behavior). Theoretically speaking, the multi-agent approach can provide flexibility, reliability, and openness of the intelligent visual surveillance systems (Vallejo, Albusac, Castro-Schez, Glez-Morcillo, & Jiménez, 2011; Ejaz, Manzoor, Nefti, & Baik, 2012; Shiang, Onn, Tee, bin Khairuddin, & Mahunnah, 2016). For instance, let us imagine that stages of the video analysis are implemented by a set of agents. Then, the visual surveillance system can be easily extended by an additional method of abnormal behavior recognition without modification of its agents and even without suspending its work. One just needs to insert into the system a new agent that can utilize results of other agents and transfer its own results to others.

In recent decade, the declarative approach to the development of the multi-agent systems is recognized as a basic idea in this research area; a set of excellent declarative multi-agent platforms and languages are developed and implemented (Baldoni, Baroglio, Mascardi, Omicini, & Torroni, 2010; Badica, Braubach, & Paschke, 2011). Unfortunately, in the framework of the intelligent visual surveillance systems, the agents are to perform very specific operations on big arrays of binary data that are out of the framework of the conventional symbolic processing operations typical for declarative languages. Thus, there is a reason for the development of new means of the multi-agent-programming for experimenting with intelligent visual surveillance systems.

A distributed version of the Actor Prolog language (Morozov, Sushkova, & Polupanov, 2017a) was developed for experimenting with the declarative agent approach to the intelligent visual surveillance.

Actor Prolog can be easily adapted to the distributed programming framework even without modifications of the syntax because this logic language is indeed an object-oriented language. The only problem is in the incorporation into the language the ability of remote procedure calls.

The term "remote procedure call" is usually associated with the OMG CORBA, Java RMI, or MS DCOM protocols. This meaning of the term is relevant to the topic because the remote predicate calls are implemented in the distributed Actor Prolog using the Java RMI protocol. At the same time, the term is linked with the general problems of the logic language design and implementation in the context of the agent logic programming.

The Actor Prolog language differs from other state-of-the-art Prolog-based agent languages like Jason (Bordini, Hübner, & Wooldridge, 2007) and 2APL (Dastani, 2008) in that it is not based on the BDI model and it does not directly offer high-level features such as planners and agent communication languages that might be expected for a multi-agent language. Actor Prolog is rather a more high-performance object-oriented logic language that is a base for implementation of real-time multi-agent application platforms.

It is known that interactions between independent agents are very hard to handle for strongly typed object systems (Odell, 2002). The main problem to be resolved in the course of adapting Actor Prolog to multi-agent paradigm was the contradiction between the strong type system of the language and the idea of independence of the software agents. The strong type system is an important feature of the language and is necessary for generation of fast and reliable executable code (Morozov & Polupanov, 2014; Morozov, Sushkova, & Polupanov, 2015b). The problem is in that one needs to transfer information about the data types between the software units to implement their link and static type-checking. This kind of information exchange between the software agents is definitely undesirable because it decreases the autonomy of the agents and complicates the agent life cycle. In the Actor Prolog language, another solution of the problem is proposed; the type system of Actor Prolog is partially softened to allow a dynamic type-checking (instead of the static one) in some restricted cases linked with the inter-agent communications.

Another problem that is close to the topic, but is still different, is a combination of the object-oriented paradigm and the strong typing. It was recognized earlier, that types are useful for formalizing and maintaining object interfaces, though types are orthogonal to objects and their integration is not a simple deed (Nierstrasz & Dami, 1995). The Actor Prolog language supports simultaneously types (domains) and classes/objects. A distinctive feature of the language is in that the "object" and "data item" notions were clearly separated in the language (Morozov, 1999). The language has the strong type system that supports various kinds of simple and composite data items like numbers, structures, lists, etc. At the same time, Actor Prolog supports classes based on the "clauses view" of the logic object-oriented programming (Davison, 1992). The instances of classes (so-called "worlds") can be processed like standard Prolog terms; they can be passed as arguments to predicates and can be included in composite terms. However, special rules are used for the unification of variables containing instances of classes and special means are to be developed for the interchange of the terms of this kind between the distributed agents.

In the Actor Prolog language, different instances of classes are always treated as different entities, that is, unification of two worlds succeeds if and only if these worlds are the same instance of the same class. The interface of the class contains all necessary information about its methods including names, arity, flow patterns, and types of arguments. The information about the determinacy/non-determinacy of the methods is also included in the interface.

The description and usage of the class interfaces are complicated a bit in Actor Prolog by the fact that the language supports concurrent processes and two different kinds of method invocations: plain and asynchronous (Morozov, 2003). The processes are a special kind of class instances; they are defined using double round brackets in the class instance constructors. The plain method invocation is a usual predicate call of standard Prolog; the predicate can be invoked in a given world using the "?" prefix. The asynchronous method invocations are indicated by special prefixes "<-" and "<<", see details in (Morozov, 2003). Only asynchronous predicate calls are applicable for the processes; an attempt to implement a plain predicate call in a concurrent process will always terminate with a failure. The *internal* keyword is introduced in the language to facilitate optimization of the logic programs. This keyword informs the compiler that a given slot of a class always contains a plain class instance, but not a process, that is important for the analysis of predicate determinacy. Obviously, one will focus on the asynchronous method invocations in this section because class instances obtained from another logic program are the processes that operate concurrently in relation to the invoking logic program.

Ordinarily, a standard static type-checking is performed in the distributed version of Actor Prolog. The dynamic type-checking is implemented only if a method (a predicate) is to be called in an object (an instance of Actor Prolog class) that is originated from another logic program and is transferred somehow to the logic program under consideration. The implementation of the remote predicate call includes the following operations:

1. One checks the name and the arity of the predicate. The predicate with the target name and arity is to be found in the object.
2. One checks the flow pattern of the predicate. The Actor Prolog language supports explicit declaration whether the argument is input or output; the flow directions of all the arguments in the predicate call are checked.
3. One checks so-called structural match of domains of all the arguments (this is a kind of dynamic type-checking).

The structural match of the domains means that the graphs representing the data structures belonging to the domains have to be equivalent, but not the names of the domains. Let us consider briefly the type system of the Actor Prolog language and the structural matching rules associated with various kinds of simple and compound data types (domains).

Actor Prolog supports the following simple data types: integer, real, symbol, and string. The difference between the integers and real numbers is in that the real numbers contain a dot. The difference between the symbols and strings is in that during the execution of the program the symbols are represented by integer codes internally, but not by the text. On the syntax level, the symbols are enclosed in single quotes (apostrophes) and the strings are enclosed in double quotes. Here is an example of using these built-in data types for the definition of user data types:

Computer Code

```
DOMAINS:
Year        = INTEGER.
Height      = REAL.
Color       = SYMBOL.
Message     = STRING.
```

During the structural matching, the integers can match only the integers, the reals can match only the reals, etc. No automatic type conversions are allowed.

Actor Prolog supports so-called numerical ranges and enumerations. A range type can be defined using the integer or real bounds, for example:

Computer Code

```
Hour          = [0 .. 24].
Angle         = [0.0 .. 360.0].
```

The procedure of structural matching checks the exact equality of the bounds of the numerical range types to be compared. The only exception is in that the real range bounds can slightly differ in accordance with the real number precision given in the translator options.

An enumeration type can be defined using a set of constants of any simple types, for instance:

Computer Code

```
Hour          = 0; 1; 2; 3; 4; 5; 6; 7; 8; 9; 10; 11; 12.
Color         = 'Red'; 'Blue'; 'Green'.
```

The structural match of two enumerations means that these types include the same sets of elements. Note that the type definition can include a set of names of other types; this is a basic difference of the Actor Prolog type system from analogous type systems in the Turbo/PDC Prolog family (Turbo Prolog Owner's Handbook, 1986). For instance, an argument of the following type can transmit both integer and real values:

Computer Code

```
Numerical     = INTEGER; REAL.
```

The type definition can refer to other types. The structural matching procedure considers all the type definitions and compares the corresponding sets of elements that can include simple domains, literals, and composite types.

There are three kinds of composite types in Actor Prolog, namely: structures, lists, and so-called underdetermined sets (Morozov, 1999). The structure type definition consists of a functor and arguments enclosed in round brackets, for example:

Computer Code

```
AppointedDate = date(Year,Month,Day).
```

The structural matching procedure checks whether two structure domain definitions contain the same functor and the same number of arguments. Then, the structural matching of the types of all corresponding arguments is implemented.

In contrast to the standard Prolog (Clocksin & Mellish, 2003), the lists are a separate type of Actor Prolog, but not a kind of the structures. The list type definition contains a name of element type and an asterisk, for instance:

Computer Code

```
Dates           = AppointedDate*.
```

The structural matching procedure checks the types of the elements of the list types.

The definition of an underdetermined set type in Actor Prolog contains an unordered set of named pairs enclosed in braces. Every pair contains the identifier of the pair and the type of the argument, for example:

Computer Code

```
Customer        = {name: STRING, birthday: Date, age: INTEGER}.
```

The structural matching procedure compares the types of all corresponding pairs in the definitions of underdetermined set types. The types must contain the pairs of the same names, but the order of the pairs is insignificant.

There are two exotic data types in Actor Prolog: so-called anonymous type "_" and so-called "any set" type "{_}". The former type indicates that a predicate accepts terms of any types; it is useful for the definition of read/write procedures, etc. The second type is used for the definition of predicates/attributes that accept terms of any types, but only in a form of an underdetermined set, for instance:

Computer Code

```
HTTP_ContentParameters = {_}.
```

By the rules of the structural matching, the anonymous type matches only the anonymous type and the "any set" type matches only the "any set" type.

All the rules described above are applicable to both the static and dynamic type-checking. A single difference relates to the structural matching data types that contain class names. The point is in that the type definition in Actor Prolog can include the name of a class enclosed in round brackets, for example:

Computer Code

```
MessageHandler = ('MyClass').
```

This type definition means that terms of the *MessageHandler* type can be instances of the *MyClass* class. The instance can be a plain world or a concurrent process of the class. The definition tells nothing about the concurrent execution of the class instance but does not prohibit this kind of class usage too. Actor Prolog considers this world data type as a simple one. The compiler of non-distributed Actor Prolog guarantees that any term of this type is an instance of the *MyClass* class or an instance of a class that inherits the *MyClass* class; this rule is softened in the distributed Actor Prolog.

The distributed Actor Prolog checks whether an instance of the class belongs to the class pointed in the type definition only if this class is defined in the same logic program (i.e., it is technically possible to check it). An instance of an external class obtained from another logical agent (program) can be freely assigned to the variable/predicate argument of any type that includes a class name. Thus, the structural matching algorithm allows matching of any world types; the names of classes in the type definitions are simply ignored when the classes are defined in different logic programs (agents).

In distributed Actor Prolog, an instance of the class can be transferred to another logic program somehow and be accepted without the check of the interface when the accepting program expects to receive an instance of some class. A real check of the class interface is to be performed when the accepting program tries to invoke a method from the external object. In this case, the structural matching procedure described above is to be performed, that can confirm the suitability of the object or yield a runtime error. Obviously, the implementation of this check requires information on the origins of all objects in the logic program. Thus, distributed Actor Prolog keeps an internal table of all class instances created during the program execution and transferred outside. Another internal table contains all objects accepted somehow from other logic programs. These tables allow Actor Prolog to distinguish clearly the instances of own and external classes and use this information in the structural matching algorithm.

Thus, the multi-agent interaction in Actor Prolog is based on the fusion of dynamic and static typing. The static type-checking and standard features of a nominative type system are implemented for all the own worlds like in the conventional Actor Prolog. At the same time, the dynamic type-checking and elements of a structural type system are implemented for all the external worlds. The authors consider the type system of Actor Prolog as a combined type system. This type system ensures the advantages of the static type-checking that are necessary for generation of high-performance code and the flexibility of the dynamic type-checking that is necessary for the multi-agent systems programming.

Let us consider an example of the remote predicate call. Suppose there are two agents: *Recognizer* and *Observer* (see Figure 9). These agents should cooperate to search and recognize people in a video scene. Suppose that *Recognizer* controls its own pan-tilt-zoom (PTZ) camera and can identify the person

Figure 9. An example of two cooperating agents

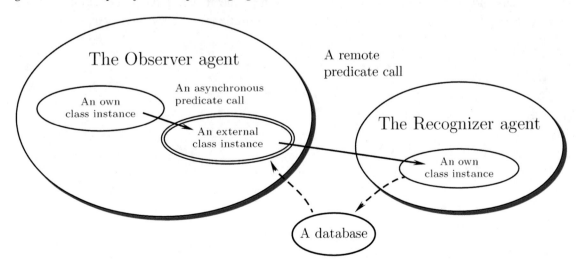

in given coordinates. The *Observer* agent can analyze the behavior of people and calculate coordinates of the persons to be identified. Suppose that these logic programs are different agents that should establish a link dynamically and exchange information to solve the problem. In Figure 9, the *Recognizer* agent publishes an instance of a class in the external database. Then, the *Observer* agent obtains this class instance and sends an asynchronous message to this class instance using Java RMI.

First, let us define a schema of the *Recognizer* logic program. The logic program below creates an instance of a class and saves it in a file to be accessible for other programs. Other programs can read this class instance and implement remote calls of predicates defined in the *Recognizer* program. Let the external program transmits coordinates of a person to be identified and the *Recognizer* logic program accepts this information and simply print it on the screen for the sake of simplicity.

In accordance with the semantics of the Actor Prolog language, the execution of the program begins with the creation of an instance of the *Main* class. In the program under consideration, the *Main* class is an instance of the *Console* built-in class that implements a text window control:

Computer Code

```
class 'Main' (specialized 'Console'):
external_file    = ('DataExchange');
[
PREDICATES:
intruder_coordinates(REAL,REAL)     - (i,i);
MODEL:
?intruder_coordinates(X,Y).
```

The *Main* class contains a single slot named *external_file*. The value of this slot is an instance of the *DataExchange* class. The *DataExchange* class implements the data exchange using a built-in database for simplicity since this is the simplest way of external file control in the Actor Prolog language.

There is a single predicate definition in the *PREDICATES* section. The *intruder_coordinates* predicate has two input real arguments. This predicate is never called directly inside the *Recognizer* logic program; that is why one should indicate in the *MODEL* section that this predicate is to be invoked with two arguments. Otherwise, the translator will discard this predicate during the optimization of the code.

The *CLAUSES* section of the *Main* class contains the definitions of the *goal* and *intruder_coordinates* predicates:

Computer Code

```
CLAUSES:
goal:-!,
      external_file ? insert(self),
      external_file ? save("e:/SharedData.db"),
      writeln("I wait for intruder coordinates...").
intruder_coordinates(X,Y):-
      writeln("X= ",X," Y= ",Y).
]
```

The *goal* predicate is called automatically during the creation of the *Main* class instance. This predicate inserts the instance of the *Main* class into the *DataExchange* database using the *insert* built-in method and the *self* keyword. Then it records the database content to the file using the *save* built-in method and writes the message on the screen: "I wait for intruder coordinates..." The *intruder_coordinates* predicate is to be invoked from outside using the remote call protocol. This predicate simply writes the coordinates on the screen.

There is yet another class definition in the text of the *Recognizer* program. The *DataExchange* class inherits methods from the *Database* built-in class that implements a simple database management system. There is a definition of the *Target* domain in the *DOMAINS* section of the *DataExchange* class. This definition is necessary in order to inform the database management system about the type of data to be stored in the *DataExchange* class instance. It is declared that the *Target* type includes instances of the *Main* class.

Computer Code

```
class 'DataExchange' (specialized 'Database'):
[
DOMAINS:
Target      = ('Main').
]
```

Let us consider the *Observer* logic program. Suppose this program should obtain an instance of an external class from the file and send to this object a message containing coordinates of a person to be identified. The *Main* class of this program inherits methods from the *Console* built-in class too.

Computer Code

```
class 'Main' (specialized 'Console'):
file        = ('InternalDatabase');
[
PREDICATES:
send_coordinates('AcceptingAgent')  - (i);
CLAUSES:
goal:-
     file ? load("e:/SharedData.db"),
     file ? find(ExternalObject),!,
     send_coordinates(ExternalObject).
send_coordinates(ExternalObject):-
     ExternalObject << intruder_coordinates(8.28,32.39),
     writeln("The information is sent...").
]
```

The *Main* class includes the *file* slot that contains an instance of the *InternalDatabase* class. There is a definition of the *send_coordinates* auxiliary predicate in the *PREDICATES* section. This predicate has one input argument that should contain an instance of a class that inherits methods from the *AcceptingAgent* interface defined below. The *goal* predicate acquires information from the external file using the *load* built-in method of the *InternalDatabase* class. Then it takes the *ExternalObject* world from the database and transmits this class instance to the *send_coordinates* predicate. The *send_coordinates* predicate implements an asynchronous predicate call in the *ExternalObject* world and writes the text message on the screen: "The information is sent..." Note that the asynchronous call is implemented using the remote call protocol because the variable *ExternalObject* contains the object that originates from another logic program. The dynamic type-checking will be implemented during the call.

The *AcceptingAgent* interface describes methods that are to be supported by the collaborator of the *Observer* agent. Note that this interface links up in no way with the classes/interfaces of the *Recognizer* agent:

Computer Code

```
interface 'AcceptingAgent':
[
PREDICATES:
intruder_coordinates(REAL,REAL)      - (i,i);
]
```

The *InternalDatabase* auxiliary class is defined in a similar way as the *DataExchange* class in the *Recognizer* agent. The single difference is in that the *Target* domain includes instances of classes that inherit the *AcceptingAgent* interface.

Computer Code

```
class 'InternalDatabase' (specialized 'Database'):
[
DOMAINS:
Target      = ('AcceptingAgent').
]
```

Let us execute the *Recognizer* logic program. The program will create the *SharedData.db* file in the "e:" hard disk and write the text on the screen:

Computer Code

I wait for intruder coordinates...

The *SharedData.db* file is text one because the *Database* class stores the information in a user-readable format. The file contains something like this (the text is reduced):

Computer Code

```
('feffo2muyuhbf2het4sx ... rlq92bop1liyzkim8pvm7q8');
```

The alphanumeric code enclosed in the apostrophes and round brackets is a usual Actor Prolog term, namely, a text representation of a class instance. On the technical level, this is an encoded instance of a Java RMI stub that refers to the class instance. Note that this format of data exchange is appropriate for any type of high-level data transfer protocols including E-Mails.

Let us launch the *Observer* agent now. This logic program will read the class instance from the *SharedData.db* external file, implement a remote predicate call in the *Recognizer* agent, and write the text message:

Computer Code

The information is sent...

The *Recognizer* agent will accept the remote predicate call and write the acquired coordinates on the screen:

Computer Code

I wait for intruder coordinates...

```
X= 8.28 Y= 32.39
```

This example illustrates the basic schema of the agent data exchange in distributed Actor Prolog using the remote predicate calls and dynamic type-checking as well as some technical details of the data encoding. The type system of Actor Prolog ensures the advantages of the static type-checking that are necessary for generation of fast executable code and the flexibility of the dynamic type-checking that is necessary for the multi-agent systems design.

Low-Level Video Processing

A typical intelligent visual surveillance program includes high-level procedures (that is, anomalous human activity recognition, etc.) and low-level video processing procedures (for instance, background subtraction, discrimination of foreground blobs, tracking the blobs over time, detection of interactions between the blobs, etc.). In the Actor Prolog language, special built-in classes supporting necessary low-level procedures are developed and implemented in Java: *ImageSubtractor*, *VideoProcessingMachine*, *FFmpeg*, *BufferedImage*, etc.

The *ImageSubtractor* class implements the following means:

1. Recognition of connected graphs of linked tracks of blob motions and creation of Prolog data structures describing the coordinates, the velocities of the blobs, and special blob motion metrics. One considers two tracks as linked if there are interactions between the blobs of these tracks. In

some applications, it is useful to eject tracks of immovable and slowly moving objects from the graphs before further analysis of the video scene.

2. Recognition and ejection of immovable and slowly moving objects. This feature is based on a simple fuzzy inference on the attributes of the tracks (the coordinates of the tracks and the average velocities of the blobs are considered).

3. Recognition of tracks of blob motions and creation of Prolog data structures describing the coordinates and the velocity of the blobs. The tracks are divided into separate segments; there are points of interaction between the blobs at the ends of the segments.

4. Recognition of moving blobs and creation of Prolog data structures describing the coordinates of the blobs in each moment.

5. Video frames preprocessing including 2D-gaussian filtering, 2D-rank filtering, and background subtraction.

The *ImageSubtractor* class is enough for starting experiments with logic programming intelligent visual surveillance applications; however, to implement an advanced video analysis the *VideoProcessingMachine* built-in class in the Actor Prolog language is to be used. The *VideoProcessingMachine* class implements a kind of a virtual machine for the low-level video processing. The principle of operation of this machine is the following one:

1. The machine keeps a sequence of commands of the low-level video processing. This sequence of commands is to be applied for every frame of the video. The loading of these commands into the machine is performed using special predicates of the *VideoProcessingMachine* class. After the loading of the commands, the sequence is repeated automatically for every incoming frame until the sequence will be updated by the programmer.

2. The machine keeps internal data arrays that are related to various sub-stages of the low-level video processing. Now the following sub-stages of the processing are implemented in the machine: pre-processing of the frame; processing of the frame in the pixel (matrix) representation; selection and processing of the foreground pixels in the frame; extraction and tracing the blobs in the sequence of the frames. Creation and conversion of the internal data arrays are implemented automatically in the course of the processing of the frame.

3. The machine supports a stack of masks of foreground pixels. This stack enables processing of different groups of blobs using different methods of image processing and blob extraction.

4. The result of the processing every frame of the video is a set of lists/graphs that contain information about the movements and other attributes of the blobs in the video scene during given time interval.

The list of the sub-stages of the low-level video processing implemented in the machine as well as the list of corresponding commands and internal data arrays are shown in Table 2.

Let us consider an example of a logic program written in Actor Prolog that extracts blobs of given types in a video (see Figure 10). The program has to create a dialog window to demonstrate the sequence of the frames and results of the detection of the blobs of the following types: electroencephalographic (EEG) cap on the head of a laboratory rat that connects the head of the rat with EEG cable, green and blue objects placed in the cage in the course of a laboratory test. In Figure 10, the logic program has

Table 2. The architecture of the video processing machine

Sub-Stage of the Processing	Examples of Commands of the Virtual Machine	Internal Arrays of the Machine Processed at this Point
Preprocessing of the frame	• Extract a sub-image. • Resize the image. • Gauss filtering.	An image (the *BufferedImage* class of the Java language)
Processing the frame in the pixel representation	• Select a channel in HSB / RGB space or the grayscale channel. • Smooth the image. • Compute gradients. • Normalize pixels.	A matrix of pixels (the int[] data type of the Java language)
Selection and processing of the foreground pixels in the frame	• Push a new mask of foreground pixels into the stack. • Pop the top mask from the stack. • Add pixels corresponding to given conditions to the mask. • Eliminate pixels corresponding to given conditions from the mask. • Subtract background using a given algorithm. • Implement rank filtering of the foreground mask. • Erode the mask of the foreground pixels. • Dilate the mask.	A mask of the foreground pixels (the boolean[] data type of the Java language)
Extraction and tracing the blobs in the frames	• Extract blobs using a given algorithm. • Fill the blobs. • Select blobs corresponding to given conditions. • Trace the movements of the blobs. • Compute color histograms of the blobs.	Lists of coordinates and other attributes of the blobs
Additional processing the tracks of the blobs	• Select tracks corresponding to given conditions.	A list of tracks

Figure 10. A laboratory rat investigates new objects in the cage

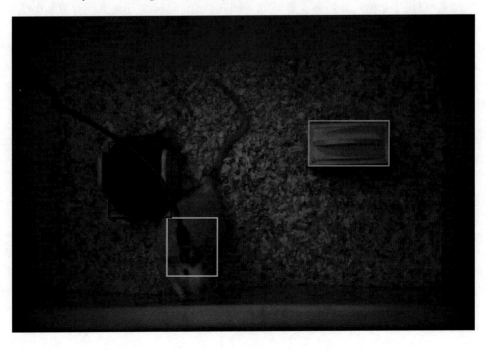

detected three blobs in the video: EEG cap that connects the head of the rat with EEG cable, a green object, and a blue object placed in the cage. The data are coming from the Institute of Higher Nervous Activity and Neurophysiology of RAS.

Let us apply the *FFmpeg* built-in class that links Actor Prolog with the FFmpeg open library to read the video file. The main class of the logic program is a descendant of the *FFmpeg* class. The *name* slot of the class contains the name of the video file. Besides, the *Main* class includes auxiliary slots *vpm*, *dialog*, *graphic_window*, *image1*, and *image2*:

Computer Code

```
class 'Main' (specialized 'FFmpeg'):
constant:
    name = "time 17-07-10 15-16-42.avi";
internal:
    vpm        = ('VideoProcessingMachine');
    dialog     = ('DemoPanel',
                    graphic_window);
    graphic_window = ('Canvas2D');
    image1     = ('BufferedImage');
    image2     = ('BufferedImage');
```

The *vpm* slot contains an instance of the *VideoProcessingMachine* class. The *dialog* class contains a dialog window. The *graphic_window* slot contains a graphics window intended for displaying video frames and blobs. The *image1* and *image2* slots contain instances of the *BufferedImage* built-in class that is intended for storing and transfer of the graphics data arrays.

The goal statement of the logic program (*goal*) loads commands to the instance of the *VideoProcessingMachine* class, then opens the dialog window, and initiates reading the video file:

Computer Code

```
CLAUSES:
goal:-!,
    store_VPM_program,
    dialog ? maximize,
    start.
```

The *store_VPM_program* predicate loads a sequence of command to the virtual machine using special predicates of the *VideoProcessingMachine* built-in class. Note that the program uses different low-level processing algorithms for extracting blobs of different types. The coordinates of EEG cap and the objects placed in the cage are detected using different sub-spaces in the Hue-Saturation-Brightness (HSB) color space.

Computer Code

```
store_VPM_program:-
    -- Suspend the video processing:
    vpm ? suspend_processing,
    -- Delete all the commands:
    vpm ? retract_all_instructions,
    -- Assign the size of blob borders:
    vpm ? blb_set_blob_borders(15,15),
    -- Create a new foreground mask:
    vpm ? msk_push_foreground,
    -- Select pixels corresponding to
    -- given interval of hue:
    vpm ? msk_select_foreground(
        'HUE',185,245),
    -- Select pixels corresponding to
    -- given interval of saturation:
    vpm ? msk_select_foreground(
        'SATURATION',50,255),
    -- Select pixels corresponding to
    -- given interval of brightness:
    vpm ? msk_select_foreground(
        'BRIGHTNESS',0,150),
    -- Erode the mask by 2 pixels:
    vpm ? msk_erode_foreground(2),
    -- Dilate the mask by 2 pixels:
    vpm ? msk_dilate_foreground(2),
    -- Extract blobs of type "cap":
    vpm ? blb_extract_blobs(
        'cap',
        'TWO_PASS_BLOB_EXTRACTION'),
    -- Select the biggest blob:
    vpm ? blb_select_superior_blob
        ('FOREGROUND_AREA'),
    -- Fill gaps in blobs:
    vpm ? blb_fill_blobs,
    -- Pop the mask from the stack:
    vpm ? msk_pop_foreground,
    -- Assign the size of blob borders:
    vpm ? blb_set_blob_borders(1,1),
    -- Create a new foreground mask:
    vpm ? msk_push_foreground,
```

```
vpm ? msk_select_foreground(
    'HUE',111,138),
vpm ? msk_select_foreground(
    'SATURATION',50,255),
vpm ? msk_erode_foreground(2),
vpm ? msk_dilate_foreground(2),
-- Extract blobs of the type:
vpm ? blb_extract_blobs(
    'green_object',
    'TWO_PASS_BLOB_EXTRACTION'),
vpm ? blb_select_superior_blob(
    'FOREGROUND_AREA'),
vpm ? blb_fill_blobs,
vpm ? msk_pop_foreground,
-- Create a new foreground mask:
vpm ? msk_push_foreground,
vpm ? msk_select_foreground(
    'HUE',147,183),
vpm ? msk_select_foreground(
    'SATURATION',103,255),
vpm ? msk_erode_foreground(2),
vpm ? msk_dilate_foreground(2),
vpm ? blb_extract_blobs(
    'blue_object',
    'TWO_PASS_BLOB_EXTRACTION'),
vpm ? blb_select_superior_blob(
    'FOREGROUND_AREA'),
vpm ? blb_fill_blobs,
vpm ? msk_pop_foreground,
-- Resume the video processing:
vpm ? process_now.
```

When a new frame is received from the video file, the *frame_obtained* predicate is invoked automatically in the logic program. The logic program informs the *FFmpeg* built-in class using the *commit* predicate that it is going to process the frame. Then it loads the content of the frame into the *image1* object using the *get_recent_image* predicate. The *process_realtime_frame* predicate of the *VideoProcessingMachine* class is used to transfer the frame to the video processing machine. Note that the arrays of the video data are encapsulated in the built-in classes all the time during the data processing that is necessary to ensure high speed of the low-level image processing.

Computer Code

```
frame_obtained:-
    commit,!,
    get_recent_image(image1),
    vpm ? process_realtime_frame(image1),
    draw_scene.
```

After that, the *draw_scene* predicate is called. This predicate accepts a list of blobs from the video processing machine and displays the blobs in the dialog window. The predicate informs the video processing machine by the *commit* predicate that it needs data structures describing the blobs in the current video frame. Then it accepts the content of the video frame using the *get_recent_image* predicate and the list of the blobs using the *get_blobs* predicate. The image and the blobs are displayed in *graphic_window* using predicates of the *Canvas2D* built-in class:

Computer Code

```
draw_scene:-
    vpm ? commit,
    vpm ? get_recent_image(image2),
    image2 ? get_size_in_pixels(IW,IH),
    vpm ? get_blobs(BlobList),
    graphic_window ? suspend_redrawing,
    graphic_window ? clear,
    graphic_window ?
        draw_image(image2,0,0,1,1),
    draw_blob_list(BlobList,IW,IH),
    graphic_window ? draw_now.
```

The list of the blobs is unrolled by the *draw_blob_list* predicate and each blob is drawn in the window as a colored rectangle:

Computer Code

```
draw_blob_list([Blob|Rest],IW,IH):-
    draw_blob(Blob,IW,IH),
    draw_blob_list(Rest,IW,IH).
draw_blob_list([],_,_).
```

The blob is described using an underdetermined set (Morozov, 1999) of Actor Prolog that contains information about the type, coordinates, width, height, and other attributes of the blob:

Computer Code

```
draw_blob(Blob,IW,IH):-
    Blob == { type:Type,x:X0,y:Y0,
              width:W1,height:H1|_},
    X2== (X0 - W1/2) / IW,
    Y2== (Y0 - H1/2) / IH,
    W2== W1 / IW,
    H2== H1 / IH,
    select_blob_color(Type,Color),
    graphic_window ?
        set_pen({color:Color,lineWidth:3}),
    graphic_window ?
        draw_rectangle(X2,Y2,W2,H2),
    fail.
draw_blob(_,_,_).
```

The *select_blob_color* predicate defines the colors of the rectangles to be used for drawing the blobs:

Computer Code

```
select_blob_color('cap','Cyan').
select_blob_color('green_object','DkGreen').
select_blob_color('blue_object','Navy').
```

This plain example illustrates both two stages of the video processing, as well as the usage of the *VideoProcessingMachine* built-in class for the extraction and analyzing blobs of various types in the video.

3D INTELLIGENT VISUAL SURVEILLANCE LOGIC PROGRAMMING

In contrast to the conventional 2D intelligent visual surveillance that cannot provide a stable recognition of the parts of the human body in most real-world applications, in 3D intelligent visual surveillance the goals of the low-level video analysis are essentially different because the methods of 3D vision can provide reliable recognition of the human body parts. For instance, the standard software of the Kinect 2 time-of-flight camera can provide ready-for-use foreground blobs, as well as up to six skeletons of people to be monitored (Chen, Wei, & Ferryman, 2013; Han, Shao, Xu, & Shotton 2013; Barmpoutis, 2013; Aggarwal & Xia, 2014; Lun & Zhao, 2015; Han, Reily, Hoff, & Zhang, 2016; Presti & Cascia, 2016). Thus, the minimal preprocessing of the data can include just a conversion of the images of the skeletons into the terms of the logic language. The high-level stage of the video processing can be implemented in accordance with the principles of conventional 2D intelligent visual surveillance, but the information about the video scene is essentially more complete, that gives an opportunity to infer much more of logical consequences on the semantics of the video scene.

Let us consider a simple logic program that analyses 3D data collected using the Kinect 2 ToF camera. The data include depth maps and images of skeletons of people, but the program has to analyze the images of the skeletons only and the depth maps are to be used only for the visualization of the video scene. Each skeleton contains 25 joints of various statuses (See Figure 11).

The statuses of joints of skeletons can be the following:

1. **Tracked:** The corresponding part of the body is directly observed by the ToF camera.
2. **Inferred:** The position of the joint is a hypothetical one.
3. **Unknown:** The device has no information about this joint.

Thus, the following set of domains (data types) can be defined in Actor Prolog to describe the data structures to be processed:

Computer Code

```
DOMAINS:
Skeletons       = Skeleton*.
```

The *Skeletons* type is a list of elements of the *Skeleton* type.

Figure 11. The images of skeletons and depth maps are the input data of the logic program. The depth maps are indicated by colors.

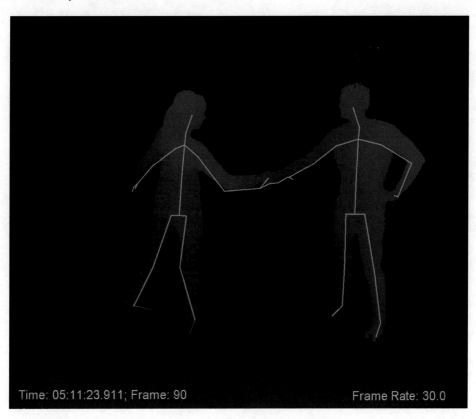

Time: 05:11:23.911; Frame: 90   Frame Rate: 30.0

Computer Code

```
Skeleton = {
        identifier: INTEGER,
        head: SkeletonJoint,
        spine: SkeletonSpine,
        left_arm: SkeletonArm,
        right_arm: SkeletonArm,
        left_leg: SkeletonLeg,
        right_leg: SkeletonLeg
        }.
```

The *Skeleton* type is an underdetermined set (Morozov, 1999) of Actor Prolog. The names of the elements of the skeleton, as well as the names of corresponding types, are obvious:

Computer Code

```
SkeletonSpine = {
        neck: SkeletonJoint,
        shoulder: SkeletonJoint,
        mid: SkeletonJoint,
        base: SkeletonJoint
        }.
SkeletonArm = {
        shoulder: SkeletonJoint,
        elbow: SkeletonJoint,
        wrist: SkeletonJoint,
        hand: SkeletonJoint,
        tip: SkeletonJoint,
        thumb: SkeletonJoint
        }.
SkeletonLeg = {
        hip: SkeletonJoint,
        knee: SkeletonJoint,
        ankle: SkeletonJoint,
        foot: SkeletonJoint
        }.
SkeletonJoint = {
        status: SkeletonJointStatus,
        position: VertexPosition3D,
        orientation: JointOrientation3D
        }.
```

The status of the joint is a symbol. Each joint is characterized by a position in 3D space, i.e. by three numbers, by orientation, i.e. by four numbers, and by two pairs of 2D coordinates that correspond to this joint and are obtained from the infrared and color images:

Computer Code

```
SkeletonJointStatus =
        'TRACKED'; 'INFERRED'; 'UNKNOWN'.
VertexPosition3D = {
        point: Point3D,
        velocity: Point3D,
        acceleration: Point3D,
        mapping1: Point2D,
        mapping2: Point2D
        }.
JointOrientation3D= q(
        Numerical,
        Numerical,
        Numerical,
        Numerical).
Point3D = p(Numerical,Numerical,Numerical).
Point2D = p(Numerical,Numerical).
```

Suppose that images of skeletons are transferred from the device in real time. Let us declare a set of predicates analyzing the 3D data:

Computer Code

```
PREDICATES:
imperative:
analyze_skeletons(Skeletons)                    - (i);
nondeterm:
a_hand_of_a_person
        (Skeletons,INTEGER,VertexPosition3D) - (i,o,o);
determ:
is_near_a_head_of_a_colleague
        (Skeletons,INTEGER,VertexPosition3D) - (i,i,i);
is_near(VertexPosition3D,VertexPosition3D)   - (i,i);
```

The *analyse_skeletons* predicate checks whether there are at least two persons in the video scene and one person punches in the face another person. If there are such persons, the predicate outputs a message on the screen (see Figure 12). Here is the definition of the predicate:

Figure 12. A punch in the face is detected by the logic program

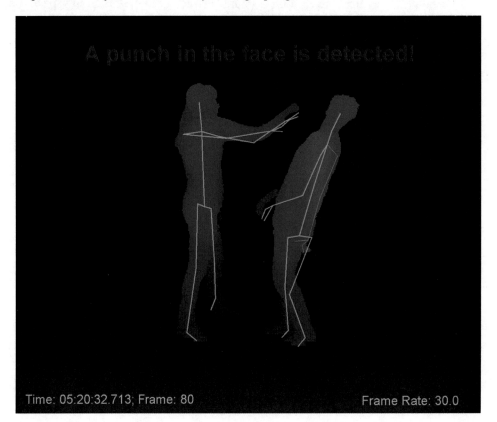

Computer Code

```
CLAUSES:
analyze_skeletons(Skeletons):-
        a_hand_of_a_person(Skeletons,Id,Position),
        is_near_a_head_of_a_colleague(Skeletons,Id,Position),!,
        graphic_window ? draw_text(
                0.93,0.08,"A punch in the face is detected!").
analyze_skeletons(_).
```

The *a_hand_of_a_person* predicate is non-deterministic. This predicate returns coordinates of hands of persons recognized in the scene.

Computer Code

```
a_hand_of_a_person([Skeleton|_],Identifier,Position):-
        Skeleton == {
                identifier:Identifier,
                left_arm:{
```

```
                    hand:{
                            status:'TRACKED',
                            position:Position|_}
                    |_}
            |_}.
a_hand_of_a_person([Skeleton|_],Identifier,Position):-
        Skeleton == {
                identifier:Identifier,
                right_arm:{
                    hand:{
                            status:'TRACKED',
                            position:Position|_}
                    |_}
            |_}.
a_hand_of_a_person([_|Rest],Identifier,Position):-
        a_hand_of_a_person(Rest,Identifier,Position).
```

The following definition is used for the detection of the punch in the face. Execution of the *is_near_a_head_of_a_colleague* predicate succeeds, if given 3D coordinates of some person's hand are close enough to the coordinates of the head of another person. The predicate checks whether the person punches in the face of another person, but not in his own face. For brevity, the velocity of the movement of the hand and the duration of the contact are not taken into consideration, but in the real intelligent visual surveillance application, these attributes are to be checked to provide acceptable specificity of the recognition procedure.

Computer Code

```
is_near_a_head_of_a_colleague([Skeleton|_],Id1,Point1):-
        Skeleton == {
                identifier:Id2,
                head:{  status:'TRACKED',
                        position:Point2|_}
                |_},
        Id1 <> Id2,
        is_near(Point1,Point2),!.
is_near_a_head_of_a_colleague([_|Rest],Id,Point):-
        is_near_a_head_of_a_colleague(Rest,Id,Point).
```

The *is_near* predicate is auxiliary. It checks whether the distance between the given coordinates is small enough. The predicate is defined using the built-in arithmetical predicates of Actor Prolog:

Computer Code

```
is_near({point:p(X1,Y1,Z1)|_},{point:p(X2,Y2,Z2)|_}):-
        Delta== ?sqrt(
                ?power(X1-X2,2) +
                ?power(Y1-Y2,2) +
                ?power(Z1-Z2,2)),
        Delta < 0.15,!.
```

For brevity, predicates implementing the data transferring from the outer world into the program are not considered here. One can find a detailed example of this kind in (Morozov, 2015). The *analyse_skeletons* predicate is to be called from time to time in the logic program to analyze images of skeletons recognized in 3D video frames transferred from the outside.

Note that the key point of this example is not recognition of the punch in the face (or recognition of a handshaking) that is a simple problem, but in that much more complex logical formulae can be automatically checked in the same way. For example, a logic program can check whether a person shakes hands with all people in the room or not, that is, the program can make some conclusions on the social status of the person or social relations between the people in the room. The same technique can be used to solve urgent problems of intelligent visual surveillance, such as recognizing armed attacks, demonstrating threatening poses, robberies, etc.

CONCLUSION

A research software platform based on the Actor Prolog concurrent object-oriented logic language and a state-of-the-art Prolog-to-Java translator was created for studying the intelligent visual surveillance. The platform includes the Actor Prolog logic programming system and an open source Java library of Actor Prolog built-in classes (Morozov, 2018). The object-oriented means of the Actor Prolog logic language allow splitting the program into interacting parallel processes that implement various stages of image processing and scene analysis, while translation to the Java language provides a high performance sufficient for analyzing real-time video, as well as reliability and stability of the work of the intelligent visual surveillance applications. This software is intended to facilitate the study of the intelligent video monitoring of anomalous people activities, the logical description and analysis of people behavior, and other intelligent video surveillance applications. One can download the software including all the examples considered in the chapter from the Web Site (Morozov & Sushkova, 2018).

Recent trends in the logic programming of the intelligent visual surveillance research are connected with the usage of 3D machine vision methods and adaptation of the distributed logic programming and multi-agent approach to the intelligent visual surveillance. The method of the object-oriented logic programming for processing the data obtained from the equipment for 3D data acquisition in the intelligent visual surveillance systems was developed. The application of methods of the object-oriented logic programming for 3D data processing gives exciting prospects in the area of intelligent visual surveillance: automatic analysis of the behavior of separate persons and groups of persons including social interactions between the people and manipulation of objects. Progress in this research area can be achieved by the development of efficient low-level 3D data processing algorithms, logic programming languages and systems for general purpose, and advanced logical methods of 3D data processing.

ACKNOWLEDGMENT

Authors are grateful to Ivan A. Kershner and Renata A. Tolmatcheva for the help in the preparation of 3D video samples and Angelos Barmpoutis for his J4K library (Barmpoutis, 2013) that was used for the data collection. Authors thank Natalia V. Gulyaeva, Ilya G. Komoltsev, Anna O. Manolova, Margarita R. Novikova, and Irina P. Levshina (IHNA and NPh RAS) for the experimental video data. Authors thank Abhishek Vaish, Vyacheslav E. Antciperov, Vladimir V. Deviatkov, Aleksandr N. Alfimtsev, Vladislav S. Popov, and Igor I. Lychkov for cooperation.

This research was supported by the Russian Foundation for Basic Research [grant number 16-29-09626-ofi_m].

REFERENCES

Aggarwal, J. K., & Xia, L. (2014). Human activity recognition from 3D data: A review. *Pattern Recognition Letters*, *48*, 70–80. doi:10.1016/j.patrec.2014.04.011

Alazrai, R., Mowafi, Y., & Lee, C. (2015). Anatomical-plane-based representation for human–human interactions analysis. *Pattern Recognition*, *48*(8), 2346–2363. doi:10.1016/j.patcog.2015.03.002

Artikis, A., Sergot, M., & Paliouras, G. (2010). A logic programming approach to activity recognition. In *International workshop on events in multimedia (EiMM 2010)* (pp. 3–8). ACM. 10.1145/1877937.1877941

Baader, F., Calvanese, D., McGuinness, D., Nardi, D., & Patel-Schneider, P. (2002). *The Description Logic Handbook*. Cambridge University Press.

Badica, C., Braubach, L., & Paschke, A. (2011). Rule-based distributed and agent systems. In *International Workshop on Rules and Rule Markup Languages for the Semantic Web* (pp. 3–28). Heidelberg, Germany: Springer.

Baldoni, M., Baroglio, C., Mascardi, V., Omicini, A., & Torroni, P. (2010). Agents, multiagent systems and declarative programming: What, when, where, why, who, how? In A. Dovier & E. Pontelli (Eds.), *A 25-year perspective on logic programming* (pp. 204–230). Heidelberg, Germany: Springer. doi:10.1007/978-3-642-14309-0_10

Banbara, M., Tamura, N., & Inoue, K. (2006). Prolog Cafe: A Prolog to Java translator system. In M. Umeda, A. Wolf, O. Bartenstein, U. Geske, D. Seipel, & O. Takata (Eds.), *Declarative programming for knowledge management* (pp. 1–11). Heidelberg, Germany: Springer. doi:10.1007/11963578_1

Baral, C., Gelfond, G., Son, T. C., & Pontelli, E. (2010). Using answer set programming to model multi-agent scenarios involving agents' knowledge about other's knowledge. In *18th International Joint Conference on Artificial Intelligence* (pp. 259–266). Toronto, Canada: Academic Press.

Barmpoutis, A. (2013). Tensor body: Real-time reconstruction of the human body and avatar synthesis from RGB-D. *IEEE Transactions on Cybernetics, 43*(5), 1347–1356. doi:10.1109/TCYB.2013.2276430 PMID:23974673

Barnum, P., Sheikh, Y., Datta, A., & Kanade, T. (2009). Dynamic seethroughs: Synthesizing hidden views of moving objects. In *Mixed and augmented reality* (pp. 111–114). ISMAR. doi:10.1109/IS-MAR.2009.5336483

Benedek, C. (2014). 3D people surveillance on range data sequences of a rotating Lidar. *Pattern Recognition Letters, 50*, 149–158. doi:10.1016/j.patrec.2014.04.010

Blunsden, S.J., & Fisher, R.B. (2010). The BEHAVE video dataset: Ground truthed video for multi-person behavior classification. *Annals of the BMVA, (4)*, 1–11.

Bordini, R. H., Hübner, J. F., & Wooldridge, M. (2007). Programming multi-agent systems in AgentSpeak using Jason. Chichester, UK: John Wiley & Sons.

Borges, P. V. K. (2013). Pedestrian detection based on blob motion statistics. *IEEE Transactions on Circuits and Systems for Video Technology, 23*(2), 224–235. doi:10.1109/TCSVT.2012.2203217

Borges, P. V. K., Conci, N., & Cavallaro, A. (2013). Video-based human behavior understanding: A survey. *IEEE Transactions on Circuits and Systems for Video Technology, 23*(11), 1993–2008. doi:10.1109/TCSVT.2013.2270402

Chaaraoui, A. A., Padilla-López, J. R., & Flórez-Revuelta, F. (2015). Abnormal gait detection with RGB-D devices using joint motion history features. In Automatic face and gesture recognition (FG) (Vol. 7, pp. 1–6). Academic Press. doi:10.1109/FG.2015.7284881

Chang, C.-L., & Lee, R. (1973). *Symbolic logic and mechanical theorem proving*. New York: Academic Press.

Chen, L., Wei, H., & Ferryman, J. (2013). A survey of human motion analysis using depth imagery. *Pattern Recognition Letters, 34*(15), 1995–2006. doi:10.1016/j.patrec.2013.02.006

Clocksin, W. F., & Mellish, C. S. (2003). *Programming in Prolog: Using the ISO standard*. Heidelberg, Germany: Springer. doi:10.1007/978-3-642-55481-0

Codognet, P., & Diaz, D. (1995). wamcc: Compiling Prolog to C. In L. Sterling (Ed.), ICLP 1995 (pp. 317–331). MIT Press.

Cook, J. J. (2004). Optimizing P#: Translating Prolog to more idiomatic C#. In CICLOPS 2004 (pp. 59–70). Academic Press.

Dastani, M. (2008). 2APL: A practical agent programming language. *Autonomous Agents and Multi-Agent Systems, 16*(3), 214–248. doi:10.100710458-008-9036-y

Davison, A. (1992, January). *A survey of logic programming-based object oriented languages* (Tech. Rep. No. 92/3). Melbourne, Australia: Department of Computer Science, University of Melbourne.

DeCamp, P., Shaw, G., Kubat, R., & Roy, D. (2010). An immersive system for browsing and visualizing surveillance video. In *International conference on multimedia* (pp. 371–380). Academic Press. 10.1145/1873951.1874002

Demoen, B., & Tarau, P. (1997). *jProlog home page.* Retrieved October 27, 2017, from https://people.cs.kuleuven.be/~bart.demoen/PrologInJava/

Diraco, G., Leone, A., & Siciliano, P. (2013). Human posture recognition with a time-of-flight 3D sensor for in-home applications. *Expert Systems with Applications, 40*(2), 744–751. doi:10.1016/j.eswa.2012.08.007

Draper, N., & Smith, H. (1998). *Applied regression analysis.* Wiley-Interscience. doi:10.1002/9781118625590

Eichberg, M. (2011). Compiling Prolog to idiomatic Java. In J. P. Gallagher & M. Gelfond (Eds.), *ICLP 2011* (pp. 84–94). Saarbrücken, Wadern: Dagstuhl Publishing.

Ejaz, N., Manzoor, U., Nefti, S., & Baik, S. (2012). A collaborative multi-agent framework for abnormal activity detection in crowded areas. *International Journal of Innovative Computing, Information, & Control, 8,* 4219–4234.

Ferryman, J., Hogg, D., Sochman, J., Behera, A., Rodriguez-Serrano, J., Worgan, S., ... Dose, M. (2013). Robust abandoned object detection integrating wide area visual surveillance and social context. *Pattern Recognition Letters, 34*(7), 789–798. doi:10.1016/j.patrec.2013.01.018

Fisher, R. (2007). *CAVIAR test case scenarios. The EC funded project IST 2001 37540.* Retrieved October 27, 2017, from http://homepages.inf.ed.ac.uk/rbf/CAVIAR/

Fisher, R. (2013). *BEHAVE: Computer-assisted prescreening of video streams for unusual activities. The EPSRC project GR/S98146.* Retrieved October 27, 2017, from http://groups.inf.ed.ac.uk/vision/BEHAVEDATA/INTERACTIONS/

Fujise, T., Chikayama, T., Rokusava, K., & Nakase, A. (1994, December). KLIC: A portable implementation of KL1. In *FGCS 1994* (pp. 66–79). Tokyo: ICOT.

Gascueña, J., & Fernández-Caballero, A. (2011). On the use of agent technology in intelligent, multi-sensory and distributed surveillance. *The Knowledge Engineering Review, 26*(2), 191–208. doi:10.1017/S0269888911000026

Gavrila, D. M. (1999, January). The visual analysis of human movement: A survey. *Computer Vision and Image Understanding, 73*(1), 82–98. doi:10.1006/cviu.1998.0716

Han, F., Reily, B., Hoff, W., & Zhang, H. (2016). *Space-time representation of people based on 3D skeletal data: A review.* arXiv preprint arXiv:1601.01006v2 [cs.CV]

Han, J., Shao, L., Xu, D., & Shotton, J. (2013, October). Enhanced computer vision with Microsoft Kinect sensor: A review. *IEEE Transactions on Cybernetics, 43*(5), 1318–1334. doi:10.1109/TCYB.2013.2265378 PMID:23807480

Hansen, D. W., Hansen, M. S., Kirschmeyer, M., Larsen, R., & Silvestre, D. (2008). Cluster tracking with time-of-flight cameras. In Computer vision and pattern recognition workshops (CVPRW'08) (pp. 1–6). Academic Press.

Haritaoglu, I., Harwood, D., & Davis, L. S. (1998, April). W4: Who? When? Where? What? A real time system for detecting and tracking people. In FG 1998 (pp. 222–227). Nara, Japan: Academic Press.

Henderson, F., & Somogyi, Z. (2002). Compiling Mercury to high-level C code. In CC 2002. Grenoble, France: Academic Press. doi:10.1007/3-540-45937-5_15

Hsieh, C.-T., Wang, H.-C., Wu, Y.-K., Chang, L.-C., & Kuo, T.-K. (2012). A Kinect-based people-flow counting system. In Intelligent signal processing and communication systems (ISPACS 2012) (pp. 146–150). IEEE. doi:10.1109/ISPACS.2012.6473470

Ibañez, R., Soria, Á., Teyseyre, A., & Campo, M. (2014). Easy gesture recognition for Kinect. *Advances in Engineering Software*, *76*, 171–180. doi:10.1016/j.advengsoft.2014.07.005

Johanna, M. (2013). *Recognizing activities with the Kinect. A logic-based approach for the support room* (Unpublished master's thesis). Radboud University Nijmegen.

Katzouris, N., Artikis, A., & Paliouras, G. (2014). Event recognition for unobtrusive assisted living. In *Hellenic conference on artificial intelligence* (pp. 475–488). Academic Press.

Kimmig, A., Demoen, B., Raedt, L. D., Costa, V. S., & Rocha, R. (2011). On the implementation of the probabilistic logic programming language ProbLog. *Theory and Practice of Logic Programming*, *11*(11), 235–262. doi:10.1017/S1471068410000566

Kowalski, R., & Sergot, M. (1986). A logic-based calculus of events. *New Generation Computing*, *1*(4), 67–96. doi:10.1007/BF03037383

Kravari, K., & Bassiliades, N. (2015). A survey of agent platforms. *Journal of Artificial Societies and Social Simulation*, *18*(1), 191–208. doi:10.18564/jasss.2661

Krishnan, P., & Naveen, S. (2015). RGB-D face recognition system verification using Kinect and FRAV3D databases. *Procedia Computer Science*, *46*, 1653–1660. doi:10.1016/j.procs.2015.02.102

Kuramochi, S. (1999). *KLIJava home page*. Retrieved October 27, 2017, from http://www.ueda.info. waseda.ac.jp/~satoshi/klijava/klijava-e.html

Lao, W., Han, J., & With, P. H. N. (2010). Flexible human behavior analysis framework for video surveillance applications. *International Journal of Digital Multimedia Broadcasting*.

Lau, T. B., Ong, A. C., & Putra, F. A. (2014, June). Non-invasive monitoring of people with disabilities via motion detection. *International Journal of Signal Processing Systems*, *2*(1), 37–41.

Lee, Y.-S., & Chung, W.-Y. (2012). Visual sensor based abnormal event detection with moving shadow removal in home healthcare applications. *Sensors (Basel)*, *12*(1), 573–584. doi:10.3390120100573 PMID:22368486

Leightley, D., Yap, M. H., Hewitt, B. M., & McPhee, J. S. (2016). Sensing behaviour using the Kinect: Identifying characteristic features of instability and poor performance during challenging balancing tasks. In Measuring behavior. Academic Press.

Lun, R., & Zhao, W. (2015, March). A survey of applications and human motion recognition with Microsoft Kinect. *International Journal of Pattern Recognition and Artificial Intelligence, 29*(5), 1555008. doi:10.1142/S0218001415550083

Machot, F., Kyamakya, K., Dieber, B., & Rinner, B. (2011). *Real time complex event detection for resource-limited multimedia sensor networks.* AMMCSS. doi:10.1109/AVSS.2011.6027378

Mastorakis, G., & Makris, D. (2014). Fall detection system using Kinect's infrared sensor. *Journal of Real-Time Image Processing, 9*(4), 635–646. doi:10.100711554-012-0246-9

Morozov, A. A. (1994). The Prolog with actors. *Programmirovanie,* (5), 66–78. (in Russian)

Morozov, A. A. (1999, September). Actor Prolog: an object-oriented language with the classical declarative semantics. In K. Sagonas & P. Tarau (Eds.), IDL 1999 (pp. 39–53). Paris, France: Academic Press.

Morozov, A. A. (2002, September). On semantic link between logic, object-oriented, functional, and constraint programming. In Proc. of MultiCPL'02 workshop (pp. 43–57). Ithaca, NY: Academic Press.

Morozov, A. A. (2003). Logic object-oriented model of asynchronous concurrent computations. *Pattern Recognition and Image Analysis, 13*(4), 640–649.

Morozov, A. A. (2007, September). Operational approach to the modified reasoning, based on the concept of repeated proving and logical actors. In V. S. C. Salvador Abreu (Ed.), CICLOPS 2007 (pp. 1–15). Porto, Portugal: Academic Press.

Morozov, A. A. (2015). Development of a method for intelligent video monitoring of abnormal behavior of people based on parallel object-oriented logic programming. *Pattern Recognition and Image Analysis, 25*(3), 481–492. doi:10.1134/S1054661815030153

Morozov, A. A. (2018). *A GitHub repository containing source codes of Actor Prolog built-in classes.* Retrieved February 12, 2018, from https://github.com/Morozov2012/actor-prolog-java-library

Morozov, A. A., & Polupanov, A. F. (2014, June). Intelligent visual surveillance logic programming: Implementation issues. In T. Ströder & T. Swift (Eds.), *CICLOPS-WLPE 2014* (pp. 31–45). RWTH Aachen University.

Morozov, A. A., & Polupanov, A. F. (2015). 5). Development of the logic programming approach to the intelligent monitoring of anomalous human behaviour. In D. Paulus, C. Fuchs, & D. Droege (Eds.), *OGRW 2014* (pp. 82–85). Koblenz: University of Koblenz-Landau.

Morozov, A. A., & Sushkova, O. S. (2016). Real-time analysis of video by means of the Actor Prolog language. *Computer Optics, 40*(6), 947–957. doi:10.18287/2412-6179-2016-40-6-947-957

Morozov, A. A., & Sushkova, O. S. (2018). *The intelligent visual surveillance logic programming Web Site.* Retrieved February 12, 2018, from http://www.fullvision.ru

Morozov, A. A., Sushkova, O. S., & Polupanov, A. F. (2015a). 8). An approach to the intelligent monitoring of anomalous human behaviour based on the Actor Prolog object-oriented logic language. In N. Bassiliades & ... (Eds.), *RuleML 2015 DC and Challenge*. Berlin: CEUR.

Morozov, A. A., Sushkova, O. S., & Polupanov, A. F. (2015b). 8). A translator of Actor Prolog to Java. In N. Bassiliades & ... (Eds.), *RuleML 2015 DC and Challenge*. Berlin: CEUR.

Morozov, A. A., Sushkova, O. S., & Polupanov, A. F. (2017a). Towards the distributed logic programming of intelligent visual surveillance applications. In O. Pichardo-Lagunas & S. Miranda-Jimenez (Eds.), *Advances in Soft Computing: 15th Mexican International Conference on Artificial Intelligence, MICAI 2016, Cancun, Mexico, Proceedings, Part II* (pp. 42–53). Cham: Springer International Publishing. 10.1007/978-3-319-62428-0_4

Morozov, A. A., Sushkova, O. S., & Polupanov, A. F. (2017b, June). Object-oriented logic programming of 3D intelligent video surveillance: The problem statement. In *IEEE 26th International Symposium on Industrial Electronics (ISIE), 2017* (pp. 1631–1636). IEEE Xplore Digital Library.

Morozov, A. A., Sushkova, O. S., & Polupanov, A. F. (2017c). Object-oriented logic programming of 3D intelligent video surveillance systems: The problem statement. *RENSIT*, *9*(2), 205–214. doi:10.17725/rensit.2017.09.205

Morozov, A. A., Vaish, A., Polupanov, A. F., Antciperov, V. E., Lychkov, I. I., Alfimtsev, A. N., & Deviatkov, V. V. (2014). Development of concurrent object-oriented logic programming system to intelligent monitoring of anomalous human activities. In A. Cliquet Jr, G. Plantier, T. Schultz, A. Fred, & H. Gamboa (Eds.), *BIODEVICES 2014* (pp. 53–62). SCITEPRESS.

Morozov, A. A., Vaish, A., Polupanov, A. F., Antciperov, V. E., Lychkov, I. I., Alfimtsev, A. N., & Deviatkov, V. V. (2015). Development of concurrent object-oriented logic programming platform for the intelligent monitoring of anomalous human activities. In G. Plantier, T. Schultz, A. Fred, & H. Gamboa (Eds.), *BIOSTEC 2014* (Vol. 511, pp. 82–97). Heidelberg, Germany: Springer. doi:10.1007/978-3-319-26129-4_6

Nierstrasz, O., & Dami, L. (1995). Component-oriented software technology. In O. Nierstrasz & D. Tsichritzis (Eds.), *Object-Oriented Software Composition* (pp. 3–28). Prentice Hall.

O'Hara, S. (2008). *VERSA – video event recognition for surveillance applications* (Unpublished master's thesis). University of Nebraska at Omaha.

Odell, J. (2002). Objects and agents compared. *Journal of Object Technology*, *1*(1), 41–53. doi:10.5381/jot.2002.1.1.c4

Ott, R., Gutiérrez, M., Thalmann, D., & Vexo, F. (2006). Advanced virtual reality technologies for surveillance and security applications. *International conference on virtual reality continuum and its applications*, 163–170. 10.1145/1128923.1128949

Patwardhan, A., & Knapp, G. (2016). *Aggressive actions and anger detection from multiple modalities using Kinect*. arXiv preprint arXiv:1607.01076.

Popa, M., Koc, A. K., Rothkrantz, L. J., Shan, C., & Wiggers, P. (2011). Kinect sensing of shopping related actions. In *International joint conference on ambient intelligence* (pp. 91–100). Academic Press.

Preis, J., Kessel, M., Werner, M., & Linnhoff-Popien, C. (2012). Gait recognition with Kinect. In *International workshop on Kinect in pervasive computing*. Newcastle, UK: Academic Press.

Presti, L. L., & Cascia, M. L. (2016). 3D skeleton-based human action classification: A survey. *Pattern Recognition, 53*, 130–147. doi:10.1016/j.patcog.2015.11.019

Raheja, J., Minhas, M., Prashanth, D., Shah, T., & Chaudhary, A. (2015). Robust gesture recognition using Kinect: A comparison between DTW and HMM. *Optik – International Journal for Light and Electron Optics, 126*(11), 1098–1104.

Rougier, C., Auvinet, E., Rousseau, J., Mignotte, M., & Meunier, J. (2011). Fall detection from depth map video sequences. In *International conference on smart homes and health telematics* (pp. 121–128). Academic Press. 10.1007/978-3-642-21535-3_16

Russell, S., & Norvig, P. (1995). *Artificial intelligence. A modern approach*. London: Prentice-Hall.

Ruta, M., Scioscia, F., Summa, M. D., Ieva, S., Sciascio, E. D., & Sacco, M. (2014). Semantic matchmaking for Kinect-based posture and gesture recognition. *International Journal of Semantic Computing, 8*(4), 491–514. doi:10.1142/S1793351X14400169

Satta, R., Pala, F., Fumera, G., & Roli, F. (2013). *Real-time appearance-based person re-identification over multiple Kinect™ cameras*. VISAPP.

Savage, R., Clarke, N., & Li, F. (2013). Multimodal biometric surveillance using a Kinect sensor. In *Proceedings of the 12th annual security conference*. Las Vegas, NV: Academic Press.

Sebe, I. O., Hu, J., You, S., & Neumann, U. (2003). 3D video surveillance with augmented virtual environments. In *First ACM SIGMM international workshop on video surveillance* (pp. 107–112). ACM.

Shen, W., Hao, Q., Yoon, H., & Norrie, D. (2006). Applications of agent-based systems in intelligent manufacturing: An updated review. *Advanced Engineering Informatics, 20*(4), 415–431. doi:10.1016/j. aei.2006.05.004

Shet, V., Harwood, D., & Davis, L. (2005). VidMAP: Video monitoring of activity with Prolog. In AVSS 2005 (pp. 224–229). IEEE.

Shet, V., Singh, M., Bahlmann, C., Ramesh, V., Neumann, J., & Davis, L. (2011, June). Predicate logic based image grammars for complex pattern recognition. *International Journal of Computer Vision, 93*(2), 141–161. doi:10.100711263-010-0343-9

Shiang, C. W., Onn, B. T., Tee, F. S., & Khairuddin, M. A. (2016). Developing agent-oriented video surveillance system through agent-oriented methodology (AOM). *CIT. Journal of Computing and Information Technology, 4*(24), 349–367. doi:10.20532/cit.2016.1002869

Silvestre, D. (2007). *Video surveillance using a time-of-light camera* (Unpublished master's thesis). Informatics and Mathematical Modelling, Technical University of Denmark.

Sinha, A., Chakravarty, K., & Bhowmick, B. (2013). Person identification using skeleton information from Kinect. In *The sixth international conference on advances in computer-human interactions (ACHI 2013)* (pp. 101–108). Academic Press.

Skarlatidis, A., Artikis, A., Filippou, J., & Paliouras, G. (2014). A probabilistic logic programming event calculus. *Theory and Practice of Logic Programming*, 1–33.

Stettner, R., Bailey, H., & Silverman, S. (2008). Three dimensional Flash LADAR focal planes and time dependent imaging. *International Journal of High Speed Electronics and Systems*, *18*(02), 401–406. doi:10.1142/S0129156408005436

Tarau, P. (2012). The BinProlog experience: Architecture and implementation choices for continuation passing Prolog and first-class logic engines. *Theory and Practice of Logic Programming*, *12*(1–2), 97–126. doi:10.1017/S1471068411000433

Turbo Prolog Owner's Handbook. (1986). Borland International.

Vallejo, D., Albusac, J., Castro-Schez, J., Glez-Morcillo, C., & Jiménez, L. (2011). A multiagent architecture for supporting distributed normality-based intelligent surveillance. *Engineering Applications of Artificial Intelligence*, *24*(2), 325–340. doi:10.1016/j.engappai.2010.11.005

Wang, C., & Liu, H. (2013). Unusual events detection based on multi-dictionary sparse representation using Kinect. In *International conference on image processing* (pp. 2968–2972). Academic Press. 10.1109/ICIP.2013.6738611

Warren, D. H. D. (1983, October). *An abstract Prolog instruction set.* Technical Note 309. Menlo Park, CA: SRI International.

Wielemaker, J., Schrijvers, T., Triska, M., & Lager, T. (2012). SWI-Prolog. *Theory and Practice of Logic Programming*, *12*(1–2), 67–96. doi:10.1017/S1471068411000494

Worch, J.-H., Bálint-Benczédi, F., & Beetz, M. (2016). Perception for everyday human robot interaction. *KI – Künstliche Intelligenz*, *30*(1), 21–27.

KEY TERMS AND DEFINITIONS

2.5D Vision: A restricted case of 3D vision. This term is applicable when one cannot determine 3D coordinates of some objects because these objects are occluded by other objects placed in front of the video scene.

3D Vision: An application of video processing based on 3D coordinates of objects in the video scene. Often this term is used as a synonym of 2.5D vision, but sometimes it is useful to distinguish 3D and 2.5D vision.

Actor Prolog: An object-oriented logic language developed in the Kotel'nikov Institute of Radio Engineering and Electronics of Russian Academy of Sciences.

Blob: A selected area of a foreground image.

Coefficient of Determination (R^2): A coefficient that indicates the proportionate amount of variation in the given response variable Y explained by the independent variable X in a linear regression model. The larger R^2 is, the more variability is explained by the linear regression model.

Intelligent Visual Surveillance: Visual surveillance implemented by a computer program. Synonyms: intelligent video surveillance, intelligent video monitoring.

Object-Oriented Logic Programming: This term is used differently by different researchers. The meanings of the term vary from solving the mathematical problem of development of a declarative semantics for basic notions of object-oriented programming to unreasoned inserting object-oriented constructs in logic languages that violate the declarative semantics of the logic programs. In the context of the chapter, an object-oriented logic language is a programming language that combines expressiveness of the logical and object-oriented approaches to the programming. In the Actor Prolog language, all object-oriented constructs have a classical declarative (model-theoretic) semantics. This means that Actor Prolog is indeed a logic language and an object-oriented language simultaneously.

Projective Transform Matrix: A matrix that allows one to calculate the pixel coordinates of an object in the video scene using its real physical coordinates and vice versa. A common approach to the problem is the usage of so-called ground plane assumption, that is, the computer program determines coordinates of body parts that are situated inside a predefined plane and this predefined plane usually is the ground. Thus, one can estimate properly the coordinates of person's feet, but a complex surface of the ground and/or presence of stairs and other objects, etc. make the problem much more complex.

RGB-D: Red, Green, Blue, Depth. This term means a description of a video scene that contains information about the colors of the objects and the distances between the camera and the points in the video scene. This description can be used as a source data for 3D/2.5D vision methods.

Skeleton: In the context of the chapter, the skeleton is a graph that describes the structure and coordinates of a human body.

Skewness: A measure of the asymmetry of the data around the mean of the sample. If skewness is negative, then the data are spread out more to the left of the mean than to the right. If skewness is positive, then the data are spread out more to the right.

Time-of-Flight (ToF): In the context of the chapter, a type of 2.5D vision hardware that determine 3D coordinates of objects in the video scene by measuring the time of movement of photons from the camera to objects and backward.

Chapter 7
Moving Object Classification Under Illumination Changes Using Binary Descriptors

S. Vasavi
V. R. Siddhartha Engineering College, India

Ayesha Farha Shaik
V. R. Siddhartha Engineering College, India

Phani chaitanya Krishna Sunkara
V. R. Siddhartha Engineering College, India

ABSTRACT

Object recognition and classification has become important in a surveillance video situated at prominent areas such as airports, banks, military installations, etc. Outdoor environments are more challenging for moving object classification because of incomplete appearance details of moving objects due to illumination changes and large distance between the camera and moving objects. As such, there is a need to monitor and classify the moving objects by considering the challenges of video in the real time. Training the classifiers using feature-based approaches is easier and faster than pixel-based approaches in object classification. Extraction of a set of features from the object of interest is most important for classification. Viewpoint and sources of light illumination plays major role in the appearance of an object. Abrupt transitions are identified using Chi-square and corners are detected using Harris corner detection. Silhouettes are captured using background subtraction and feature extraction is done using ORB. k-NN classifier is used for classification.

INTRODUCTION

Now-a-days objects such as human beings, animals, buildings, vehicles recognition and classification have become important in video surveillance system. Classifying moving objects with in a video sequence is challenging in outdoor environments because of incomplete appearance details, occlusions,

DOI: 10.4018/978-1-5225-5751-7.ch007

dynamic background and illumination conditions. Recorded videos cannot be analyzed manually and as such requires a robust system that can monitor and classify the moving objects by considering the challenges of video in the real time.

Motivation

Visual surveillance and monitoring moving objects is required to identify suspicious activities at public places such as shopping malls, airports, railway stations, bus junctions, banks and military applications. Human operators monitoring manually for long durations is infeasible due to monotony and fatigue. As such, recorded videos are inspected when any suspicious event is notified. But this method only helps for recovery and does not avoid any unwanted events. "Intelligent" video surveillance systems can be used to identify various events and to notify concerned personal when any unwanted event is identified. Such a system requires algorithms that are fast, robust and reliable during various phases such as detection, tracking, classification etc. This can be done by implementing a fast and efficient technique to classify the objects that are present in the video in the real time.

Problem Statement

Basic video analysis operations such as object detection, classification and tracking require scanning the entire video. But this is a time consuming process and hence we require a method to detect and classify the objects that are present in the frames extracted from a real time video. Our earlier work on moving object classification is done by extracting texture, color and structural features, Zernike moments. It was noticed that efficiency of classification is dependent on how far an object is detected or objects appearance in the video frame. Object detection varies because of illumination changes. This chapter is on Moving object classification under illumination variations and abrupt changes by extracting features from key frames that are robust to illumination.

Background

The field of computer vision requires understanding of key terms color spaces, histogram comparison, background modeling, key frame extraction, scene detection, silhouette extraction, feature extraction and classification as described in the following sections.

Color Spaces

Color is a phenomenon that relates to the physics of light, chemistry of matter, geometric properties of object and human visual perception. Human eye can distinguish thousands of colors with different shades and intensities. Attributes that define color are Hue(set of pure colors: green, red, magenta, orange, yellow, blue that range from 0 to 359 degrees), Tint (amount of white added to the original color), Shade(amount of black added to the original color),Tone (amount of black and white added to the original color), Saturation(percentage of original color that ranges from 100% to 0%), Lightness(percentage of original color from 0% to 100%), Chromacity (intensity that contributes to color perception), Luminance (describes the perceived brightness of a color). Different types of color spaces are used to represent an image such as Red Green Blue (RGB), Cyan Magenta Yellow Key/Black (CMYK), Hue Saturation

Value (HSV), Hue Saturation Intensity (HSI). Out of which RGB color space and HSI color space are most commonly used.

RGB Color Space

In the RGB model, each color appears in its primary spectral components of red, green, and blue. RGB model is based on a Cartesian coordinate system. RGB image is an image in which each of the red, green, and blue images are an 8-bit image. The term full-color image is used often to denote a 24-bit RGB color image as shown in figure 1.

HSI Color Space

The HSI (hue, saturation, intensity) color space as shown in figure 2 can approximate the way humans perceive and interpret color. In this, Hue component describes the color in the range between [0,360] degrees, Saturation component specifies the mixture of white color with original color in the range of [0, 1] and intensity in the range of [0, 1] where 0 means black and 1 means white.

Histogram Comparison

Histogram of an image presents frequency of pixel intensity values. These geomtrical clues helps for image analysis such as to find duplicate frames. There are several ways to perform histogram comparision (bin-bin comparision and cross-bin comparision, correlation, chi-square, intersection and battacharya distance, histogram intersection algorithm, Kolmogorov-Smirnov test). Euclidean distance, Manhattan distance and Chebyshev distance between pixels of two frames can also be used for histo-

Figure 1. RGB color space (Poorani, Prathiba, & Ravindran, 2013)

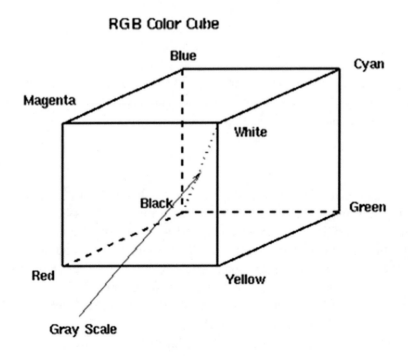

Figure 2. HSI Color Space ("HSI Color Space - Color Space Conversion",2016)

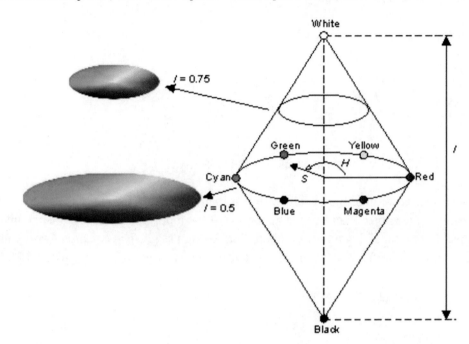

gram comparison. Correlation method uses array of bins (each of the bin contains color samples). A chi-square (χ^2) method performs statistical hypothesis test to conclude on pixel values are equal or not. Pearson Product Moment Coefficient (PPMC) is used to determine whether to accept or to reject initial hypothesis. It calculates difference between the expected frequency and the observed frequency of all the features. Intersection method calculates similarity of two histograms in the range of [0, 1], where 0 specifies no overlap and 1 as similar. Bhattacharyya coefficient can be used to determine how far the two histograms are close to each other. Hence in order to say that two objects are same, we can compute similarity between their histograms. Let I represent pixel intensity, g is the number of pixels in the first image having pixel intensity I, denoted by $H_1(I)$. Now two histograms say H_1 and H_2 can be compared using the following four measures:

1. **Correlation Method:** For two histograms H_1, H_2, Histogram correlation can be defined as given in equation 1 and equation 2 ["Histogram Comparison", (2016)].

$$d(H_1, H_2) = \frac{\sum_I (H_1(I) - \bar{H}_1)(H_2(I) - \bar{H}_2)}{\sqrt{\sum_I (H_1(I) - \bar{H}_1)\sum_I (H_2(I) - \bar{H}_2)}} \tag{1}$$

where

$$\bar{H}_k = \frac{1}{N}\sum_J H_k(J) \tag{2}$$

N is the total number of histogram bins.

2. **Chi-Square Method:** In chi-square, goodness of fit test is used to test the hypothesis, that data comes from a normal hypothesis. It is calculated using equation 3 ("Histogram Comparison", 2016)

$$d(H_1, H_2) = \sum_I \frac{(H_1(I) - H_2(I))^2}{H_1(I)} \tag{3}$$

Where H1 and H2 are two histograms and I is the intensity.

3. **Intersection Method:** Histogram intersection is mainly used when color of the object is the means for identifying the object. Let I be the histogram of the input frame/image, M is the histogram of the object frame containing n bins, computes intersection as given in equation 4 with output value between 0 (no overlap) and 1 (identical distributions) ("Histogram Comparison", 2016)

$$d(H_1, H_2) = \sum_I \min(H_1(I), H_2(I)) \tag{4}$$

Where min function returns smallest of the two parameters, that represents, number of pixels with the same color in the input image. When we divide $d(H_1, H_2)$ with number of pixels, the result is normalized in the range between 0 and 1.

4. **Bhattacharya distance Method:** Bhattacharyya coefficient is a distance measure between two histograms that gives a measure of relative closeness of the two statistical samples being considered. Given two histograms H_1, H_2, N is the mean, equation 5 calculates Bhattacharyya measure value: ("Histogram Comparison", 2016)

$$d(H_1, H_2) = \sqrt{1 - \frac{1}{\sqrt{\bar{H}_1 \bar{H}_2 N^2}} \sum_I \sqrt{H_1(I).H_2(I)}} \tag{5}$$

Background Modeling

Moving objects in a video sequence can be identified using background subtraction method. For this each of the video frames should be compared with respect to a fixed initial frame, in the case of dynamic background modeling, this initial frame is updated as and when new objects that occur in the video becomes static. In some cases background modeling is still difficult because of illumination changes, objects being added or removed quite frequently. Hence background modeling techniques can be said as non predictive modeling (it creates dynamic model by analyzing whether a pixel belongs to foreground or background) and predictive modeling (creates a statistical model) (Xu, Dong, Zhang, & Xu, 2016). Background modeling techniques can further be classified as recursive and non recursive, out of which recursive methods may pose errors even though it requires less memory for computation. Examples of background modeling techniques are: Frame differencing, Average filtering, median filtering, Gaussian mixture model, minimum-maximum filtering, linear predictive filter, approximated median filter, running Gaussian average method, mixtures of Gaussians, kernel density estimation, Harris corner based modeling. Frame differencing method detects foreground objects by taking the difference between current frame and the previous frame. But this method is sensitive to noise and illumination changes. Also efficiency is reduced when size of the foreground object is large. Average of the frames for a particular period of time is used to create background model in average filtering method. Efficiency of this method is reduced when object size is large, illumination changes occur, very fast or very slow objects appear. Median of all the pixel values is computed in median filtering method to find background model. Similarly, if Medoid is computed, it finds color and sharper background estimation. Advantage of median filter is that, it can detect edges and background without any blurring. Linear predictive filter such as Wiener filter, Kalman filter predicts background model based on k-pixel values. These filters are not suited in real time environment because computing the coefficients for each frame takes significant amount of time. Gaussian Mixture Model (GMM) models the background as a weighted sum of M Gaussian densities as shown in equation 6 (Xu, Dong, Zhang, & Xu, 2016).

$$P(x_t) = \sum_{i=1}^{K} \omega_{i,t} \cdot \eta(x_t, \mu_{i,t}, \sum_{i,t}) \tag{6}$$

where K is the number of Gaussian distributions, $\eta(x_t, \mu_{i,t}, \sum_{i,t})$ is the *i-th* Gaussian probability density function, $\omega_{i,t}$ is its weight at time t, $\mu_{i,t}$ is the mean value of the *i-th* Gaussian in the mixture at time t, and $\sum_{i,t}$ is the covariance matrix of the *i-th* Gaussian in the mixture at time t.

Non parametric method such as kernel density estimator (KDE), uses kernel function to model the distribution. Let the N recent sample of intensity values are $\{x_1, x_2..., x_N\}$ to model the background distribution, then Kernel density estimation of intensity for every pixel x_t at time t is estimated using equation 7 (Xu, Dong, Zhang, & Xu, 2016).

$$\Pr(x_t) = \frac{1}{N} \sum_{i=1}^{N} \prod_{j=1}^{d} \frac{1}{\sqrt{2\pi\sigma_j^2}} e^{-\frac{1}{2}\frac{(x_{t_j} - x_{i_j})^2}{\sigma_j^2}} \tag{7}$$

where N represents number of samples, d is the number of channels and σ is the kernel function bandwidth for every color channel that is estimated by equation 8 (Xu, Dong, Zhang, & Xu, 2016). $\sigma = \dfrac{m}{0.68\sqrt{2}}$ (8) where m is the median absolute deviation over the sample for consecutive values of the pixel. Clustering based background modeling is called as codebook method. In this method, for each pixel, a series of key color values (codewords) are stored. Adaptive GMM reconstructs the background and adapts to the scene by adjusting the parameters and choosing number of components for each pixel. Let data sample x_t at time t, then parameters $\omega_{i,x,t}$, $\mu_{i,x,t}$ and $\sigma_{i,x}$ *are updated* recursively using equation 9, equation 10 and equation 11 (Xu, Dong, Zhang, & Xu, 2016).

$$\omega_{i,x,t} = \omega_{i,x,t} + \alpha(o_{i,x,t} - \omega_{i,x,t}) \tag{9}$$

$$\mu_{i,x,t} = \mu_{i,x,t} + o_{i,x,t}(\alpha \: / \: \omega_{i,x,t})\overline{o}_{i,x,t} \tag{10}$$

$$\sigma^2_{i,x,t} = \sigma^2_{i,x,t} + o_{i,x,t}(\alpha \: / \: \omega_{i,x,t})(\delta^T_{i,x,t}\delta_{i,x,t} - \sigma^2_{i,x,t}) \tag{11}$$

Few other methods that perform background modeling are Consensus-based method (SACON) that builds statistical background model, a self-organizing background subtraction method (SOBS) that creates adaptive background modelling and robust to illumination changes, a universal background subtraction algorithm (Vibe) that estimates the background with the help of first frame, Pixel-based adaptive segmenter (PBAS) that uses past image values for foreground segmentation proved to be better in generating precise results.

Harris corner-detector is based on sparse feature set of detected corners in each of the video frame. Corner is a point where directions of two edges can be changed and such variation can be used to detect it. In this, selected corners are used for modeling as given in equation 12 (Lim, Ramesh, Yang, Xiang, Gao, & Lin, 2017).

$$E(u,v) = \sum_{x,y} w(x,y)[I(x+u, y+v) - I(x,y)]^2 \tag{12}$$

where w (x, y) can be a window (rectangular or Gaussian) that gives weight to the surrounding pixels, E is the difference between the original and the moved window, u is the window's displacement in the x direction, v is the window's displacement in the y direction, I is the intensity of the image at a position (x, y), I(x+ u, y+ v) is the intensity of the moved window, I(x, y) is the intensity of the original [Poonam, Mind, & Gumaste, 2017).

Key Frame Extraction

A key frame in a video can be defined as starting and ending points that includes any transition such as shadow, lighting conditions, illumination changes, weather conditions. Key frame algorithms can be broadly classified into five categories (Dang & Radha,2014) namely: Shot boundary based methods(frames are extracted from a fixed position and then chooses the first frame, middle frame and the last frame as key frames), Visual content based approach (key frames are detected using features such as edge, color and texture changes), Clustering based approach (calculates distance from each frame to the existing clustering center and if the distance is less than a threshold value, then classified into the smallest distance cluster, or creates a new cluster), Motion analysis based approach (key frames are detected using movement information of objects in the video), Compressed video stream extraction based approach (uses features of the compressed video data). Other methods such as pixel based approach (each pixel is compared to understand significant changes, not suitable for real time environments as it takes lot of time for computation), template matching (a fixed pattern is compared to detect key frame) and histogram based method (compares two histograms to detect changes). Sequential based approaches (uses visual features and temporal information) and cluster based (frames of a shot are clustered). Shot boundary based key frame extraction method for segmenting user generated videos on video sharing websites using visual content is proposed by (Kelm, Schmiedeke, &Sikora, 2009). Image epitome based framework for extracting key frame from consumer video that uses image epitome to measure dissimilarity between frames of the input video and does not require shot detection, segmentation is proposed by Dang, Kumar, and Radha (2012). Shao and Ji (2009) proposed key frame extraction based on intra-frame and inter-frame motion histogram analysis where entropy values of the consecutive frames are used to conclude on motion objects. Dang and Radha (2014) use heterogeneity image patch (HIP) index of a frame in a video sequence to judge on key frame. Huang, Xia, Zhang, and Dong (2013) proposed a hybrid approach that uses both shot boundary and visual content based methods for key frame extraction.

Scene Detection

A scene is a collection of adjoining shots that represents points of interest (El-Qawasmeh, 2003). The following figure 3 presents structure and various terms that defines a video.

Scene change can be abrupt or gradual(dissolve, fade, wipe). Scene change detection algorithms can be classified as top down approach and bottom up approach. Top down approach is based on explicit modelling where as bottom up apporaches work on both compressed and uncompressed video data. Several methods have been proposed for scene detection such as pixel-difference method as shown in figure 4 (computes the difference between pixels belonging to consecutive frames and not suited for real time environments because of intensive computations), likelihood ratio (first frames are divied into blocks and comptation is performed based on blocks of pixels, not suited for real time environments becuase of intensive computations), histogram comparision (bin wise difference is calculated for concluding upon scene), motion vector scheme(motion vector counts are used for detecting scene), Kolmogorov-Smirnov Test (pixel luminance for two consecutive frames is calculated for detecting scene), edge change ratio (intensity of edges between consecutive frames is computed to identify scene).

Figure 3. Structure of a video (Eyas EL-QAWASMEH, 2003)

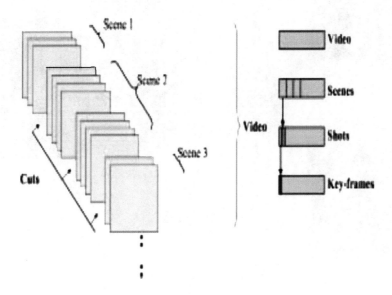

Ford, Robson, Temple, and Gerlach (1997) presented five metrics for scene change detection in video sequences. Hierarchical scene change detection is proposed by Shin, Kim, Lee, and Kim (1998) for MPEG-2 compressed video. When compared to pixel domain processing, their method reduced computational requirement. Fernando, Canagarajah, and Bull(2000) proposed scene change detection approach that works for uncompressed and compressed video. Even though their approach detects scene changes such as abrupt transitions and gradual transition accurately, but could not work well for camera movement detection within the same framework. Ji Zhang, Feng Liu, Hui Shao, and Wang (2007) presented an effective error concealment framework for H.264 decoder based on video scene change detection. Their framework has better robustness and can efficiently improve the visual quality and PSNR of the decoded video. Drawback is that work cannot be applied for scene change detection of B frames. Sakarya and Telatar(2010) presented a video scene detection using graph based representations and unsupervised clustering procedure. Their graph based video scene boundary detection method is evaluated and compared with the existing graph based video scene detection methods. Chauhan, Parmar, Parmar, and Chauhan (2013) presented hybrid approach called video compression based scene change detection along with block based motion estimation algorithms (BME). Chhasatia, Trivedi, and Shah (2013) presented a person localization system for video segmentation, scene detection and object localization. Test results showed that their algorithm worked well in different conditions such as object's motion, illumination changes, and different poses. Shukla, Mithlesh, and Sharma (2013) presented a detailed survey on existing scene change detection algorithms.

Figure 4. Frame differencing method (Shukla, Mithlesh, & Sharma, 2015)

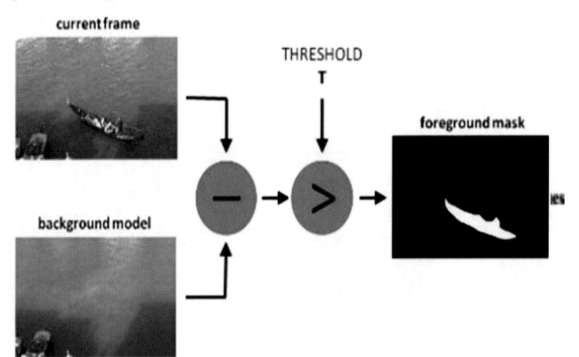

Silhouette Extraction

Suppose E(u, v) is the eye vector, let σ(u, v) is a point on a surface and surface normal is N(u, v), is a silhouette point if E(u, v)·N(u, v) = 0, that is, the angle between E(u, v) and N(u, v) is 90 degrees (Gooch,2017). Figure 5 presents silhouette example for a polygon surface.

N · E < 0 means polygon is front-facing, N · E > 0 means polygon is back-facing and if N · E = 0 then the polygon is perpendicular to the view direction.

Silhouette can be extracted by modelling in object space (computes edges and curves) or in image space (uses image processing techniques). Object space algorithms such as brute force approach (every edge is verified whether it belongs to the point of interest or not), edge buffer(facets are used instead of edges), probabilistic (start with some initial edges and verifies whether these seed edges belong to point of interest or not), Gauss map arc hierarchy (a tree structure is created to store angles of arcs between front and back facing polygons), normal cone hierarchy (a tree structure is created to store Polygon normal's that are grouped into cones), implicit surfaces (ray tests are used to find points on the silhouette curve), NURBS surfaces (marching cube algorithm is used to find Silhouette curves). Image space silhouette algorithms such as two pass methods (uses Back facing and front facing polygons), environment map (shading maps are used), one pass method (two cube maps are created during pre process), model augmentation (every edge is checked whether it is a silhouette edge or not), depth discontinuity methods (depth difference between two pixels is calculated to conclude on silhouette). An algorithm is developed in Sulaiman, Hussain, Tahir, Samad, and Mustafa(2008) to extract human silhouette by separating foreground pixels from its background. Methods commonly used for silhouette extraction are background subtraction (difference between the current frame and initially assumed background image is computed to detect moving object), statistical methods (this approach is robust to noise, shadow, illumination changes, detects changed regions with the help of pixel statistics and concludes on foreground pixel), temporal differencing method (This method is good for dynamic scene changes but cannot extract

Figure 5. Example for 2D silhouette example (Gooch,2017).

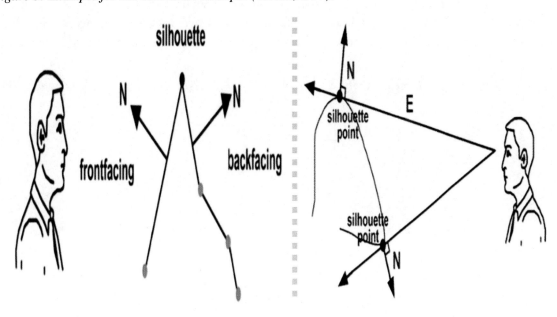

foreground object that moves slowly or when an object stops, consecutive frames pixel difference is used to detect moving region in a video). Optical flow (basing on the pixel direction and time, a particular region of a frame is estimated), Thresholding (a simplest technique where a threshold value is used to conclude on foreground and background pixel). Silhouette based object classification consists of two steps: Offline and Online step. In offline phase, a sample template database is created with some know object silhouettes as shown in figure 6 (24 templates, 14 for human, 5 for human and 5 for vehicles). In online step, in each of the frame of the given video sequence, silhouette is extracted for each of the detected object, compare it with the existing ones in real time.

Feature Extraction

Feature extraction is the process where we retrieve the content of an object that is useful for further processing such as image amalysis. Features such as color, texture, stuctural, wavelet transformation, local binary pattern(LBP), corners, edges, ridges, invariant features, blobs, Scale invariant features (SIFT) (Lowe, 1999), speededup robust features (SURF)[Herbert Bay,Andreas Ess, Tinne Tuytelaars, LucVan Gool(2008)], principal component analysis (PCA) (Pearson,1901), features from accelerated segment test (FAST), Oriented FAST and Rotated BRIEF (ORB) (Rublee, Rabaud, Konolige, & Bradski, 2011). Feature extraction method should be robust so that it can identify features during the conditions such as noise, illumination and scale invariant. Other features such as Histogram of Gradients (HOG) (McConnell, 1986), Luminace symmetry, central moments, ART moments, cumulants, horizontal and vertical projection, morphological features can also be extracted for moving object classification. The following figure 7 presents taxonomy of descriptors based on computational and storage requirements.

A study among various methods such as Harris corner detection, SIFT (Scale Invariant Feature Transform) and SURF is carried out by [Kartikay Lal, Khalid Mahmood Arif (2016)] and are evaluated, compared on various performance measures such as robustness, amount of points detected, time complexity. Surveillance system for detecting moving objects classification is proposed in Mustafah, Zainuddin, Rashidan, Aziz and Saripan (2017). Their method extracts low level features such as color, edge and texture. Algorithm performs adaptive background modeling and could handle illumination and adverse weather. Cars, motorcycles and pedestrians are classified. Figure 8 presents details of background subtraction algorithm where video frames are divided into patches and patches are analyzed for

Figure 6. Sample object silhouette template database (Dedeoglu, Toreyin, Gudukbay, & Cetin, 2006)

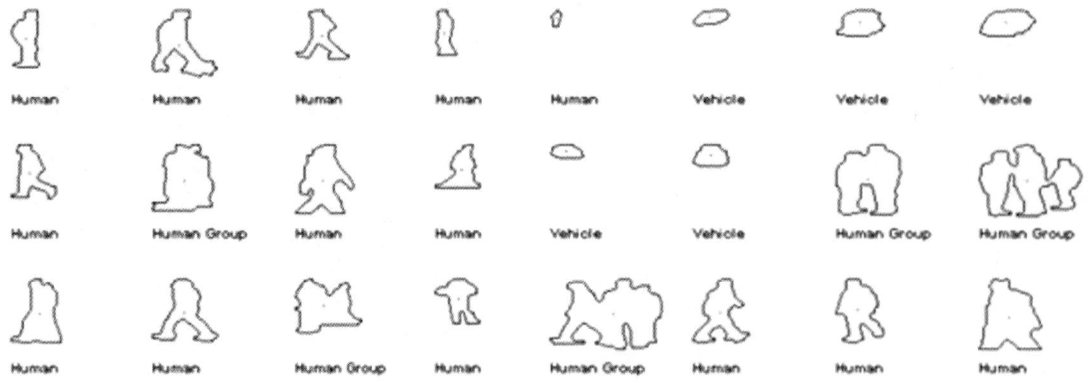

Figure 7. A brief taxonomy of descriptors based on computational and storage requirements (Heinly, Dunn, & Frahm,2012)

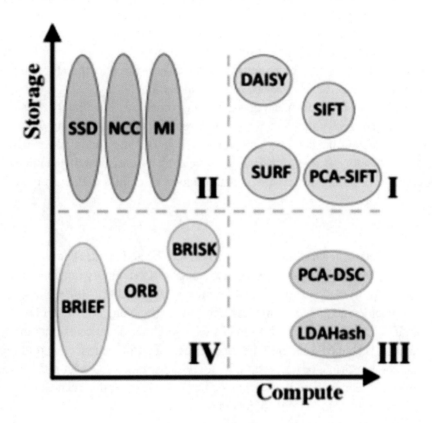

Figure 8. Overview of Background modelling (Mustafah, Zainuddin, Rashidan, Aziz & Saripan, 2017)

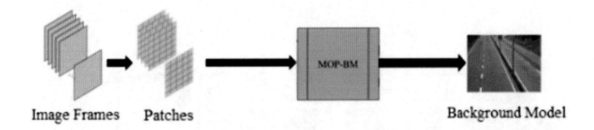

background reconstruction. This system involves patch computation for every frame in the video and can face time complexity problem in real-time.

An integrated system is proposed in Liang and Juang (2015) for classifying objects such as pedestrians, cars, motorcycles, and bicycles. Shadow removal technique is used to improve the performance of classification. A two-level Haar wavelet transform is used to extract features such as shape features, HOG features. Performance of this method is compared with existing approaches and it proved to be better. But the method cannot work in the presence of illuminations changes and invariant features.

Vanithamani (2017) presents moving object classification where for each object features such as top left and bottom right of the rectangle, that indicates X and Y coordinates of the contour points, motion features such as current speed, acceleration & deceleration value, change in size, velocity are extracted and classified into classes such as static to static, static to moving, moving to moving, moving to static, sudden stop, sudden speed. Classification results and performance measures are not mentioned in this work. Oh and Kang (2017) presents object-detection and classification method for a multi-layer LiDAR and a CCD sensor. Figure 9 presents overview of their method.

It can be observed that results from unary classifiers are combined to accurately classify object on each sensor modality. Also red arrows shown in the figure 9 represents the processing of unary classifier for each sensor and green arrows indicate the fusion processing. This method involves high computational cost. Drawback of this method is that, if there is a failure at unary classifier, it can reduce the classification results. Advantage of binary descriptors is that, each bit associates with exactly one comparison.

Classification

Moving objects can be classified basing on shape, color, motion and texture features. Yet again classification is done based on pixel wise or from prediction image or from binary image. Image Classification methods can be roughly divided into two broad families of approaches: (a) Learning based classifiers also known as parametric methods require learning phase. (b) Nonparametric classifiers require no learning and classification decision is based on Nearest-Neighbor distance estimation of the data. Yet another categorization is shape based classification (uses spatial information) and motion based classification (use temporal information). Few authors used dispersedness, aspect ratio, area, perimeter as a classification metric for shape based classification. Temporal properties such as optic flow, velocity, speed are used by few authors for moving object classification.

k-Nearest Neighbor algorithm (k-NN) is a non-parametric method used for classification and also for regression. The training samples are stored as n- dimensional numeric attributes. When a test sample is given, the k-nearest neighbor classifier searches the k training samples which are closest (Euclidean

Figure 9. Overview of the system for classifying objects on each sensor modality (Oh & Kang, 2017)

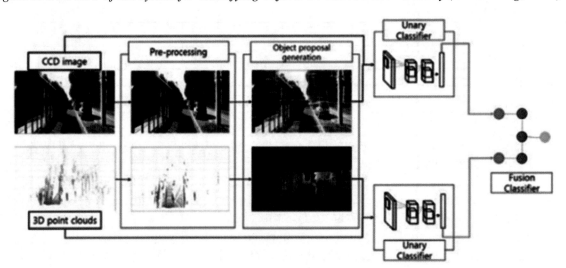

distance) to the unknown sample. Euclidean distance between two points $X_1(x_1,x_2,....x_n)$ and $X_2(x_1,x_2,....x_n)$ is given in equation 13.

$$\text{dist } (X_1,X_2) = \sqrt{\sum_{i=1}^{n} (x1i - x2i)^2} \tag{13}$$

The main objective of this chapter is to propose a system that can classify the objects into human and vehicle object class in a video taken from the stationary camera under various illumination changes such as change in the background light, intensity variations in light, abrupt changes such as sudden adding of new/old objects in to a frame. The process of moving object classification for a stationary camera video sequence can be used in surveillance systems and also to detect abnormal object behavior.

1. Frame Extraction: Frames are extracted from the video at the rate of 30 frames per second.
2. Frame Conversion: Frames thus obtained are in RGB color space model should be converted into HSI color space model.
3. Histogram Analysis: Histogram values are derived for each frame under illumination changes using the Chi Square test.
4. Discovery of Abrupt Transitions: Comparing the obtained Chi Square test value with a threshold value of 0.9.
5. Application of Background modeling: Harris Corner based background modeling technique is used for corner detection.
6. Key frame extraction from the scene and Meta data generation: Metadata (Frame number and the timestamp at which the frame was captured) of the extracted key frames is stored in a file.
7. Capture Silhouette and feature extraction: Silhouettes of the extracted key frames are generated and the features are extracted using ORB feature detector that can efficiently compute oriented BRIEF features.
8. Classification of an object under Illumination changes: Objects are classified using k-NN classifier based on the features extracted. Classification is done in R programming.

LITERATURE REVIEW

Image object classification has its significance in video surveillance systems. Classification can be on both static and moving objects. This requires robust methods for extracting features and also representing them.

Segmentation of Moving Objects Using Background Subtraction Method in Complex Environments (Kumar & Yadav, 2016)

In this, authors addressed the problem of detecting an object by under various spatiotemporal conditions such as illumination changes, low resolution, water ripples. Background pixels are updated by calculating current spatial variance. Their work is compared with other background subtraction methods and proved to better. Their work comprises of two stages, in the first stage, background modeling and its updation is done, where as in the second stage, regions are processed for extracting connected components.

Advantages

1. No need of allocating memory buffers for generation of background model.
2. Background is characterized basing on spatial and temporal features
3. Background model adaption during temporal changes in the environment, aperture distortion, ghost effect and over segmentation error.

Disadvantages

This approach does not give good results in the following conditions.

1. If the scene contains many, slowly moving objects(mean and median)
2. If the objects are fast and frame rate is slow
3. If general lighting conditions in the scene change with time.

Detection of Scene Change in Video (Mithlesh & Shukla, 2016)

Authors of this work presented a system on video shot boundary detection using color histogram. Histogram difference between two consecutive frames is used to detect shot. The scene change detection system is developed using MATLAB. The following figure 10 presents block diagram of their system.

Advantages

1. This system provides an easy way to determine threshold using scaled frame difference
2. Provides stable Scene change detection in the presence of abrupt transition.
3. Authors claim on low false alarms and robust to gradual transitions

Disadvantages

1. Complexity of the proposed system is high. Scene Change Detection technique using Graph based is complex when compared to Histogram and adaptive thresholding method.
2. No results are presented to verify precision, recall and PSNR(Peak Signal to Noise Ratio).

Comparison of Histograms in Physical Research (Bityukov, Maksimushkina, & Smirnova, 2016)

Histogram is a simple method for many of the image processing applications to represent the given object. This work has used methods such as statistical histogram comparison method (SCH), Kolmogorov–Smirnov (KS) method and Anderson–Darling (AD) and results are assessed. When random variables have differences in mathematical expectation then Anderson–Darling and Kolmogorov–Smirnov proved to be better. Also, when random variables have differences in the widths of distributions then statistical comparison of histograms proved to be better.

Figure 10. Block diagram for detection of scene change (Mithlesh & Shukla, 2016)

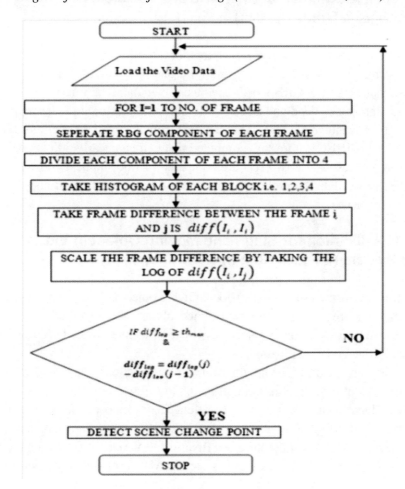

Human Action Recognition From Multiple Views Based on View-Invariant Feature Descriptor Using Support Vector Machines (Sargano, Angelov, & Habib, 2016)

Feature descriptor is proposed in this work for multi-view human action recognition. Region-based features are extracted from the human silhouette and SVM is used for action classification. Their work is compared with existing works and results proved to be better. This work is aimed at reducing classification cost for feature extraction. This system works in various stages. In the first step, silhouette is preprocessed. Then silhouette is divided into radial bins, and features such as geometrical and Hu moments are extracted. Finally SVM is used for classification. IXMAS Dataset is used for experimentation. Even though this method works well for 2D methods, in real time videos, silhouette extraction becomes complex.

ORB: An Efficient Alternative to SIFT or SURF (Rublee, Rabaud, Konolige, & Bradski, 2011)

Current methods for feature matching use descriptors for detection and matching that are complex and time consuming. Authors proposed a binary descriptor based on BRIEF, called ORB, that is rotation invariant and also handles noise. ORB is validated using two datasets: images with synthetic in-plane rotation with Gaussian noise added and a real-world dataset of textured planar images captured from different viewpoints. Advantage of ORB is that it is free from licensing restrictions of SIFT and SURF. oFAST keypoints and rBRIEF features are computed that extracts nearly 500 keypoints per image for training and testing images. These two sets are compared to find the best correspondence. ORB has outperformed SIFT and SURF. But problem with ORB is that recovery of variance, which makes nearest neighbor more efficient.

A Study on Classification for Static and Moving Object in Video Surveillance System (Mishra, Kumar, & Saroha, 2016)

Authors of this work performed survey on various moving object classification systems that are based on motion and shape features. Figure 11 presents object classification hierarchy.

This paper outlined various phases in object classification such as object detection, feature extraction, dimensionality reduction and classification.

Many other works are reported for moving object classification. Work of Boukhriss, Fendri, and Hammami (2016) is about moving object classification in Infrared and Visible spectra. Features based on shape, texture and motion are used for classification. Their proposed method is on visible spectrum or infrared spectrum depending on weather conditions such as snow, fog, rain, sunny and also on the recorded time of video. Another work reported by Bhanse and Vasekar (2017) is on recognizing moving objects that helps in Advanced driver assistance systems (ADAS) as shown in figure 12.

Figure 11. Overview of object classification (Mishra, Kumar, & Saroha, 2016)

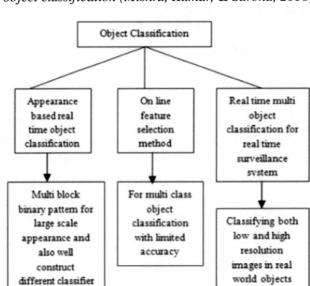

Figure 12. Architecture for tracking task with simultaneous confinement and mapping (SLAM) and detects and tracks the moving objects(DATMO)

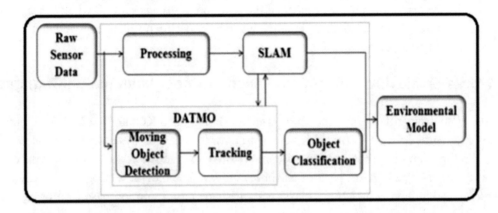

Authors of Higashia, Fukuib, Iwahoria, Adachia, and Bhuyanc (2016) treated shadows as illumination changes and proposed a new feature for handling shadows. This feature is constructed using L*a*b* components, Peripheral Increment Sign Correlation and Normalized Vector Distance that are robust to illumination changes. Shadows are detected using histograms and finally SVM is used for classifying scenes such as Room, Campus, highway, Hallway, Lab. Authors explained various features that are robust to illumination changes such as Normalized Vector Distance(NVD) and Peripheral Increment Sign Correlation(PISC). NVD is a distance between the normalized irradiances of the background image and that of the observed image. NVD at a pixel, x = (x, y), is calculated by the equations 14,15 and 16 (Higashia, Fukuib, Iwahoria, Adachia, & Bhuyanc, 2016)

$$NVD_{x,\,y} = \left| \frac{I_{x,\,y}}{\left| I_{x,y} \right|} - \frac{B_{x,y}}{\left| B_{x,y} \right|} \right| \tag{14}$$

$$\left| I_{x,\,y} \right| = \sqrt{\sum_{i=-N_b/2}^{N_b/2} \sum_{j=-N_b/2}^{N_b/2} \left| I_{x-i,\,y-j} \right|^2} \tag{15}$$

$$\left| B_{x,\,y} \right| = \sqrt{\sum_{i=-N_b/2}^{N_b/2} \sum_{j=-N_b/2}^{N_b/2} \left| B_{x-i,\,y-j} \right|^2} \tag{16}$$

where NVDx,y means the NVD value at x, and Ix,y and Bx,y mean the data at x in the observed image and the background image, respectively. Nb represents the block size.

PISC is a correlation of the Local Binary Pattern (LBP) of a background image and that of an observed image. PISC is obtained by equation 17 (Higashia, Fukuib, Iwahoria, Adachia, & Bhuyanc, 2016)

$$PISC(x) = \frac{1}{N} \sum_{k=1}^{N} \left(f_k(x) \cdot b_k(x) + (1 - f_k(x)) \cdot (1 - b_k(x)) \right) \tag{17}$$

where PISC(x) means the PISC value at x, $f_k(x)$ and $b_k(x)$ represents the sign of the difference of the pixel value between x and its k-th neighbor pixel for the observed image and that for the background image, respectively, and N means the number of the neighbors. $f_k(x)$ is set to 0 when the sign is minus. Otherwise, it is set to 1. $b_k(x)$ is also set as the same manner. Experiments results proved that their method is better than existing works.

Other Works on Moving Object Classification Under Illumination Changes

MOG based Visual surveillance system is proposed in Mangal, and Kumar (2017). Their system works in both color and gray level. CAVIAR database is used for testing their system. Objects such as human, animals and vehicles are classified. Their work cannot deal with shadows and also time complexity is high.

Authors Joshi and Patel (2016) performed survey on various object detection (frame difference, optical flow and background subtraction), classification(shape based, motion based, color based, texture based) and tracking (matching based, Region based, feature based such as color, gradient, edge, biological, spatio temporal, deformable template based tracking, model based, filtering based such as kalman filter, particle filter, class based tracking, fusion based tracking) algorithms.

In the work given by Kavitha and Kavitha (2017), authors proposed a system to detect abnormal moving objects under illumination changes both in indoor and outdoor. Classification is done using sparse graph based k-NN so that nearest neighbor classification can approximate real-time moving object classification. Experimental results proved that their method outperforms all other existing works.

Drawbacks of Existing Methods

1. Object classification is done only on the fixed classes and gives accurate results only when the classes are more specific.
2. HOG features did not perform well and is less discriminating for recognizing object classes.
3. RGB and Ohta histograms are severely degraded when the illumination source is not kept constant.
4. Most of the existing methods work only under indoor environments.
5. More redundancy in the case of Hu's moments. Also accuracy is low for Hu's moments. Hu's moments are more complex.
6. Visual appearance of the moving object varies basing on time and camera distance and viewing angle.
7. Moving object classification system should work for spatio temporal data that varies with time.

The proposed approach considers the training dataset consisting of various types of image samples and the features are extracted from those samples, which is used to train the classifier. To classify the objects, video is divided into frames. Abrupt transitions are identified using GLCM, Chi-square test and corners are detected using Harris corner detection. Meta data is generated for the Key Frames which can be used for further processing such as image indexing, object querying and retrieval. Silhouettes are captured using background subtraction and feature extraction is done using ORB. Finally k-NN classifier is used for classification.

PROPOSED SYSTEM

This section presents the detailed functionality of the proposed system to perform moving object classification in a video sequence. Our proposed method gives solution to the drawbacks that are already mentioned in the previous section. The main functional requirements are:

1. The video can be taken from a stationary camera or can be uploaded from device.
2. Given a video, the software must be able to extract the frames out of it.
3. Given two successive frames, the software should find the similarity between them.
4. Given a processed frame, the software must be able to detect the corners accurately.
5. Given the abrupt frames, the software must generate object silhouettes.
6. Given the features of training and testing silhouettes, the software must classify the objects present.

Non-functional requirements of our system are as follows:

1. Once the user runs the program, (extraction of frames in HSI format) the image file must be saved immediately. It should just need a 'one-click' effort.
2. The system must be interactive. It should provide flexibility.

Methodology for Processing Spatial Queries

Figure 13 presents flowchart of the proposed system.

- **Step 1:** Read a Local Video.
- **Step 2:** Converting the video into frames, converting from RGB to HSI.
- **Step 3:** Detection of foreground scene using Chi-square test
- **Step 4:** Background modeling of the scene using Harris Corner Based Background modeling technique.
- **Step 5:** Key frame extraction from the scene and Meta data generation
- **Step 6:** Capture silhouette and Feature extraction
- **Step 7:** Classification of object under illumination changes.

Detailed functioning of each step is discussed in the following paragraphs. In the first step as shown in figure 14, a video frame is extracted and converted from RBG to HSI. Reason for choosing HSI color space is that, Hue, Saturation and Intensity components reflects the way a human visual system will perceive color through their eyes. Also RGB color space poses difficult description, redundant and correlated (luminance information in all channels), that can reduce efficiency. So we have chosen HSI color model which is designed to approximate the way humans perceive and interpret color. These HSI frames are used for further processing.

Gray Level Cooccurrence Matrix is computed for each frame (for texture analysis) and similarity between consecutive frames is found out using chi-square test (histogram values) . 0.9 threshold is taken to classify the possible abrupt transition. If abrupt transition is detected, we treat the frame as keyframe. Figure 15 presents flowchart of histogram analysis.

Figure 13. Flowchart of the proposed system   *Figure 14. RGB to HSI conversion*

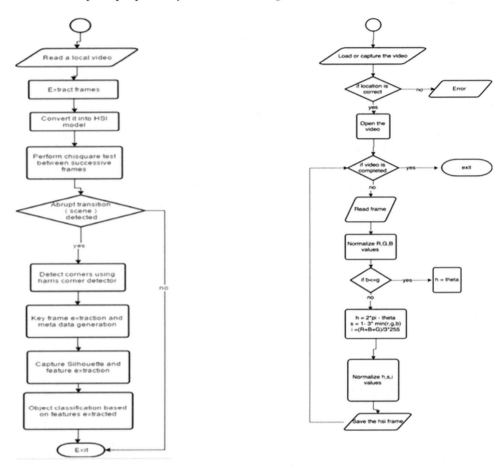

Corner based background modelling that applies correlation (mean and variance-adaptive threshold) between 2 corners to say they belong to foreground or background (faster than pixel based methods) is the next step. Harris corner detection is used to identify static and motion corners as shown in figure 16.

The next step is meta data generation for the extracted keyframes. Figure 17 presents the flowchart for meta data generation.

The following figure 18 presents Meta data file that stores key frame numbers and corresponding locations of the object within that frame. For example first row indicates that in frame number 1; object was located with coordinates (x, y) (a, b) (p, q) and (r, s) and so on.

The next step in moving object classification is to capture silhouette. Figure 19 presents flowchart of this process.

For each of the silhouette, 26 features are extracted and ORB descriptor is created. This process is shown in figure 20.

The final step is to classify the object into predefined class such as human or vehicle. This process is shown in figure 21.

Figure 15. Chi-square Test for histogram analysis

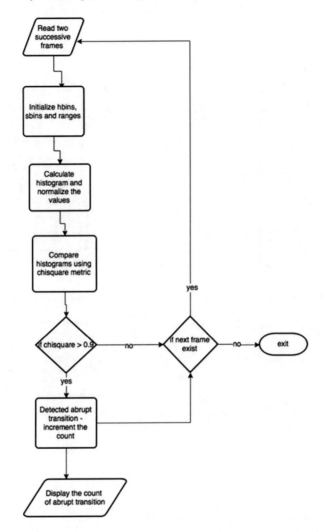

Algorithms

This section presents various algorithms that describe the detailed working of the proposed approach. Algorithm1 presents the conversion of video from RGB to HSI color space mode and extracting the frames, Algorithm2 is used for finding the similarity between frames and abrupt transitions, Algorithm3 describes the background modeling of the scene. Algorithm 4 algorithm 5 algorithm 6 and algorithm 7

Converting the Video From RGB to HSI Color Space Model

Algorithm1: Converting the given video from RGB to HSI color space model and extracting the frames

Input: Sample video (uploading a file or captured from camera)
Output: HSI frames for the given video

Figure 16. Detecting corners using Harris Corner Detector

Figure 17. Meta data generation for the extracted key frames

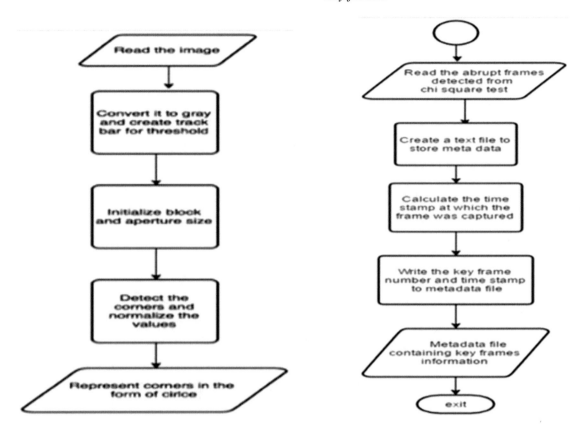

Figure 18. Meta data format

Frame No	Coordinates
1	(x, y) (a, b) (p, q) (r, s)
2	(c, d) (m, n) (t, u) (i, j)
...	....

1. Read the video
2. Extract the frames
3. For each frame perform the steps 4 through 6
4. Beginning with normalizing RGB values using equations 18,19 and 20 (Gonzalez & Woods, 1992)

$$r = R/(R + G + B) \tag{18}$$

$$g = G/(R + G + B) \tag{19}$$

Figure 19. Flowchart for capturing silhouette

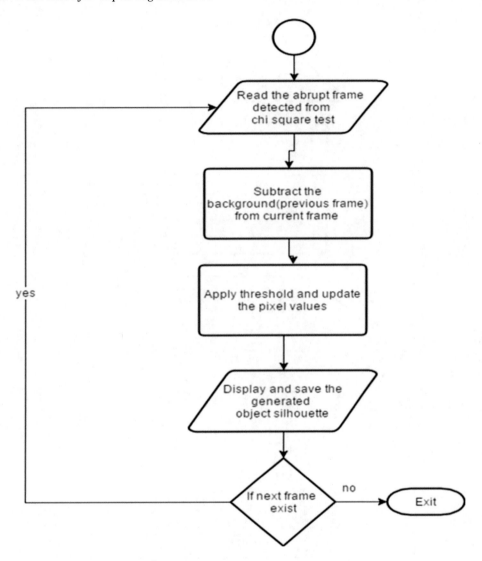

$$b = B\big/(R + G + B) \tag{20}$$

5. Normalized H, S and I components are then obtained by using equations 21, 22, 23 and 24.(Gonzalez & Woods, 1992)

$$h = \cos^{-1}\left\{\frac{0.5[(r-g) + (r-b)]}{\sqrt{[(r-g)^2 + (r-b)(g-b)]}}\right\} \tag{21}$$

h belongs to [0,π] for b≤g

Figure 20. Feature extraction using ORB

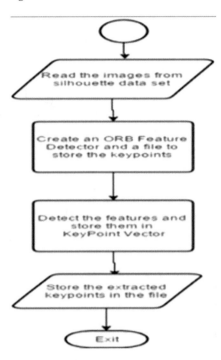

Figure 21. Object classification

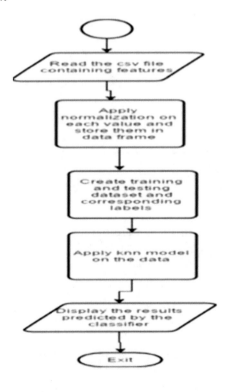

$$h = 2\pi - \cos^{-1}\left\{ \frac{0.5[(r-g)+(r-b)]}{\sqrt{[(r-g)^2 + (r-b)(g-b)]}} \right\} \tag{22}$$

h belongs to [π,2π] for b > g

$$s = 1 - 3 \cdot \min(r,g,b) \text{ s belongs to } [0.1] \tag{23}$$

$$i = (R+B+G)\big/(3*255) \text{ where i belongs to } [0.1] \tag{24}$$

6.　h, s and i values are converted in the ranges of [0,360], [0,100], [0,255] respectively as given in equation 25 and equation 26.

$$H = h*180\big/\Pi \tag{25}$$

$$S = s*100 \tag{26}$$

$$I = i*255 \tag{27}$$

Histogram Comparison Using Chi-Square Test and Detection of Abrupt Transitions

Data is divided into k bins (16 in our experimentation) and test statistic is applied. It is used as goodness of fit test. Histograms are collected *counts* of data organized into a set of predefined *bins. Each of the histogram has the following components such as dims (that specifies nu*mber of parameters), Bins(number of subdivisions in each dim), Range(limits for the values to be measured). For our experimentation, we calculated based on hue and saturation. Dims=2, we have considered 50 bins for hue and 60 bins for saturation.

Range for hue as [0,180] and saturation as [0,255] and finally normalized into the range [0,1].

From the four different metrics, we have chosen chi-square metric for finding the similarity between successive frames as it produces accurate results when compared to others.

The chi square statistic is defined as follows:

$$\chi^2 = \sum_i \frac{(O_i - E_i)^2}{E_i} \tag{28}$$

O_i: The observed number of cases in category i
E_i: The expected number of cases in category i.

This chi square statistic is obtained by calculating the difference between the observed number of cases and the expected number of cases in each category. This difference is squared and divided by the expected number of cases in that category. These values are then added for all the categories, and the total is referred to as the chi squared value.

$$d(H_1, H_2) = \sum_I \frac{(H_1(I) - H_2(I))^2}{H_1(I)} \tag{29}$$

Algorithm2: Finding similarity between successive frames using chi-square metric and Detection of abrupt transitions
Input: Successive Frames extracted from the video
Output: Abrupt transitions

1. Load the two frames
2. Calculate the H-S histogram for all the frames and normalize them in order to compare them.
3. Compare the histogram of the frames using chi square metric as given in equation 28. If the chi square value is less than 0.9, then we move to next frame if there are some more frames exist Else if the chi square value is greater than or equal to 0.9, then we detect the frame as abrupt transition frame (keyframe), increment the count of abrupt transitions and move to next frame if any.
4. Display the count of abrupt transitions detected.

Harris Corner Based Background Modeling

Even though GMM does not require storage of input data in the running process, it is not suited for moving object scenario, because a series of (K is 3, 4 or 5) training frames without moving objects will be given to train the GMM. Similarly KDE depends on pixel values that are observed earlier, but retaining these N frame pixels is time consuming especially when N is large. Inspite that codewords can be updated during training time, it is not suited for dynamic background modelling. A corner is a point where a gradient will have two or more dominant directions and such corners will be continuously detected under illumination changes. Hence Corner based background modelling that applies correlation (mean and variance-adaptive threshold) between 2 corners to say they belong to foreground or background (faster than pixel based methods) is used in the proposed system inspite it generates large of corners. Algorithm starts finding corners in frames so as to estimate motion of the objects between consecutive frames.

Algorithm 3: Detection of corners in the given frame
Input: KeyFrames whose chi-square satisfies the threshold
Output: Corners plotted on the image (side bar to adjust the threshold)

1. Read the abrupt transition frame
2. Create a track-bar for adjusting the threshold value
3. Initialize detector parameters, blockSize as 2 and apertureSize as 3.
4. Detect corners using Harris detector using equation 30, 31 and equation 32 (Nelli,2017).

$$E(u, v) = \sum_{x,y} w(x, y)[I(x + u, y + v) - I(x, y)]^2 \tag{30}$$

$$E(u,v) \approx \begin{bmatrix} u\, v \end{bmatrix} M \begin{bmatrix} u \\ v \end{bmatrix} \tag{31}$$

Where

$$M = \sum_{x,y} w(x,y) \begin{bmatrix} I_x^2 & I_x I_y \\ I_x I_y & I_y^2 \end{bmatrix} \tag{32}$$

Ix and Iy are the derivatives of the image in the x and y.

5.   5. Represent the corners in the form of circles as given in equation 33 (Nelli,2017).

$$R=\det(M)-k(\text{trace}(M))^2 \tag{33}$$

Where $\lambda1$ and $\lambda2$ are the Eigen values of M, det(M)=$\lambda1\lambda2$, trace(M)=$\lambda1+\lambda2$

From this we can determine whether a given region is a corner, an edge or flat as follows:

If |R| is small it is considered as flat, if R<0 then it is edge and if R is large then it is considered as corner.

Metadata Generation for the Extracted Key Frames

A separate file is created to store text information of a video frame. Such file is called as metadata. Metadata is also defined as data about data i.e. information about a video, image, frame. Such information is required in many image processing applications such as video tagging, summarization, video indexing, contextualizing visual information, security applications. Metadata can be of three types [Margaret Rouse (2016)]: technical metadata (brand and model of the camera, time and date of frame, GPS location, aperture, shutter speed, ISO number, focal depth, dots per inch pertaining to camera details and settings), descriptive metadata (image creator name, image keywords, comments, titles, captions), administrative metadata (owner information, usage, copyrights, restrictions on reuse). Technical metadata is generated automatically where as descriptive and administrative metadata are created manually. In the proposed system, after performing chi-square test, we have got the frames where the movement of objects is detected. The phenomenon of detecting the movement of object is termed as abrupt transition (scene) and the frames in which the abrupt transitions took place are termed as Key Frames. These frames represent the important information of the video.

Algorithm4: Generation of metadata for the extracted key frames
Input: Key frames
Output: Metadata file
1.   Read the key frames extracted
2.   Create a text file to store metadata
3.   Calculate the timestamp at which the frame was captured using the frame number as given in equation 34, 35 and 36. Since, the frames are extracted 30fps, we have:

Hour = (frame_number / 30) / 3600 (34)

Minute = ((frame_number/30) % 3600) / 60 (35)

Second = ((frame_number/30) % 3600) % 60 (36)

4.   Store the key frame number and timestamp information in the metadata file as follows:

outputfile << "Key frame number" << "\t" << "Timestamp" << "\n\n";

Capturing Object Silhouette

A silhouette can be treated as a solid shape with same color having edges matching to some of the objects such as a person, vehicle, building, and animal. Silhouette can be extracted from object (edges or curves for a given viewpoint) and screen space (image processing algorithms). The target object in each of the key frame is first segmented from its background. In the proposed system silhouettes are extracted using background subtraction followed by morphological operations.

Algorithm5: Generation of object silhouette
Input: Key Frame extracted
Output: Object Silhouettes
   1.   Read the abrupt frames
   2.   For each frame, perform the following steps
        a.   Subtract the background (previous frame) from the current frame. Only pixels in both frames that haven't changed will result in zero. This is done using equation 37. (Hong, Lee, Jung, & Oh, 2008)

$$S(x,y) = \begin{cases} 1, if \left| I_c(x,y) - I_b(x,y) \right| > thresho \\ 0, otherwise \end{cases}$$ (37)

        b.   Segment the resulting difference image by applying threshold. This will result in a binary image, ideally with zero where the image was static, and 1 or 255 where motion was present.
        c.   Update the pixel values and save the generated silhouette.

Feature Extraction Using ORB Feature Detector

Feature is a property that describes a real world entity. Image object poses two types of features such as spatial features (shape, size, texture, color, coordinates) and temporal features (time, movement). Feature extraction and description plays important role in moving object classification. Feature extraction computes image information by analyzing every image point to be an image feature. This step should be robust to noise, illumination changes, invariance and affine transformations. Even though SIFT proved to better in object recognition but computational complexity is high and hence cannot be suited for real time applications. Simplified version of SIFT called as SURF can perform faster than SIFT and can detect points well, but in intensity change situations it is not as good as SIFT. Another alternative for SIFT called as Robust Independent Elementary Features (BRIEF), is less complex than SIFT and with similar matching performance. Another feature detector developed alternative to SIFT is Oriented FAST and Rotated BRIEF (ORB) that is fast, suitable in real time and efficient in computation. FREAK's methods, based on ORB is coarse-to-fine ordering.

Standard silhouettes of human and vehicle are taken for training the classifier. The silhouettes given as input and output is the .csv file containing the features. The features of both the training silhouettes and testing silhouettes (silhouettes generated in the previous step) are extracted.

ORB starts with FAST that operates on a circle of sixteen pixels around the corner candidate p. Let I_p is the intensity at point p. Now p is classified as a corner, if there exists a set of twelve contiguous pixels whose intensity is between $I_p - t$ or $I_p + t$, for a threshold value t.

Algorithm6: Extracting the features
Input: Object Silhouette
Output: CSV File containing the features of moving object
    1.   Read the object silhouette image
    2.   Create an ORB Feature Detector and a csv file to store key points
    3.   For each silhouette image, perform the following steps

Detect the feature points (Rublee, Rabaud, Konolige, & Bradski, 2011)

a.   Moment is calculated using equation 38.

$$m_{pq} = \sum_{x,y} x^p y^q I(x,y) \tag{38}$$

Centroid is calculated using equation 39 .

$$C = \left(\frac{m_{10}}{m_{00}}, \frac{m_{01}}{m_{00}} \right) \tag{39}$$

Compute the descriptors for each keypoint (Rublee, Rabaud, Konolige, & Bradski, 2011)

b.   A binary test for a patch p is defined as shown in equation 40.

$$\Gamma(p; x, y) = \begin{cases} 1; p(x) < p(y) \\ 0; p(x) \geq p(y) \end{cases} \tag{40}$$

Where p(x) is the intensity of p at a point x.

Each feature is defined as a vector of n binary tests as given in equation 41. (Rublee, Rabaud, Konolige, & Bradski, 2011)

$$f_n(p) = \sum_{1 \le i \le n} 2^{i-1} \Gamma(p; x_i, y_i) \tag{41}$$

Create an object with the feature descriptors as shown in equation 42. (Rublee, Rabaud, Konolige, & Bradski, 2011)

a.

$$S = \begin{pmatrix} x_1, ..., x_n \\ y_1, ..., y_n \end{pmatrix} \tag{42}$$

Derive steered(rotated) version

S_θ using equation 43 and equation 44. (Rublee, Rabaud, Konolige, & Bradski, 2011)

$$S_\Theta = R_\Theta S \tag{43}$$

$$g_n(p, \Theta) = f_n(p) \mid (x_i, y_i) \in S_\Theta \tag{44}$$

b.　Create Bag of Words K-Means Trainer and cluster the features extracted into a dictionary

keypoints.append(temp_feature)
descriptors.append(temp_descriptor)

c.　Store them in the .csv file

Moving Object Classification Using k-NN Algorithm

During feature extraction ORB will use FAST to detect keypoints. A total of 32 keypoint features are extracted. Now Harris corner measure is used to find top N points among them. In order to get invariant features, we compute centroid of the silhouette from whose direction we can get invariant features. These features are given to k Nearest Neighbor classifier. k-NN classifier is a simplest algorithm that depends on limited number of neighboring samples for performing classification and proved to better than other classifiers. Another advantage is that, we need not have a separate training set for and can be done in constant amount of time. training, and the time complexity of training is a constant. k-NN classification uses class membership function such as majority vote to classify an object. There are many studies in the literature to conclude on the value of k. Heuristic techniques can be used to compute the value of k for a given data set. Min-Mix Normalization as given in equation 45 is used for transforming original data in to a range 0 to 1.

$$Zi = xi - min\left(x\right) \ / \ max\left(x\right) - min\left(x\right) \tag{45}$$

where x=(x$_1$,..., x$_n$) and z$_i$ is ith normalized data

Algorithm7: Object Classification using k-NN
Input: Training and Testing datasets
Output: Assigns a class label C $\in$ C1,C2
    1.   Read the training and testing datasets

Train the classifier using f1 //f1 – training data

Upload f2 for testing //f2 – testing data

    2.   Normalize the values using min-max normalization
    3.   Set parameters of k-NN classifier

for 1…i do //i – no of rows (data samples) in f1

for 1…j do //j – no of rows (data samples) in f2

Find similarity between instances (max k instances) using Euclidean distance

Collect the k neighbours

if (features in f1 is nearly $\in$features in f2) then

Assigns (1…m) $\in$ C;

    4.   Display the results predicted by the classifier

Implementation and Results

Dataset Used

Benchmark datasets have been taken and tested.
    The following table 1 presents the benchmark datasets used for testing our proposed system.

Table 1. Benchmark datasets

S.No	Name of the Dataset	From University	Time of Video in Minutes
1	1b video	University Of Michigan	1.45
2	2b video	University Of Michigan	1.00
3	3b video	University Of Michigan	2.49
4	4b video	University Of Michigan	2.30
5	5b video	University Of Michigan	3.23
6	6b video	University Of Michigan	2.45
7	conv4a additional bag	PIROPO Database	0.01
8	conv4a illumination	PIROPO Database	1.57
9	Conv4a seat1	PIROPO Database	1.42
10	Conv4a test2	PIROPO Database	2.06
11	Conv4a test3	PIROPO Database	1.51
12	Conv4a test4	PIROPO Database	1.05
13	Conv4a test5	PIROPO Database	1.55
14	Conv4a test6	PIROPO Database	1.19
15	Conv4a test7	PIROPO Database	1.33
16	Conv4a test8	PIROPO Database	01.01
17	Conv4a test9	PIROPO Database	02.17
18	Conv4a test10	PIROPO Database	01.01
19	Conv4a test11	PIROPO Database	01.31
20	Con4a testing	PIROPO Database	00.14
21	Michigan	University Of Michigan	00.46
22	PNNL-parking-LOT(1)trim	Massachusetts Institute Of Technology	00.27
23	PNNL-parking-LOT(2)trim	Massachusetts Institute Of Technology	00.45
24	Road crossing	University Of Florida	00.35
25	Road2	University Of Florida	00.25
26	Walkbyshop1cor	Massachusetts Institute Of Technology	01.34

Implementation Overview

Frames are extracted from the video at the rate of 30 frames per second so that the total frames extracted are 35*30=1050 frames.

These frames which are in RGB color space model are then converted into HSI color space model as shown in figure 22. Frames are stored in both RGB and HSI format for further processing.

Histogram values are derived for each frame under illumination changes using the Chi-Square test as shown in figure 23.

For consecutive frames Chi-Square test is applied and compared with a threshold of 0.5 as shown in figure 24.

Figure 22. Frames conversion

Figure 23. Histogram analysis

Abrupt frames are saved which are having the chisquare value greater than 0.5. Key frames. Harris Corner based background modeling technique is used for corner detection. In this step, corners have been detected for the abrupt frames generated as shown in figure 25.

Figure 24. Discovery of abrupt transitions

Figure 25. Detected corners

Metadata is generated for the extracted key frames and is stored in a text file as shown in figure 26.

Human and vehicle silhouettes are taken for training the classifier as shown in figure 27 which is termed as training silhouettes. Silhouettes are generated for the extracted key frames which are termed as testing silhouettes.

17 silhouettes were generated for this video which was given for testing. Features are extracted by ORB feature detector for both the training and testing silhouettes and stored in a .csv file. k-NN classifier is used for classification. Also Min-max normalization is applied. Figure 28 presents a sample snapshot for the results obtained on a sample video.

Figure 26. Metadata file

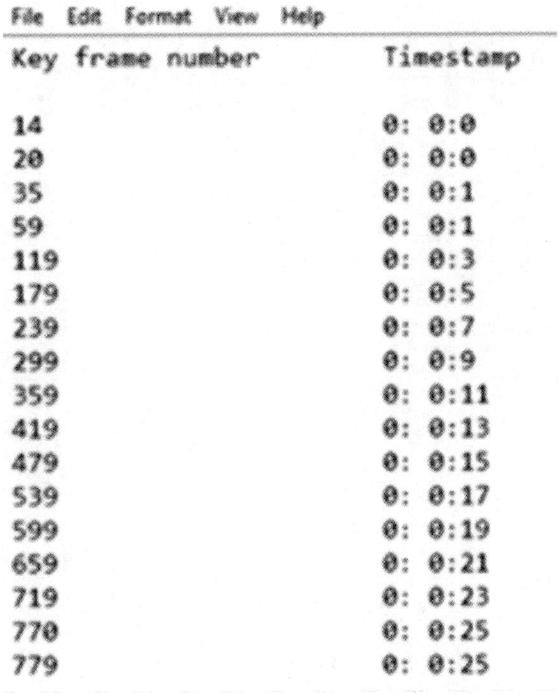

File	Edit	Format	View	Help

Key frame number	Timestamp
14	0: 0:0
20	0: 0:0
35	0: 0:1
59	0: 0:1
119	0: 0:3
179	0: 0:5
239	0: 0:7
299	0: 0:9
359	0: 0:11
419	0: 0:13
479	0: 0:15
539	0: 0:17
599	0: 0:19
659	0: 0:21
719	0: 0:23
770	0: 0:25
779	0: 0:25

Figure 27. Training silhouettes

Figure 28. Predicted results by k-NN classifier

```
> library(class)
> prc24_test_pred9<-knn(train=prc24_train,test=prc24_test,cl=prc24_train_labels,k=9)
> prc24_test_pred9
 [1] h h h h h h h h v h h h h h h
Levels: h v
> library(caret)
Loading required package: lattice
Loading required package: ggplot2
> conf_mat249<-confusionMatrix(prc24_test_pred9,prc24_test_labels)
> conf_mat249
Confusion Matrix and Statistics

          Reference
Prediction  h  v
         h 12  4
         v  1  0

               Accuracy : 0.7059
                 95% CI : (0.4404, 0.8969)
    No Information Rate : 0.7647
    P-Value [Acc > NIR] : 0.8085

                  Kappa : -0.1039
 Mcnemar's Test P-Value : 0.3711

            Sensitivity : 0.9231
            Specificity : 0.0000
         Pos Pred Value : 0.7500
         Neg Pred Value : 0.0000
             Prevalence : 0.7647
         Detection Rate : 0.7059
   Detection Prevalence : 0.9412
      Balanced Accuracy : 0.4615

       'Positive' Class : h

> |
```

Performance Evaluation

This section presents results of Performance is calculated for the classification algorithms k-NN for various values of k 7, 8, 9 depending on the parameters given in table 2.

1. Confusion matrix: Also known as contingency table or error matrix. It is a specific table that allows the visualization of the classification performance of the proposed algorithm as mentioned in the table 2. In the confusion matrix, '**a**' and '**b**' are class labels and
    a. TP indicates True Positives
    b. FP indicates False Positives

Table 2. Confusion Matrix (Han & Kamber,2006)

	a	B
actual a=0	TP	FN
actual b=1	FP	TN

c. TN indicates True Negatives

d. FN indicates False Negatives

2. Precision: Ratio of the number of matched objects retrieved to the total number of irrelevant and relevant matched objects retrieved as given in equation 46. (Han & Kamber,2006)

$$Precision = \frac{tp}{tp + fp} *100\%$$ (46)

Where, tp is the number of relevant matched objects retrieved

fp is the number of irrelevant matched objects retrieved

3. Recall: Ratio of the number of relevant matched objects retrieved to the total number of relevant matched objects present as given in equation 47. (Han & Kamber,2006)

$$Recall = \frac{tp}{tp + fn} *100\%$$ (47)

Where, tp is the number of relevant matched objects retrieved

fn is the number of relevant matched objects not retrieved

4. Error rate: It measures the average magnitude of the error.

Classification Accuracy Rate (CAR or Accuracy): It is measured using all the prediction values that are correct as given in equation 48. (Han & Kamber,2006)

$$accuracy = \frac{tp + tn}{tp + tn + fp + fn} *100\%$$ (48)

5. F-Measure: It is the harmonic mean of precision and recall as given in equation 49. (Han & Kamber,2006)

$$F = 2\left(\frac{precision * recall}{precision + recall}\right)$$ (49)

Experimental Setup in R Language for Classification

Need to install two packages namely 'class' and 'caret'. Class library consists of the k-NN function definition which we use to predict the class to which the objects belong to. Caret library is used to generate confusion matrix. Figure 29 presents snapshot of experimental setup required for performing classification.

Figure 29. Experimental setup for classification

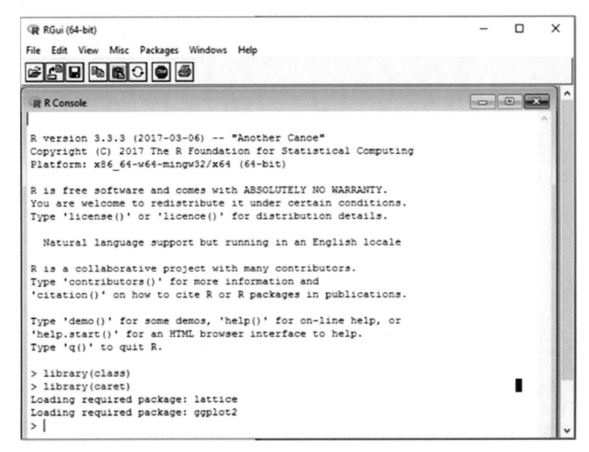

Comparison

The following tables from table 3 to table 5 present the different classification results for a sample video listed in table 1. CAR, Precision and Recall values for k-NN Classifier when k=7, 8 and 9 for a sample Video 1 listed in table 1.

Table 6 presents F-Measure value when k=7,8,9.

The following table 7 and table 8 presents performance measure values summarized for all 26 videos listed in table 1.

Table 3. CAR, Precision and Recall values for k-NN Classifier when k=7

Accuracy =21.43%			
	True Human	**True Vehicle**	**Class Precision**
Pred Human	3	0	100
Pred Vehicle	11	0	0
Class Recall	21.43	0	-

Table 4. CAR, Precision and Recall values for k-NN Classifier when k=8

Accuracy =57.14%			
	True Human	**True Vehicle**	**Class Precision**
Pred Human	8	0	100
Pred Vehicle	6	0	0
Class Recall	57.14	-	-

Table 5. CAR, Precision and Recall values for k-NN Classifier when k=9

Accuracy =100%			
	True Human	**True Vehicle**	**Class Precision**
Pred Human	14	0	100
Pred Vehicle	0	0	-
Class Recall	100	-	-

Table 6. F_Measure Values for various k values

	k=7	**k=8**	**k=9**
Human	**35.29**	**72.72**	**100**
Vehicle	**0**	**0**	**0**

Table 7. Summary of precision and recall for all the 26 videos

	k=7		**k=8**		**k=9**	
	Precision	**Recall**	**Precision**	**Recall**	**Precision**	**Recall**
Human	79.545	45.61	98.63	63.97	98.189	95.20
Vehicle	14.357	83.33	16.43	75	-	-

Table 8. Summary of F-Measure for all the 26 videos

	k=7	**k=8**	**k=9**
Human	**55.44**	**75.705**	**96.181**
Vehicle	**23.86**	**25.45**	**-**

A total of 26 benchmark datasets have been taken. Moving objects have been classified as human or vehicle based on the features extracted. For 4 videos, no abrupt changes were detected which says that there are no moving objects. For the remaining 22 videos, k-NN classifier is applied for various values of k say 7, 8 and 9. Precision, Recall and Accuracy has been determined. Using these values, F_Measure has been determined. Results show that, the classifier produces more accurate results at k=9.

CONCLUSION AND FUTURE WORK

In this internet of everything (IoE) era, video data is generated very fast by many sources. Such big data requires insight analysis such as moving object classification. Main purpose of moving object classification is to classify the different types of objects that are present in a video while monitoring the video in the important areas Even though many works are reported, they could classify moving objects into predetermined classes such as pedestrians, animals, buildings and vehicles. Some works used frames as the basic unit of processing, other works on scene or shot based. Very less works are reported to recognize objects, even in various lightening conditions, camera angle and distance. Better performance should be achieved under the illumination variations.

This chapter proposed a system for moving object classification under illumination changes by considering scenes in a video sequence. Our proposed approach uses frame extraction from a video, conversion from RGB to HSI model, finding similarity (dissimilarity) between consecutive frames using chi-square test. Then we detect feature points using a Harris corner detector and represent them as ORB descriptors. Generation of Meta data from the Key Frame, silhouette capturing and feature extraction is done to achieve classification of the moving object in a video sequence under illumination changes. Even though vector based descriptors such as SURF, SIFT proved to be good for invariance, but are not suited in low power device environment. As such binary descriptors such as ORB are used. The k-NN classification algorithm compares features of the silhouettes of the detected moving objects and concludes on human, vehicle class (predefined). For this we created a sample template database with known object silhouettes.

A total of 26 datasets are used and accuracy of 46.63% is achieved when k=7 and 65.5195% when k=8 and 90.691% when k=9.

The objects have been detected and classified successfully. But, when the object tends to move then the coordinates of that object changes with respect to the frame, then it is considered as a different object. So some techniques, such as remembering all the previous objects and their occurrence in the future frames could be detected and can be used for false abrupt change, so that detected objects are not detected again. Camera calibration can be done as the videos recorded by the camera vary in the speed of the movement of the objects.

REFERENCES

Bay, H., Ess, A., Tuytelaars, T., & Van Gool, L. (2008, June). Speeded-Up Robust Features (SURF). *Computer Vision and Image Understanding, 110*(3), 346–359. doi:10.1016/j.cviu.2007.09.014

Bityukov, S. I., Maksimushkina, A. V., & Smirnova, V. V. (2016). Comparison of histograms in physical research. *Nuclear Energy and Technology, 2*(2), 108–113. doi:10.1016/j.nucet.2016.05.007

Boukhriss, Fendri, & Hammami. (2016). Moving object classification in infrared and visible spectra. *Proc. SPIE 10341, Ninth International Conference on Machine Vision (ICMV 2016).*

Chauhan, Parmar, Parmar, & Chauhan. (2013). Hybrid Approach for Video Compression Based on Scene Change Detection. *IEEE International conference on Signal Processing, Computing and Control (ISPCC)*, 1-5.

Chhasatia, N. J., Trivedi, C. U., & Shah, K. A. (2013). Performance Evaluation of Localized Person Based Scene Detection and Retrieval in Video. *IEEE International Conference on Image Information Processing (ICIIP)*, 78 – 83. 10.1109/ICIIP.2013.6707559

Dang, C. T., Kumar, M., & Radha, H. (2012). Key frame extraction from consumer videos using epitome. *19th IEEE International Conference on Image processing (ICIP)*, 93 – 96. 10.1109/ICIP.2012.6466803

Dang, C. T., & Radha, H. (2014, June). Heterogeneity Image Patch Index and its Application to Consumer Video Summarization. *IEEE Transactions on Image Processing*, *23*(6), 2704–2718. doi:10.1109/TIP.2014.2320814 PMID:24801112

Dedeoglu, Y., Toreyin, B. U., Gudukbay, U., & Cetin, A. E. (2006). Silhouette Based Method for Object Classification and Human Action Recognition in Video. *European Conference on Computer vision*, 64 - 77. 10.1007/11754336_7

Eyas, E. (2003). Scene Change Detection Schemes for Video Indexing in Uncompressed Domain. *INFORMATICA*, *14*(1), 19–36.

Fernando, W. A. C., Canagarajah, C. N., & Bull, D. R. (2000). A Unified Approach To Scene Change Detection In Uncompressed And Compressed Video, Digest of Technical Papers. *International Conference on Consumer Electronics*, 350-351.

Ford, R. M., Robson, C., Temple, D., & Gerlach, M. (1997). Metrics for Scene Change Detection in Digital Video Sequences. *IEEE International Conference on Multimedia Computing and Systems*, 610 - 611. 10.1109/MMCS.1997.609780

Gonzalez, W. (1992). *Derivation of HSI to RGB and RGB to HSI conversion equations, Digital Image Processing* (1st ed.). Addison Wesley.

Gooch, B. (2017). *Silhouette Extraction*. Retrieved from https://www.cs.rutgers.edu/~decarlo/readings/gooch-sg03c.pdf

Han & Kamber. (2006). Data Mining: Concepts and Techniques (2nd ed.). Morgan Kaufmann Publishers.

Higashia, K., Fukuib, S., Iwahoria, Y., & Adachia, Y. (2016). New Feature for Shadow Detection by Combination of Two Features Robust to Illumination Changes. *Procedia Computer Science*, *96*, 896–903. doi:10.1016/j.procs.2016.08.268

Histogram Comparison. (2016). Retrieved from http://docs.opencv.org/2.4/doc/tutorials/imgproc/histograms/histogram_comparison/histogram_comparison.html

Hong, K., Lee, C., Jung, K., & Oh, K. (2008). Real-time 3D Feature Extraction without Explicit 3D Object Reconstruction. *International Journal of Computer, Electrical, Automation. Control and Information Engineering*, *2*(8), 2613–2618.

HSI Color Space - Color Space Conversion. (2016). Retrieved from https://www.blackice.com/color-spaceHSI.htm

Huang, M., Xia, L., Zhang, J., & Dong, H. (2013). An Integrated Scheme for Video Key Frame Extraction. *2nd International Symposium on Computer, Communication, Control and Automation*, 258-261. 10.2991/3ca-13.2013.64

Joshi, U., & Patel, K. (2016). Object tracking and classification under illumination variations. *International Journal of Engineering Development and Research, 4*(1), 667–670.

Kavitha & Kavitha. (2017). Abnormal Moving Object Detection Using Sparse Based Graph K Nearest Neighbour (SGk-NN). *International Journal of Innovative Research in Computer and Communication Engineering, 5*(4), 7901–7906.

Kelm, P., Schmiedeke, S., & Sikora, T. (2009). Feature-based video key frame extraction for low quality video sequences, Image Analysis for Multimedia Interactive Services. *10th Workshop on Image Analysis for Multimedia Interactive Services*, 25 – 28.

Lal, K., & Arif, K. M. (2016). Feature extraction for moving object detection in a non-stationary background. *12th IEEE/ASME International Conference on Mechatronic and Embedded Systems and Applications (MESA)*, 1-6. 10.1109/MESA.2016.7587172

Liang, C.-W., & Juang, C.-F. (2015). Moving object classification using local shape and HOG features in wavelet-transformed space with hierarchical SVM classifiers. *Applied Soft Computing, 28*, 483–497. doi:10.1016/j.asoc.2014.09.051

Lim, R., & Yang, X. Gao, & Lin. (2017). *Real-Time Optical flow-based Video Stabilization for Unmanned Aerial Vehicles*. Cornel University Library. arXiv:1701.03572 [cs.CV]

Lowe, D. G. (1999). Object recognition from local scale-invariant features. *Proceedings of the Seventh International Conference on Computer Vision*, 1150–1157. 10.1109/ICCV.1999.790410

Mangal, S., & Kumar, A. (2017). Real time moving object detection for video surveillance based on improved GMM. *International Journal of Advanced Technology and Engineering Exploration, 4*(26), 17–22. doi:10.19101/IJATEE.2017.426004

McConnell, R. K. (1986). *Method of and apparatus for pattern recognition*. U.S. Patent No. 4,567,610.

Mishra, P. K., & Saroha, G. P. (2016). A Study on Classification for Static and Moving Object in Video Surveillance System. *International Journal of Image, Graphics & Signal Processing, 8*(5), 76–82.

Mithlesh, C. S., & Shukla, D. (2016). Detection of Scene Change in Video. *International Journal of Science and Research, 5*(3), 1187–1201.

Mustafah, Y. M., Zainuddin, N. A., Rashidan, M. A., Aziz, N. N. A., & Saripan, M. I. (2017). Intelligent Surveillance System for Street Surveillance. *Pertanika Journal of Science & Technology, 25*(1), 181–190.

Nelli, F. (2017). *OpenCV & Python – Harris Corner Detection – a method to detect corners in an image*. Retrieved from http://www.meccanismocomplesso.org/en/opencv-python-harris-corner-detection-un-metodo-per-rilevare-i-vertici-in-unimmagine/

Oh, S.-I., & Kang, H.-B. (2017). Object Detection and Classification by Decision-Level Fusion for Intelligent Vehicle Systems. *Sensors (Basel)*, *17*(1), 207. doi:10.339017010207 PMID:28117742

Pearson, K. (1901). On Lines and Planes of Closest Fit to Systems of Points in Space. *Philosophical Magazine*, *2*(11), 559–572.

Poonam, D. (2017). Image Stitching Based on Corner Detection. *International Journal on Recent and Innovation Trends in Computing and Communication, Volume*, *5*(4), 351–354.

Poorani, M., Prathiba, T., & Ravindran, G. (2013). Integrated Feature Extraction for Image Retrieval. *International Journal of Computer Science and Mobile Computing*, *2*(2), 28–35.

Rouse, M. (2016). *Image metadata*. Retrieved from http://whatis.techtarget.com/definition/image-metadata

Rublee, E., Rabaud, V., Konolige, K., & Bradski, G. (2011). ORB: An efficient alternative to SIFT or SURF. *Proceedings of the 2011 International Conference on Computer Vision*, 2564-2571. 10.1109/ICCV.2011.6126544

Sakarya, U., & Telatar, Z. (2010). Video scene detection using graph-based representations. *Signal Processing Image Communication*, *25*(10), 774–783. doi:10.1016/j.image.2010.10.001

Sargano, A. B., Angelov, P., & Habib, Z. (2016). Human Action Recognition from Multiple Views Based on View-Invariant Feature Descriptor Using Support Vector Machines. *Applied Sciences*, *6*(10), 309. doi:10.3390/app6100309

Satrughan, K. U. M. A. R., & Jigyendra Sen, Y. A. D. A. V. (2016, June). Segmentation of Moving Objects using Background Subtraction Method in Complex Environments. *Wuxiandian Gongcheng*, *25*(2), 399–408.

Shao, L., & Ji, L. (2009). Motion histogram analysis based key frame extraction for human action/activity representation. *2009 Canadian Conference on Computer and Robot Vision, IEEE CRV'09*, 88 – 92.

Shin, T., Kim, J.-G., Lee, H., & Kim, J. (1998). Hierarchical Scene Change Detection In An Mpeg-2 Compressed Video Sequence. *IEEE International Symposium on Circuits and Systems*, 4, 253 - 256.

Shukla, Mithlesh, & Sharma. (2015). A Survey on Different Video Scene Change Detection Techniques. *International Journal of Science and Research*, 214 – 219.

Sulaiman, S., Hussain, A., Tahir, N., Samad, S. A., & Mustafa, M. M. (2008). Human Silhouette Extraction Using Background Modeling and Subtraction Techniques. *Information Technology Journal*, *7*(1), 155–159. doi:10.3923/itj.2008.155.159

Bhanse & Vasekar. (2017). Survey on Classification for Moving Object Detection and Tracking Using Multiple Sensor Fusion Architecture. *International Journal of Innovative Research in Computer and Communication Engineering*, *5*(2), 2054–2060.

Vanithamani, S. (2017). Vehicle classification and analyzing motion features. *Indian Journal of Engineering*, *14*(36), 89–94.

Xu, Y., Dong, J., Zhang, B., & Xu, D. (2016, January). Background modeling methods in video analysis: A review and comparative evaluation. *CAAI Transactions on Intelligence Technology*, *1*(1), 43–60. doi:10.1016/j.trit.2016.03.005

Zhang, J., Liu, F., Shao, H., & Wang, G. (2007). An Effective Error Concealment Framework For H.264 Decoder Based on Video Scene Change Detection. *Fourth International Conference on Image and Graphics*, 285 - 290. 10.1109/ICIG.2007.174

Chapter 8
Machine Vision Application on Science and Industry:
Machine Vision Trends

Bassem S. M. Zohdy
Institute of Statistical Studies and Research, Egypt

Mahmood A. Mahmood
Institute of Statistical Studies and Research, Egypt

Nagy Ramadan Darwish
Institute of Statistical Studies and Research, Egypt

Hesham A. Hefny
Institute of Statistical Studies and Research, Egypt

ABSTRACT

Machine vision studies opens a great opportunity for different domains as manufacturing, agriculture, aquaculture, medical research, also research studies and applications for better understanding of processes and operations. As scientists' efforts had been directed towards deep understanding of the particular material systems or particular classes of types of specific fruits, or diagnosis of patients through medical images classification and analysis, also real time detection and inspection of malfunction piece, or process, as various domains witnessed advancement through using machine vision techniques and methods.

INTRODUCTION

Machine vision studies open a great opportunity for different domains as manufacturing, agriculture, aquaculture, medical research, also research studies and applications for a better understanding of its processes and operations. As scientists efforts had been and still directed towards the understanding of materials, systems or/and specific classes of types of fruits, or diagnosis of patients through medical images classification, and analysis, and also real-time detection and inspection of malfunction piece,

DOI: 10.4018/978-1-5225-5751-7.ch008

or process, as various domains witnessed advancement through using machine vision techniques and methods. Researchers in material sciences achieved very important findings in understanding and analyzing microstructural images, as the aim is to produce a common method to extract meaningful features regarding micrographs. Also image analysis main aim is to extract meaningful data included in images to be analyzed, the image analysis contains many processes including image registration and image fusion, image registration uses two images, the first one is called the reference image, while the second one called the sensed image, the sensed image is aligned to the reference image, the result or output of this process is used in another process is called image fusion, image fusion is the process of combining the data contained in two images in one image, as the resulting image will be more informative than the input images, mentioned processes, image registration, and image fusion are very helpful in many practical and application domains, as industrial, medical, aquaculture, and agriculture and other areas that use machine vision technologies and applications (DeCost, 2015;El-Gamal, 2016;Saberioon, 2016;Benalia, 2016).

MACHINE VISION RECENT STUDIES

Industrial Studies

Studies revealed many types of images or video streams that could be used in machine vision applications and techniques, as in Aldrich (2010) the main concern of the mineral processing industry is to extract features from froth images, this features could be obtained through employing color feature extraction of red, green, blue (RGB) to better determine the minerals captured by pictures.

In Liu (2017) a research study proposes that the Internet of Thinks (IoT) makes possible the use of intelligent system for mechanical products assembly for (IIASMP), the proposed system make use of a mechanical product assembly process to evaluate the characteristics of IoT-based manufacturing systems, among the characteristics of the proposed system is the self-regulation and self-organization, meaning that the system can monitor itself through collecting information about its status and analyze this data to enhance or react to current status, as computer vision systems are embedded in material handling robot, smart controller, actuator, and a training information consisting of image database, image processing algorithm and encoding rules, as the system recognizes material types through capturing images of materials and then processing it. The framework proposed by this study consists of five layers, first is the sensing assembly layer, this layer employs resources identification technology along with multiple source and sensor data acquisition to obtain the state of the assembly resource, then a feedback formulates instructions in order to adjust the assembly process, the second layer is the net layer that could convert the protocol, store, route, transfer the sensing data, the third layer is the where the sensor data is fusion, in this layer the sensing data map the performance of the resources assembly, transforming the data into information in order to extract and combine this data, the forth layer is where the decision and applications is made, the extracted information from the last layer could be audited and analyzed to support the intelligent management control, the fifth layer is system service part providing resource configuration and data security, the sixth and last layer is interface layer that enables methods of exchanging data.

In Dcost (2015) previously established computer vision methods are used to define quantitative microstructure descriptors and methods to identify classes of microstructures, from a diverse collection recorded in databases. The results are computed in real-time, capturing the most important characteristics details from microstructural images without fine-tuning from human experts.

The author of Di Leo (2016) proposed a vision system for online quality of industrial manufacturing, with a reference values the proposed measurement system detects the defects of electromechanical parts, using components include two cameras to produce image for quality monitoring purposes, as the images produced by each camera go for two different procedures, first is the top image processing, and the objective is to measure the length for design specification purposes, as this procedure adopt NI vision by National Instrument, the other procedure obtains images for parts that could not be seen from the top angle, this procedure uses backlight illuminator to create a binary image with 75% threshold.

A comparison study in Rashidi (2016) assess various machine learning techniques to detect three categories of construction; concrete, red bricks and Oriented Strand Boards (OSBs), the study compares Radial Basis Function (RBF), Multilayer Perceptron (MLP), and Support Vector Machine (SVM), according with the results, SVM shown better performance than the other two techniques, for the three kind of materials their surface textures were accurately detected.

In Lapray (2016) is proposed HDR-ARtiSt (High Dynamic Range Adaptive Real-time Smart camera) which is complete FPGA-based smart camera architecture that obtains a real-time video stream with a high dynamic range from multiple acquisitions, this study produces uncompressed B&W 1, 280*1,024-pixel HDR live video at 60 fps, displaying the HDR live video on a LCD monitor with the help of an embedded DVI controller.

A study conducted by authors in Gade (2014) to review the thermal cameras and its applications, illustrated that thermal cameras are classified as passive sensors, these sensors detect the infrared radiation emitted by bodies with a temperature above absolute zero. Thermal camera was developed for the military purpose, as a surveillance and night vision tool; nowadays the price has decreased, opening up a broader field of applications. These types of cameras overcome the illumination problems in grayscale and RGB images. As the produced images from thermal cameras represented as grayscale images with a depth of 8 to 16 bit per pixel as images can be compressed with standard JPEG and video can be compressed with H264 or MPEG.

Agriculture Studies

A review study conducted by Di Leo (2016) in research domain of computer vision in agriculture assures that using computer vision techniques in evaluating the quality of fruits and vegetables will decrease time wasted in human check for fruits and vegetables, as it provides powerful tools for external quality assessment and evaluation of fruits and vegetables. Also, a review paper by Tscharke (2016) investigates the application of machine vision systems to recognize and monitor the behavior of animals in a quantitative manner.

In Di Leo (2016) the author reviews the three main types of machine vision systems used in automated inspection of fruits and vegetable qualities, the traditional computer vision systems deals with (RGB) color cameras that simulate the human eyesight, by capturing images with the three filters centered red, green, and blue (RGB). Hyperspectral machine vision systems unlike the traditional systems, as it combines in one system a spectroscopic with imaging techniques to obtain set of monochromatic images. In multispectral machine vision systems, they are different from the traditional and hyperspectral systems in the number of monochromatic images in the spectral domain, as the main advantage of this system is the number of wavelengths of monochromatic images captured can be chosen freely by using narrowband filters.

A review describes and assesses the most recent technologies of various optical sensors and its suitability for fish farming management. Measurement and prediction of fish products quality are performed (Saberioon, 2016) the major areas of optical sensors applications in aquaculture are discussed in this review: pre-harvesting, during cultivation and post-harvesting. Machine Vision Systems (MVSs) and optical sensors are excellent options for real-world application based on digital camera development thanks to the increasing speed of computer-based processing.

A survey study conducted by Cubero (2016) displays the recent researches that use color and non-standard computer vision systems for the inspection of citrus in an automated way. The existing technologies to acquire fruits images and their use for the non-destructive inspection of internal and external characteristics of these fruits are explained. Machine vision has been proved to be a very useful and practical tool for automatic purposes during the fruit inspection. Indoors and outdoors inspections have shown in excellent performance and accuracies.

In Benalia (2016) the authors proposed a system for real-time and in-line sorting of dried figs through color assessment measured by chroma meter CR-400; which is handheld, portable measurement instrument designed to evaluate the color of objects, particularly with smoother surface conditions or minimal color variation, and CIE illuminant D65 and the 10o observer standard that is a commonly used standard illuminant defined by the International Commission on Illumination (CIE). It is part of the D series of illuminants that try to portray standard illumination conditions at open-air in different parts of the world.

A Canon EOS 550D digital camera was used for the fig image acquisition, which captured images of 2592* 1728 pixels size, with a resolution of 0.06 mm/pixel. Eight fluorescent tubes (BIOLUX 18 W/965, 6500 K, OSRAM, Germany) were used to lighting placing them on the four sides of a square inspection chamber in a 0°/45° configuration.

The results from Chroma meter and image analysis made possible a complete classification between high quality and deteriorated figs, by the evaluation of the figs color attributes.

In Beltrán Ortega (2016) the study to evaluate the quality of virgin olive oil during its manufacturing in-line process and storage is presented. The study list the sensing technologies used in the olive oil industry. For tasks that allow an effective process supervision and the control of the virgin olive oil production. Electronic noses pose good classification abilities to discriminate between olive oils in terms of quality and origin and even quantify different olive pastes. Valuable information is extracted online at different stages. Polyphenols are measure through the use of Electronic tongues that are also able to mimic the human sense of taste. Also the dielectric spectroscopy is able to mimic the human sense of taste, mainly used for the detection of water content in olive oil.

Medical Studies

The research by Norouzi (2014) classifies the segmentation methods of image processing regarding medical imaging into four categories, first region based methods, then clustering methods, then classifier methods, and finally hybrid methods. In region-based methods, which considered as the bases of most image segmentation methods, and the most popular region-based methods are thresholding and region growing, as the second method is classification, that uses training data as sample images to extract features for future detection and classification of medical images, as the algorithms used in this method are k-nearest neighbor, and maximum likelihood. The third method is clustering methods, this method is the classification method but it does not use training data but rather using statistical techniques to extract the features of data, it employs k-means, fuzzy C-mean and expectation maximization algorithms, the last method is the hybrid method, which combines boundary and regional information in order to segment medical images.

As Gomes (2011) the computer integrated surgery employs two dimensional (2D) or three dimensional (3D) medical imaging for preoperative planning, in order to collect information about the patient, these images combined together with the human anatomy to produce a computer model that is employed in surgical planning, using intraoperative sensing to register patient data.

The robotic surgical systems rely on 3D imaging, as the vision system uses high-resolution 3D endoscope and image processing equipment, also Praxiteles robotic system developed by Praxim Medivision presented a bone-mounted guide positioning to align a cutting guide in image-free total knee arthroplasty (TKA), used during surgeon process to perform the planar cuts manually using the guide.

In Havaei (2016) the author proposed a deep neural network method (DNN) to segment brain tumor images using Magnetic resonance (MR) images. The study proposes a convolutional neural network (CNN) adapting Deep Neural Network (DNN). As the results revealed that the proposed architecture is 30 times faster than currently published studies. The study uses Pylearn 2 library that is a deep learning open source library, the library supports GPUs usage that increases the speed of deep learning algorithms, as convolutional neural network (CNN) learns the data features, the proposed method uses minimal pre-processing, the preprocessing follows 3 steps, first removing 1% high and low intensities, then applying N4ITK bias correction, then subtracting the channel's mean and dividing by the channel's standard deviation in order to normalize the data, connected components simple method is employed for post-processing.

The study (Escalera, 2016) proposed technique for detecting multiple sclerosis in the brain using stationary wavelet transform (SWT), also the study uses three machine learning classifiers, namely; decision tree, k-nearest neighbors, and support vector machine, the technique implies using two-level stationary wavelet entropy (SWE), first; it applies stationary wavelet transform (SWT) on a given image, second; the randomness of the sub-bands is measured by the Shannon entropy A, defined as follows:

$$A = -\sum_i x_i^2 \log x_i^2 \tag{1}$$

Here A represents the entropy and X_i represents the ith element of a given sub-band. Note that for an *m-level* decomposition, there are in total *(3m+ 1)* sub-bands, and thus a vector of *(3m+ 1)* elements was formed.

Then this methodology applied three classifiers as mentioned above, and the results showed that k-nearest neighbors outperform the other algorithms.

An experiment study performed in Tajbakhsh (2016) considered four medical imaging applications involving classification, detection, and segmentation from various different imaging modality systems, the experiment showed that the use of pre-trained CNNs with fine tuning outperformed the trained CNNs from scratch.

Ophthalmic imaging provides a way to diagnose and objectively assess the progression of a number of pathologies including neo-vascular age-related macular degeneration and diabetic retinopathy. In De Fauw (2017) a study counts on digital fundus photographs digital Optical Coherence Tomography (OCT) images using DeepMind (Beattie, 2016).

MACHINE VISION TECHNIQUES AND ALGORITHMS

Feature extraction in mineral processing industries as reported in the review study conducted by Aldrich (2010) categorized in three categories; first the physical features to detect the bubble size and shape through the usage of the edge detection method, which detects the gradients of the pixel intensities between bubbles in froth images by using the valley edge detection and valley edge tracing methods to segment froth images, in valley edge detection the concern is to detect the valley edges between the bubbles, images then filtered to extract noise and then image pixels are assessed to promote possible edge candidates, then the next step is to cleanup based on valley edge tracing to ensure the removal of gaps between valley edges.

Another technique of image segmentation used in the mineral processing industry, which employs two-stage procedure, through identifying the local minima in pixel intensities, then calculating the bubble diameter in order to border thinning, as reported by Aldrich (2010) that this approach is more accurate for larger bubble size. Also, another algorithm of physical feature extraction in the mineral processing industries is a watershed algorithm which is a formation approach relies on simulation of water rising from a set of markers; the approach relies on identifying the minima and regional maxima by locating trends in pixel intensities along different scan lines. Another approach is froth color, which extracts the color features of minerals loaded, red, green, and blue (RGB), also hue, saturation, intensity or values (HSI) or (HSV).

Other approaches used in the industrial domains are the statistical approaches, which used in the extraction of features from images, as fast Fourier Transforms (FFT), wavelet transforms, Fractal descriptors, co-occurrence matrices and their variants, texture spectrum analysis, latent variables which implies; principal component analysis, Hebbian learning, multilayer perceptrons, cellular neural networks.

It is noteworthy to mention that other methods used specifically in mineral processing industries are dynamic features; which uses descriptors designated specifically to capture the behavior of froth, which implies mobility techniques; that refers to the speed and direction of movement, mobility techniques that include; bubble tracking, block matching, cluster matching and pixel tracing. Besides the mobility techniques, there is a stability technique that reveals the appearance and disappearance of bubbles in the froth.

In Lapray (2016) was used the multiple exposure control (MEC) algorithm for low and high exposure images, as it is more close to Kang's exposure control and Alston's multiple exposure systems, as the proposed study the exposure system is alternate between two values which is continuously adapted to reflect changes in the scene.

Authors in Beltrán Ortega (2016) reviewed four types of medical image segmentation using region-based, classification, clustering, and hybrid algorithms, first the region-based methods categorized in two algorithms, first is thresholding, in this technique the image is formed with different gray levels, therefore,

$$g(x,y) = \begin{cases} foreground\ if\ f(x,y) \geq T \\ background\ if\ f(x,y) < T \end{cases} \tag{2}$$

Where $f(x, y)$ is the pixel intensity in the (x, y) position and T is the threshold value. An inappropriate threshold value leads to poor segmentation results.

For an image that does not have the constant background and has a diversity across the object, the local thresholding method is used to divide images into sub-images, to posteriorly calculate the threshold value for each part. The thresholding results for each part of an image are merged. The image is divided into vertical and horizontal lines, each part includes a region of both the background and the object. Finally, an interpolation is made to obtain appropriated results.

In the discipline of machine learning, SVM is classified as a supervised learning model with associated learning algorithms capable of recognizing patterns in the data analyzed. The simple task of SVM is predicts information from an input data set, for each given input, and two possible classes form the output, making it a no probabilistic binary linear classifier. An SVM model is a representation of the samples as points in feature space, mapped so that the samples of the separate categories are divided by an hyper-plane that is as far as possible from marginal samples of each category. Figure 1 shows the distribution of samples related to an example problem in a 2D space. Additionally, it shows the resulted hyper-plane (a line in 2D space) and margins by SVM. SVMs can efficiently perform non-linear classifications using different kernel functions. Kernel functions map inputs into higher-dimensional feature spaces. By following this process, the problem will be reformulated as a linear problem; thus, the ordinary SVM can perform linear classification in the new feature space (Gutschoven, 2000).

Figure 1. RBF network

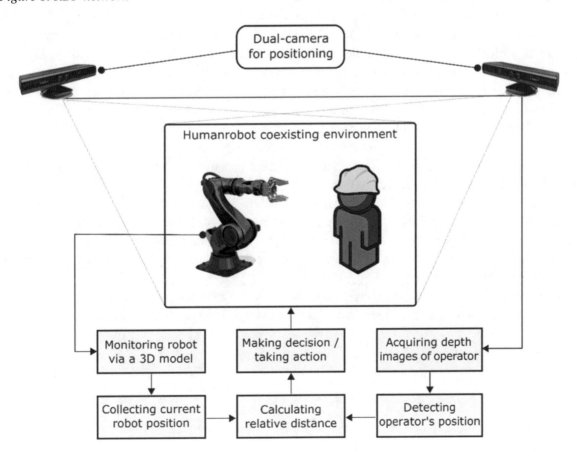

In the study Di Leo (2016) the authors review the various computer vision systems that are used in agriculture research domain, the systems could be categorized in three main types; the traditional, the hyper-spectral, and the multispectral computer vision systems.

SVM was trained to classify microstructures into one of seven groups with greater than 80% accuracy over 5-fold cross-validation (Dcost, 2015), as SVM can classify microstructures into groups automatically and with high accuracy, even using relatively small training data sets. In addition, the feature histogram can provide the basis for a visual search engine that finds the best matches for a query image in a database of microstructures. This automatic and objective computer vision system offers a new approach to archiving, analyzing, and utilizing microstructural data.

The methodology proposed by Tseng (2015) for e-quality control using support vector machine starts with embedding different test samples, and then a remote inspection process, followed by data acquisition, in order to prepare the training data, then selecting the model, then a sensitivity analysis is conducted with different C values and comparing results with different kernels, this leads to identifying optimal values. In training SVMs, the study selects a kernel and set a value to the margin parameter C. To develop the optimal classifier, the proposed methodology needs to determine the optimal kernel parameter and the optimal value of C. A k-fold (a value of k=10) cross-validation approach is adopted for estimating the value of training parameter. The minimum value considered was 0.1 and the maximum value was 500. The value of C with the high-level training accuracy percentage was identified as the optimal value (highlighted in bold characters). The performance of the classifiers is evaluated by using different kernel functions in terms of testing accuracy, training accuracy, a number of support vectors, and validation accuracy. Four different kernel functions are identified for this research. They include (1) Linear Kernel, (2) Polynomial Kernel, (3) Radial Basis Function (RBF) Kernel, and (4) Sigmoid Kernel. Polynomial and RBF kernels are by far the most commonly used kernels in the research world.

In the study Tscharke (2016) machine vision system is used in behavior recognition process to determine the welfare of a livestock, and this process has four main parts depicted in Figure 2, first is the initialization, which leads to calibration process that requires data represented as model constraints, next software variables, and hardware camera variables, second element is tracking through segmentation that classifies object and backgrounds, that leads to the third element of pose estimation through a predictive model, and finally results or the recognition.

Another machine learning technique is MLP (Jazebi, 2013) that is derived from the artificial neural networks to solve more sophisticated nonlinear problems, each MLP encompasses a number of basic neurons, which are organized in three layers: the input layer, the hidden layer, and the output layer as depicted in Figure 1. An MLP is mainly a feedforward network, which implies the error back propagation concept for the training purposes.

RBF (Man, 2013) is a one hidden layer neural network, which may use several forms of radial functions as the activation function. The most common one is the Gaussian function defined by:

$$f_j\left(x\right) = exp\left(\frac{x - \mu_j^{\,2}}{2\sigma_j^2}\right) \tag{3}$$

Where σ is the radius parameter, μ is the vector determining the center of basis function f_j and x is the d-dimensional input vector. In an RBF network as in Figure 1, a neuron of the hidden layer is acti-

Figure 2. The recognition process

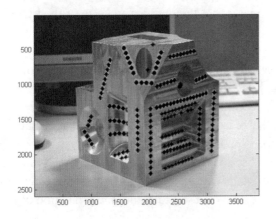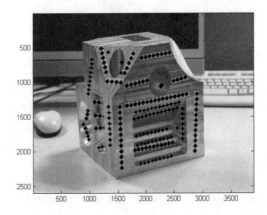

Figure 3. Multilayer perceptron (MLP)

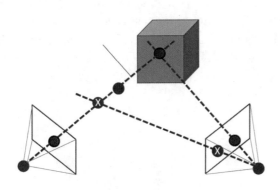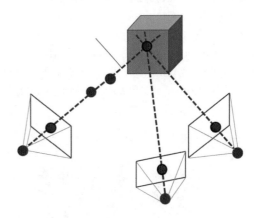

vated whenever the input vector is close enough to its center vector μ. There are several techniques and heuristics for optimizing the basis functions parameters and determining the number of hidden neurons needed to achieve the optimal classification. The second layer of the RBF network, which is the output layer, comprises one neuron to each class. The output is the linear function of the outputs of the neurons in the hidden layer and is equivalent to an OR operator. The final classification is given by the output neuron with the greatest output. An RBF neural network with one output neuron is depicted in Figure 1.

Authors in Havaei (2016) applied deep neural networks, especially CNN has recently many studies in the medical domain, as CNN consists of a succession of layers, which perform operations on the input data. Convolutional layers (symbol C^k_s) convolve the images I_{size} presented to their inputs with a pre-defined number (k) of kernels, having a certain size s, and are usually followed by activation units which rescale the results of the convolution in a nonlinear manner. Pooling layers (symbol P^{stride}_{size}) reduce the dimensionality of the responses produced by the convolutional layers through downsampling, using different strategies such as average pooling or max pooling. Finally, fully connected layers ($F_{\#neurons}$) extract compact, high-level features from the data. The kernels belonging to convolutional layers as well as the weights of the neural connections of the fully connected layers are optimized during training through back-propagation. The user specifies the network architecture, by defining the number of layers, their kind, and the type of activation unit. Other relevant parameters are the number and size of the kernels employed during convolution, the number of neurons in the fully connected part and the downsampling ratio applied by the pooling layers.

Machine vision techniques employ technologies for robots as three-dimensional perception that considered one of the technologies used to aid in recognizing the surroundings and navigation in order to accomplish tasks in fully understood environment, as controlling and operating robots requires the ability to visualize the surroundings in a human way, for that reason robot needs to gain a 3D information.

Mathematical models in Salvi (2002) can give the corresponding point in an image given a point in the site of view, as determining a set of parameters that describes the mapping between the 2D image coordinates and 3D point in world coordinate system, in other words, this situation called the camera calibration. Pinhole camera model presented in Faugeras (1993) is used to model the projection of the world coordinates into images.

As depicted in Figure 4, the image of the 3D point P, is generated by optical ray passing by the optical center and intersecting the image plane, resulting p' in the image plane, that is located at a distance focal length f behind the optical center (Usamentiaga, 2014), in order to describe the projection of 3D points on the 2D image plane mathematically is the transformation of the real coordinate system to W to the camera coordinate system C, this transformation is done using the equation (4), so the camera coordinates of point $P_c=(x_c, y_c, z_c)^T$ calculated from the world coordinates $P_w=(x_w, y_w, z_w)^T$ using rigid transformation $H_{w \to c}$, The transformation $H_{w \to c}$ include three translations (T_x, T_y, T_z) and three rotations

Figure 4. Pinhole Camera Model

(α, β, γ), using equation (5) the projection in the camera C into the image coordinate system is calculated.

$$\begin{pmatrix} P^c \\ 1 \end{pmatrix} = Hw \rightarrow c \begin{pmatrix} p^w \\ 1 \end{pmatrix} \tag{4}$$

$$\begin{pmatrix} u \\ v \end{pmatrix} = \frac{f}{z^c} \begin{pmatrix} x^c \\ y^c \end{pmatrix} \tag{5}$$

Figure 5a shows the point projection in the real worldview on an image, although the point in the image cannot directly be obtained by the original point as it is not one to one relationship but one to many so the inverse problem is weakly stated, as different points could be on the same pixel, effectively dealing with the inverse problem by forming a straight line contains formed by all the points of the same pixel of the image as depicted in Figure 5b.

Another technique defined as passive techniques like stereo-vision that requires only ambient light deals with this issue through locating the same point and calculating the intersection of projection lines of multiple images. Other techniques employ an infrared pattern to estimate the depth information from the returning time. As active vision techniques like laser triangulation, structured light, and light coding use different illumination method as the more appropriate features, the object has will increase the accuracy when passive vision is used. Unlike active vision techniques and passive vision, techniques require multiples cameras for the sake of 3D reconstruction as it depends on the application.

Stereo-vision and photogrammetry is a 3D reconstruction technique that uses conventional 2D imaging, measuring real dimensions of an object from an image (Eisenbeiss, 2005). Stereo-vision and photogrammetry calculate the intersection of the projection line as depicted in Figure 6 through obtain-

Figure 5. From 3D to 2D, (a) Direct problem, (b) Inverse problem

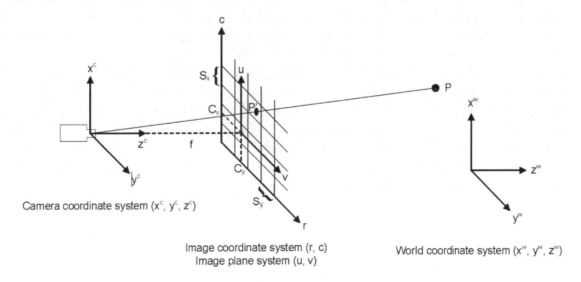

Camera coordinate system (x^c, y^c, z^c)

Image coordinate system (r, c)
Image plane system (u, v)

World coordinate system (x^w, y^w, z^w)

Figure 6. From 2D to 3D, (a) Homologous points, (b) Intersection of project lines

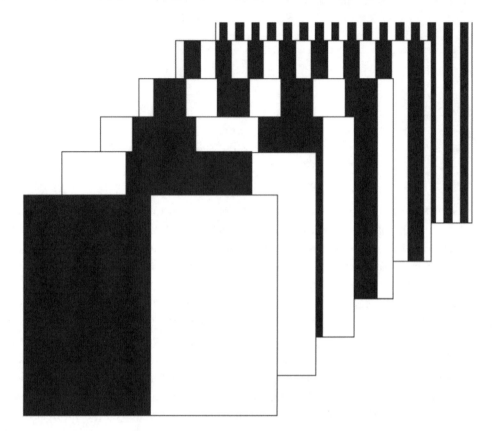

ing the same point in other images preferably in three other images in order to improve the accuracy as to get the accurate 3D position homologous points must be selected.

Ensuring the detection could be obtained through the usage of laser points or other techniques of physical marks with high contrast, as shown in Figure 7 physical marks taken and processed, in order to calculate the 3D position there are many factors that must be considered, these factors are the calibration parameters and the spatial position of the cameras, as the marks could be paired using epipolar geometry and the intersection lines of the projection lines (Luhmann, 2006).

Markerless stereo-vision (Canny, 1987) uses feature-tracking algorithm to find, extract, and match characteristics of objects between similar images abandoning physical marks as depicted in Figure 8.

Time of flight (ToF) (Kolb, 2009) as an active vision technique projects infrared pattern to get 3D data depicted in Figure 9, as this technique employs a Tof camera that uses light pulses as its illumination, it is on for a very short time, the resulting light pulse is projected on the objects and reflected as the camera captures the reflected light into the sensor plane. A delay of returning light could be caused by the distance and could be calculated by equation (6).

$$t_D = 2 \cdot \frac{D}{c})$$

Figure 7. Detection of marks from several images

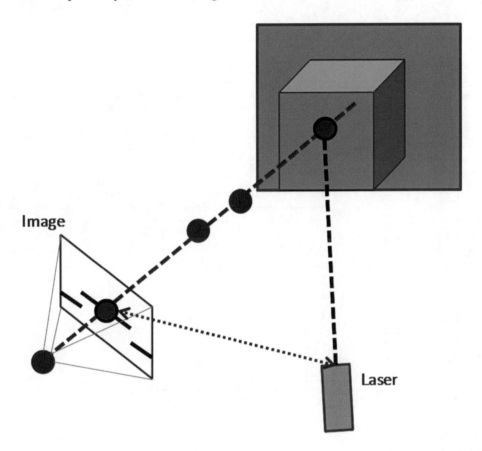

Figure 8. Feature tracking algorithm

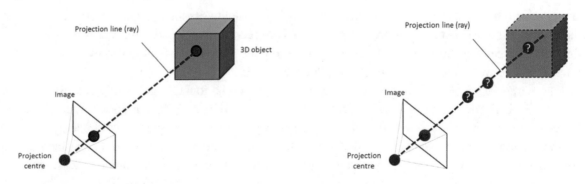

Where t_D is the delay, D is the distance to the object and c is the speed of light.

Structured light techniques have two main categories of techniques, time-multiplexing techniques and one-shot techniques (Chen, 2000), in time-multiplexing techniques there is no limit for the number of patterns, so the number of correspondences and larger resolution could be obtained, but all factors like camera, projector, and the object have to remain fixed and unchanged. In one-shot techniques moving

Figure 9. Projecting a pattern on the object

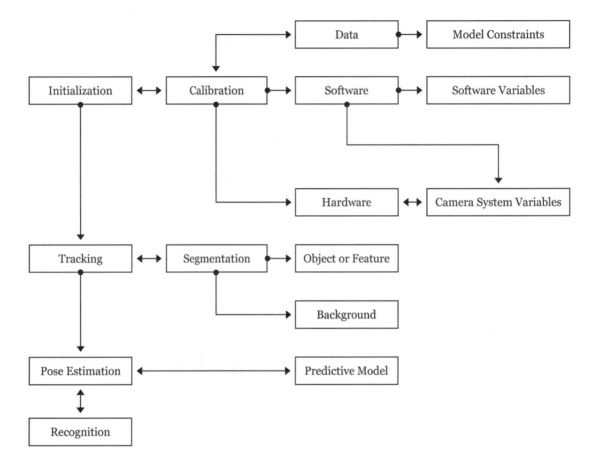

camera could be used, as each line or point could be identified by its local neighborhood, unlike laser the projected light is not harmful to human, the camera records a stripped light that was originally a light transformed from its original status as shown in Figure 10, as depth is obtained by calculating the intersection between lines and planes (Pages, 2006; Salvi, 1998).

The stereo-vision and photogrammetry (Rusu, 2011) could be found in automotive industry for the sake of car body deformation measurements, adjustment of tools and rigs, as in aerospace industry for measurement and adjustment for mounting rigs, alignment between parts, wind energy systems for deformation measurements and production control, as for water dams, tanks and plant facilities in construction, the stereo-vision and photogrammetry offers accuracy and highly precise measurements. There are three main different approaches to photogrammetric solutions (Hefele, 2001). First, relies on two or more static cameras observing target points called forward intersection observing moving targets. In the second approach, one or more cameras are placed to observe the fixed target, this approach called resection. The third approach, called bundle adjustment combines the first and second approach.

In laser triangulation technique (Mahmud, 2011), a laser emitter, the camera, and the point form a triangle as shown in Figure 11. The shape and size of the triangle are known by the distance between the camera and the laser emitter, the angle of the laser emitter corner is known, and camera corner determined by the location of the laser dot corner of the triangle.

Figure 10. Structured light pattern

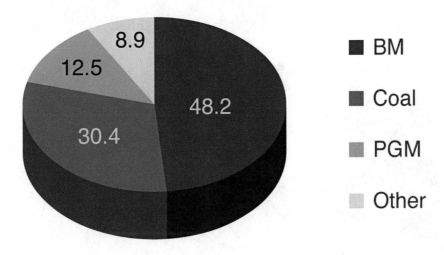

Figure 11. Laser triangulation

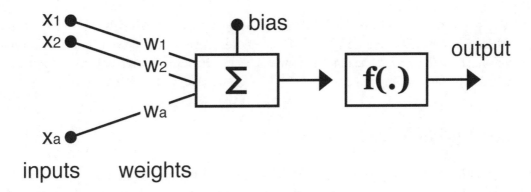

Structured light supply visual features not depending on the site, in Claes (2007) a solution is presented, shown in Figure 12, where only one camera is attached to the end effector, and a static projector used to illuminate the work in hand. Reported that the accuracy is 3 mm, and this is appropriate for concrete implementation. As Pages (2006) proposed coded structured light, a robust visual feature is provided by the coded light pattern. Even in the existence of occlusions, placing the robot with reference to planar object yields well-accepted results.

Light coding techniques in Patra (2012) and Susperregi (2013) have many industrial and robotic applications, also used for people tracking in video games. As it is a great innovation offered at a very low cost. But has some shortcomings, as it provides a noisy point cloud. One of the solutions is to combine HD camera with light coding sensors to obtain high-resolution point cloud. Experimental results show that this approach enhances both indoor and outdoor sites with a significant increase in the resolution of the point cloud. In order to improve the detection of people on mobile robots, thermal sensors could be placed on mobile platforms and use classifiers of supervised learning, since humans could be detected through their thermal characteristics distinguished from other objects, as the experimental results show

Figure 12. Robot positioning using structured light

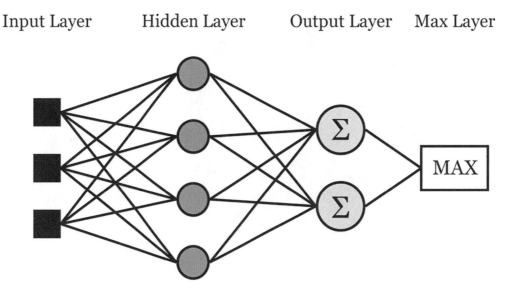

decreased false positive rate. Wang (2013) Shows that using sensor data and virtual 3D robot models for collision detection depicted in Figure 13, shop floor environment displayed as 3D models connected to motion sensors, used to simulate the real environment, as unstructured foreign objects including mobile operators added by light coding sensors.

MACHINE VISION APPLICATIONS

Applications of machine vision systems could be exploited in different and various domains, in Aldrich (2010) a comprehension of procedures happening in the froth floatation systems has long been held key to understanding of the overall behavior of froth floatation systems, as humans cannot diagnose and predict the status related to more advanced structure in the froth, the developed techniques of machine

Figure 13. Sensor data and 3D robot models

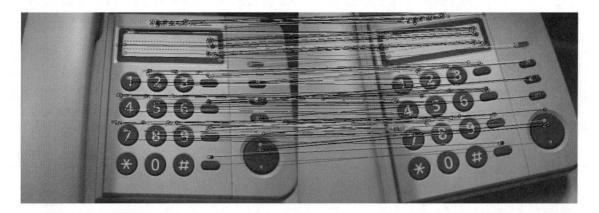

vision enables such industries to extract features of images, the Figure 14 below gives indication of the application of froth imaging systems in the mineral processing industries.

Figure 14 shows that the majority of applications (approximately 48.2%) have been reported in the base metals (BM) industry, which mostly includes copper, lead, and zinc, with a few papers related to nickel, magnesium, and tin. Application in the coal industries (30.4%), particularly in China, is second, followed by application (12.5%) in the platinum group metal (PGM) industry, mostly in South Africa. The balance of the applications reported in the literature (8.9%) is associated with oxides, such as P_2O_5, SiO_2, and CaO.

In Gomes (2011) the authors review surgical robotics field and how to use the computer vision techniques like image processing to mimic the human ability in different operations, as the key player of such systems is the da Vinci robotics system and others, as the accuracy have been accomplished through the use of image processing and image registration to robotic system.

In Lapray (2016) the implementation of HDR-2 video system has been first prototyped on the HDR-ARtiST platform with the limitation *P=2*, using only two frames to generate each HDR frame. The authors decided to continuously update these of exposure times from frame to frame to minimize the number of saturated pixels by instantaneously handling any change of the light conditions. The estimation of the best exposure times is computed from the 64-level histogram *(q)* provided automatically by the sensor in the data-stream header of each frame. For each low-exposure frame (I_L) and each high exposure (I_H),

Figure 14. Application of machine visions systems in the metallurgical industries

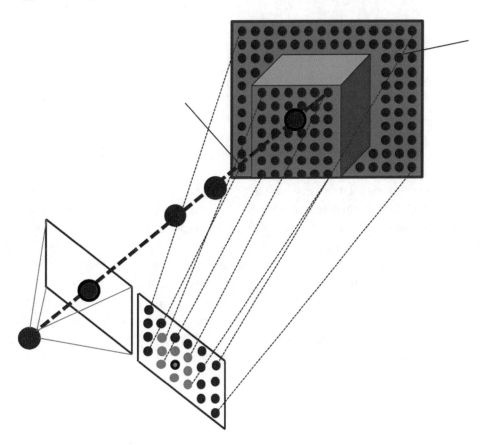

evaluating Q_L and Q_H that are, the ratio of pixels in the four lower levels and the four higher levels of the histogram:

$$Q_L = \sum_{h=1}^{h=4} \frac{q(h)}{N} \quad Q_H = \sum_{h=60}^{h=64} \frac{q(h)}{N} \tag{7}$$

Where $q(h)$ is the number of pixels in each histogram category h and N the total number of pixels. A series of decisions can be performed to evaluate the low-exposure time $(Dt_{L;t,l})$ and the high-exposure time $(Dt_{H;t,l})$ at the next iteration $(t + 1)$ of the acquisition process.

$$\Delta t_{L,t+1} = \begin{cases} \Delta t_{L,t} + 10x & if \, Q_L > Q_{L,req} + thr_{Lp} \\ \Delta t_{L,t} - 10x & if \, Q_L > Q_{L,req} - thr_{Lp} \\ \Delta t_{L,t} + 1x & if \, Q_L > Q_{L,req} + thr_{Lm} \\ \Delta t_{L,t} - 1x & if \, Q_L > Q_{L,req} - thr_{Lm} \end{cases}$$

$$\Delta t_{H,t+1} = \begin{cases} \Delta t_{H,t} + 10x & if \, Q_H > Q_{H,req} + thr_{Hp} \\ \Delta t_{H,t} - 10x & if \, Q_H > Q_{H,req} - thr_{Hp} \\ \Delta t_{H,t} + 1x & if \, Q_H > Q_{H,req} + thr_{Hm} \\ \Delta tH_{,t} - 1x & if \, Q_H > Q_{H,req} - thr_{Hm} \end{cases} \tag{8}$$

where, x is the integration time of one sensor row. $Q_{L,req}$ and $Q_{H,req}$ are, respectively, the required number of pixels for the low levels and the high levels of the histogram. To converge to the best exposure times as fast as possible, we decide to use two different thresholds for each exposure time: thr_{Lm} and thr_{Lp} are thresholds for the low-exposure time whereas thr_Hm and thr_{Hp} are for the high-exposure time. In our design, values of $Q_{L,req}$ and $Q_{H,req}$ are fixed to 10% pixels of the sensor. Values of thr_Lm and thr_Hm are fixed to 1% whereas thr_{Lp} and thr_{Hp} are fixed to 8%.

Detecting temperature of specific scene which is produced by thermal cameras could have many advantages (Gade, 2014), in order to detect some objects by their temperature may lead to a wide research and applications, this application may give clues to health, type of object or material, as in animal and agriculture detecting disease in animals could be easily detected by thermal cameras. The stress level of animals before slaughtering is important to the meat quality. The stress level is correlated with the blood and body temperature of the animal. It is therefore important to monitor and react to a rising temperature. Another application is the detection of heat loss in buildings, as many application reviewed and conducted for human in detecting facial expressions and medical analysis using thermal cameras.

Detecting and evaluating the quality in industrial, medical, agriculture and aquaculture domains as reviewed by Tseng (2015), has a lot of applications with the aid of computer vision systems, automated diagnosis, automatic target defect identification, intelligent faults diagnosis, intelligent quality management, on-line dimensional measurement, optimization, quality control, quality monitoring as for e-quality control.

In Rashidi (2016) is proposed an approach to detect images of building materials, this approach contains two steps, first is feature extraction, and second is supervised learning algorithm specifically classification. First, the technique detects the feature set. Then the classifier takes the combined feature set in order to classify the objects, as a result, the detection or in other words, the job is done using machine learning supervised learning algorithms especially RBF, SVM, and MLP. The model with the best results and performance is chosen.

REFERENCES

Aldrich, C., Marais, C., Shean, B. J., & Cilliers, J. J. (2010). Online monitoring and control of froth flotation systems with machine vision: A review. *International Journal of Mineral Processing, 96*(1), 1–13. doi:10.1016/j.minpro.2010.04.005

Beattie, C., Leibo, J. Z., Teplyashin, D., Ward, T., Wainwright, M., Küttler, H., . . . Schrittwieser, J. (2016). *Deepmind lab.* arXiv preprint arXiv:1612.03801

Beltrán Ortega, J., Gila, M., Diego, M., Aguilera Puerto, D., Gámez García, J., & Gómez Ortega, J. (2016). Novel technologies for monitoring the in-line quality of virgin olive oil during manufacturing and storage. *Journal of the Science of Food and Agriculture, 96*(14), 4644–4662. doi:10.1002/jsfa.7733 PMID:27012363

Benalia, S., Cubero, S., Prats-Montalbán, J. M., Bernardi, B., Zimbalatti, G., & Blasco, J. (2016). Computer vision for automatic quality inspection of dried figs (Ficus carica L.) in real-time. *Computers and Electronics in Agriculture, 120*, 17–25. doi:10.1016/j.compag.2015.11.002

Canny, J. (1987). A computational approach to edge detection. In Readings in Computer Vision (pp. 184-203). Academic Press. doi:10.1016/B978-0-08-051581-6.50024-6

Chen, F., Brown, G. M., & Song, M. (2000). Overview of 3-D shape measurement using optical methods. *Optical Engineering (Redondo Beach, Calif.), 39*(1), 10–23. doi:10.1117/1.602438

Claes, K., & Bruyninckx, H. (2007, August). Robot positioning using structured light patterns suitable for self calibration and 3D tracking. *Proceedings of the 2007 International Conference on Advanced Robotics.*

Cubero, S., Lee, W. S., Aleixos, N., Albert, F., & Blasco, J. (2016). Automated systems based on machine vision for inspecting citrus fruits from the field to postharvest—a review. *Food and Bioprocess Technology, 9*(10), 1623–1639. doi:10.100711947-016-1767-1

De Fauw, J., Keane, P., Tomasev, N., Visentin, D., van den Driessche, G., Johnson, M., ... Peto, T. (2017). Automated analysis of retinal imaging using machine learning techniques for computer vision. *F1000 Research, 5*. PMID:27830057

DeCost, B. L., & Holm, E. A. (2015). A computer vision approach for automated analysis and classification of microstructural image data. *Computational Materials Science, 110*, 126–133. doi:10.1016/j.commatsci.2015.08.011

Di Leo, G., Liguori, C., Pietrosanto, A., & Sommella, P. (2016). A vision system for the online quality monitoring of industrial manufacturing. *Optics and Lasers in Engineering*, *89*, 162–168. doi:10.1016/j.optlaseng.2016.05.007

Eisenbeiss, H., Lambers, K., Sauerbier, M., & Li, Z. (2005). Photogrammetric documentation of an archaeological site (Palpa, Peru) using an autonomous model helicopter. In CIPA 2005 (pp. 238-243). Academic Press.

El-Gamal, F. E. Z. A., Elmogy, M., & Atwan, A. (2016). Current trends in medical image registration and fusion. *Egyptian Informatics Journal*, *1*(17), 99–124. doi:10.1016/j.eij.2015.09.002

Escalera, S., Athitsos, V., & Guyon, I. (2016). Challenges in multimodal gesture recognition. *Journal of Machine Learning Research*, *17*(72), 1–54.

Faugeras, O. (1993). *Three-dimensional computer vision: a geometric viewpoint*. MIT Press.

Gade, R., & Moeslund, T. B. (2014). Thermal cameras and applications: A survey. *Machine Vision and Applications*, *25*(1), 245–262. doi:10.100700138-013-0570-5

Gomes, P. (2011). Surgical robotics: Reviewing the past, analysing the present, imagining the future. *Robotics and Computer-integrated Manufacturing*, *27*(2), 261–266. doi:10.1016/j.rcim.2010.06.009

Gutschoven, B., & Verlinde, P. (2000, July). Multi-modal identity verification using support vector machines (SVM). In *Information Fusion, 2000. FUSION 2000. Proceedings of the Third International Conference on* (*Vol. 2*, pp. THB3-3). IEEE.

Havaei, M., Davy, A., Warde-Farley, D., Biard, A., Courville, A., Bengio, Y., ... Larochelle, H. (2016). Brain tumor segmentation with deep neural networks. *Medical Image Analysis*, *35*, 18–31. doi:10.1016/j.media.2016.05.004 PMID:27310171

Hefele, J., & Brenner, C. (2001, February). Robot pose correction using photogrammetric tracking. In *Machine Vision and Three-Dimensional Imaging Systems for Inspection and Metrology* (Vol. 4189, pp. 170–179). International Society for Optics and Photonics. doi:10.1117/12.417194

Jazebi, F., & Rashidi, A. (2013). An automated procedure for selecting project managers in construction firms. *Journal of Civil Engineering and Management*, *19*(1), 97–106. doi:10.3846/13923730.2012.738707

Kolb, A., Barth, E., Koch, R., & Larsen, R. (2009, March). Time-of-Flight Sensors in Computer Graphics. In Eurographics (STARs) (pp. 119-134). Academic Press.

Lapray, P. J., Heyrman, B., & Ginhac, D. (2016). HDR-ARtiSt: An adaptive real-time smart camera for high dynamic range imaging. *Journal of Real-Time Image Processing*, *12*(4), 747–762. doi:10.100711554-013-0393-7

Liu, M., Ma, J., Lin, L., Ge, M., Wang, Q., & Liu, C. (2017). Intelligent assembly system for mechanical products and key technology based on internet of things. *Journal of Intelligent Manufacturing*, *28*(2), 271–299. doi:10.100710845-014-0976-6

Luhmann, T., Robson, S., Kyle, S. A., & Harley, I. A. (2006). *Close range photogrammetry: principles, techniques and applications*. Whittles.

Mahmud, M., Joannic, D., Roy, M., Isheil, A., & Fontaine, J. F. (2011). 3D part inspection path planning of a laser scanner with control on the uncertainty. *Computer Aided Design*, *43*(4), 345–355. doi:10.1016/j.cad.2010.12.014

Man, Z., Lee, K., Wang, D., Cao, Z., & Khoo, S. (2013). An optimal weight learning machine for hand-written digit image recognition. *Signal Processing*, *93*(6), 1624–1638. doi:10.1016/j.sigpro.2012.07.016

Norouzi, A., Rahim, M. S. M., Altameem, A., Saba, T., Rad, A. E., Rehman, A., & Uddin, M. (2014). Medical image segmentation methods, algorithms, and applications. *IETE Technical Review*, *31*(3), 199–213. doi:10.1080/02564602.2014.906861

Pages, J., Collewet, C., Chaumette, F., & Salvi, J. (2006, June). A camera-projector system for robot po-sitioning by visual servoing. In *Computer Vision and Pattern Recognition Workshop, 2006. CVPRW'06. Conference on* (pp. 2-2). IEEE. 10.1109/CVPRW.2006.9

Patra, S., Bhowmick, B., Banerjee, S., & Kalra, P. (2012). High Resolution Point Cloud Generation from Kinect and HD Cameras using Graph Cut. *VISAPP*, *12*(2), 311–316.

Rashidi, A., Sigari, M. H., Maghiar, M., & Citrin, D. (2016). An analogy between various machine-learning techniques for detecting construction materials in digital images. *KSCE Journal of Civil Engineering*, *20*(4), 1178–1188. doi:10.100712205-015-0726-0

Rusu, R. B., & Cousins, S. (2011, May). 3d is here: Point cloud library (pcl). In *Robotics and automa-tion (ICRA), 2011 IEEE International Conference on* (pp. 1-4). IEEE.

Saberioon, M., Gholizadeh, A., Cisar, P., Pautsina, A., & Urban, J. (2016). Application of machine vision systems in aquaculture with emphasis on fish: State-of-the-art and key issues. *Reviews in Aquaculture*.

Salvi, J. (1998). *An approach to coded structured light to obtain three dimensional information*. Uni-versitat de Girona.

Salvi, J., Armangué, X., & Batlle, J. (2002). A comparative review of camera calibrating methods with accuracy evaluation. *Pattern Recognition*, *35*(7), 1617–1635. doi:10.1016/S0031-3203(01)00126-1

Susperregi, L., Sierra, B., Castrillón, M., Lorenzo, J., Martínez-Otzeta, J. M., & Lazkano, E. (2013). On the use of a low-cost thermal sensor to improve kinect people detection in a mobile robot. *Sensors (Basel)*, *13*(11), 14687–14713. doi:10.3390131114687 PMID:24172285

Tajbakhsh, N., Shin, J. Y., Gurudu, S. R., Hurst, R. T., Kendall, C. B., Gotway, M. B., & Liang, J. (2016). Convolutional neural networks for medical image analysis: Full training or fine tuning? *IEEE Transac-tions on Medical Imaging*, *35*(5), 1299–1312. doi:10.1109/TMI.2016.2535302 PMID:26978662

Tscharke, M., & Banhazi, T. M. (2016). A brief review of the application of machine vision in livestock behaviour analysis. *Journal of Agricultural Informatics, 7*(1), 23-42.

Tseng, T. L. B., Aleti, K. R., Hu, Z., & Kwon, Y. J. (2015). E-quality control: A support vector machines approach. *Journal of Computational Design and Engineering*, *3*(2), 91–101. doi:10.1016/j.jcde.2015.06.010

Usamentiaga, R., Molleda, J., & Garcia, D. F. (2014). Structured-light sensor using two laser stripes for 3D reconstruction without vibrations. *Sensors (Basel)*, *14*(11), 20041–20063. doi:10.3390141120041 PMID:25347586

Wang, L., Schmidt, B., & Nee, A. Y. (2013). Vision-guided active collision avoidance for human-robot collaborations. *Manufacturing Letters*, *1*(1), 5–8. doi:10.1016/j.mfglet.2013.08.001

Section 3
Machine Vision for Structural Health Monitoring

The most recent and successful experiences in the application of machine vision-based techniques for structural health monitoring are described in this section. Chapter 9 introduces an unmanned bridge inspection and evaluation plane system, a railway foreign body intrusion recognition system, and a concrete crack tracking and evaluation system. Chapter 10 is devoted to the analysis of seismic testing using an integration of 3D motion capture methodologies with motion magnification analysis. Chapter 11 describes different methods and devices that can be used in optical scanning systems for noise reduction.

Chapter 9
Application of Computer Vision Technology to Structural Health Monitoring of Engineering Structures

X. W. Ye
Zhejiang University, China

T. Jin
Zhejiang University, China

P. Y. Chen
Zhejiang University, China

ABSTRACT

The computer vision technology has gained great advances and applied in a variety of industry fields. It has some unique advantages over the traditional technologies such as high speed, high accuracy, low noise, anti-electromagnetic interference, etc. In the last decade, the technology of computer vision has been widely employed in the field of structure health monitoring (SHM). Many specific hardware and algorithms have been developed to meet different kinds of monitoring demands. This chapter presents three application scenarios of computer vision technology for health monitoring of engineering structures, including bridge inspection and evaluation with unmanned aerial vehicle (UAV), recognition and surveillance of foreign object intrusion for railway system, and identification and tracking of concrete cracking. The principles and procedures of three application scenarios are addressed following with the experimental study, and the possibilities and ideas for the application of computer vision technology to other monitoring items are also discussed.

INTRODUCTION

A computer vision system is a system which uses vision information to understand scenes and objects, including shape, size and color. It grabs images by computer vision device (CMOS and CCD) and delivers them to a processing unit and determines the parameters by their pixel distribution, brightness, color,

DOI: 10.4018/978-1-5225-5751-7.ch009

etc. The concept of computer vision system was emerged around the 1950s and developed rapidly after 1990s. This technology is featured by advantages like high speed, high accuracy, low noise, excellent anti-interference ability, convenience, long-term monitoring duration, etc. So far, the computer vision technology has been widely applied in the semiconductor industry, electron industry, aviation industry and quality inspection.

In the community of civil engineering, the structural health monitoring technology has been widely used in monitoring different kinds of structures (Chong et al., 2003; Ye et al., 2012). Due to its unique advantages, the computer vision technology is gradually applied in structural health monitoring. Nowadays, structural health monitoring based on computer vision technology for the tunnel surface inspection (Huang et al., 2017), surface crack feature identification (Schwitzke et al., 2010), and structural deformation monitoring (Fukuda et al., 2010) are under development and improvement.

This chapter is aimed to introduce three structural health monitoring techniques based on computer vision technology. They are automatic bridge inspection and evaluation system, railway foreign body intrusion recognition system, and concrete crack tracking and evaluation system.

AUTOMATIC BRIDGE INSPECTION AND EVALUATION SYSTEM

Traditional ways of inspecting bridges mainly depend on manual works or complicated and expensive sensors (Hu et al., 2013). During the inspection, bridge inspection vehicles or vessels are usually applied which is costly and has an adverse impact on traffic. The automatic bridge inspection and evaluation system is a technique combining the unmanned aerial vehicle (UAV) technique with the computer vision technique (Zhang et al., 2012; Ellenberg et al., 2016). This technique aims at solving the problems encountered by the traditional bridge inspection methods including difficulty of operation, existence of blind zones, high cost of inspection, massive time-consumption, and adverse impact on traffic.

During the operation of monitoring, this system obtains images of bridges by visual sensors on-board (Morgenthal et al., 2014), analyzes the images by computer software, and determines accordingly if there are bridge diseases such as cracks, damages, exposure of reinforcing bars, and separation of bearings. Also, it determines the condition of a bridge by the changes of displacement between different parts of the bridge. For example, corrosion of steel bridge can be evaluated by color lump distributions on the bridge.

Brief Introduction of the System

The automatic bridge inspection and evaluation system is consisted of the UAV system and the vision system. The UAV system carries the visual sensors to target positions and makes the cameras to grab the images for analysis. The computer vision system is mainly consisted of high-definition cameras and relevant software. After reaching the target position, the camera starts to grab images of the bridge, and then the system deals with the images using image identification technology for recognizing cracks, damages, exposure of reinforced bars, separation of bearings or other bridge illnesses. To illustrate how the system works, crack identification is demonstrated as below. The main procedure is shown in Figures 1 and 2.

System Test

Figure 3 shows a typical three-span simply supported bridge. This kind of structure is widely used in practice which takes a great proportion in small and medium bridges. The visual inspection of this kind of bridge is divided into four parts: the bridge deck system inspection, the upper structure inspection, the support system inspection, and the substructure inspection.

Figure 1. Main procedure of the system

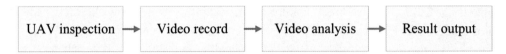

Figure 2. Image processing procedure

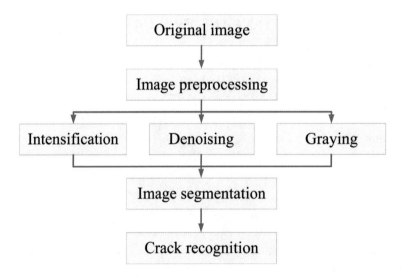

Figure 3 Three-span simply supported bridge

Nowadays, commercial drones are steady and mature. They have great flying capability so that high resolution cameras can be equipped, which will facilitate the possibility of combination between computer vision technology and UAV technology. With the help of the computer vision UAV system, the inspection of bridge will be much faster. The main procedure is illustrated in Figure 4. The UAV for the system test is a Phantom 4 drone with a 12 megapixel camera on board. The critical point is the UAV must reach a certain altitude in order that the camera on board can cover the entire bridge. When the UAV flies along the bridge, it is supposed to keep the same altitude over the bridge to get a steady video. After retrieving the UAV, video records can be obtained. The video is to be processed by corresponding computer vision software which will read a frame at certain time interval and analyze the image for recognition of diseases. If a disease is found, the software will evaluate the disease, as shown in Figure 5.

Figure 4. Procedure for scanning damages on bridge deck

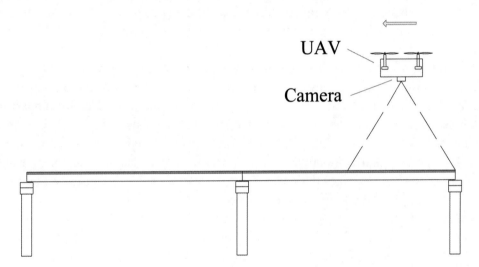

Figure 5. Field test result

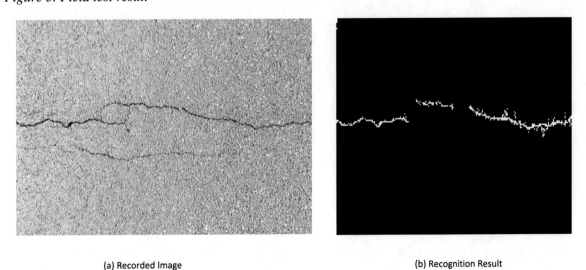

(a) Recorded Image   (b) Recognition Result

RAILWAY FOREIGN BODY INTRUSION RECOGNITION SYSTEM

With the development of railway traffic, many extremely long railroads have been built. However, in recent years, some reports show that several railway accidents were caused by foreign body intrusion. This typical accident endangers the security of railway system and the passengers. The railway foreign body intrusion and recognition system is based on computer vision technology. Several algorithms are involved in the system including multi-Gaussian possibility background modeling algorithm, frame difference method, and tracking algorithm.

Brief Introduction of the System

The railway foreign body intrusion and recognition system consists of distributed stereo image sensors and relevant image processing software. Stereo cameras are deployed in critical sections around the railway in order to obtain images of target sections. The software system will analyze the images from the inspection cameras and determine if there is any foreign body on the railroad. The main procedure is shown in Figure 6.

The recognition of the foreign body is implemented with the help of foreground and background segmentation algorithm. There are several possible algorithms for the segmentation task, such as frame-differencing, single Gaussian background model, and Gaussian mixture model. In comparison with the first two choices, Gaussian mixture background model has advantages that it can handle small motions like a waving plant caused by the wind. Additionally, most pixels in high dynamic scenes do not typically comply with single Gaussian distributions. Thus, the mixture of the Gaussian can improve the accuracy. The Gaussian mixture model takes (basically 3 to 5) Gaussian models to describe the features of the pixel sets in the images. We use the method proposed by Zoran Zivkovic to determine the number of Gaussian models. When a new frame of image is obtained, the Gaussian mixture models are updated. The present pixel sets are matched with the Gaussian mixture models, and then the specific pixel sets are judged as background or the sets will be judged as foreground.

The Gaussian mixture model takes a Gaussian distribution to define a corresponding state in k states, one part represents the gray value of the background, and the other part represents the gray value of the moving introduction foreign body. Assuming that the gray value of each pixel sets is described by X, the distribution density function will able to be described by k Gaussian functions:

Figure 6. Main procedure of the system

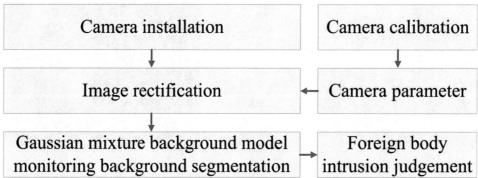

$$f(X_i = x) = \sum_{i=1}^{k} w_{ij} \bullet P(x, \mu_{ij}, \sum ij) \tag{1}$$

where $P(x, \mu_{ij}, \sum ij)$ is the ith Gaussian distribution at time t, the mean value is μ_{ij}, the covariance is $\sum ij$, the weight of the ith Gaussian distribution is w_{ij}, and $\sum_{i=1}^{K} w_{ij} = 1$. $P(x, \mu_{ij}, \sum ij)$ is expressed by:

$$P(x, \mu_{ij}, \sum ij) = \frac{1}{(2\pi)^{d/2} \left| \sum ij \right|^{1/2}} e^{-\frac{1}{2}(X_i - \mu_o)^T \sum_{ij}^{-1} (X_i - \mu_{ij})} \tag{2}$$

For the Gaussian distribution that matches the current pixel sets, the structure will be changed as:

$$w_{ij}' = (1 - \alpha)\mu_{ij}' + \rho x_{t+1} \tag{3}$$

$$\sum_{ij}^{t+1} = \alpha \eta(x_{t+1}, \mu_{ij}^t, \sum_{ij}^t) \tag{4}$$

System Test

In order to simulate the situation of a real intrusion incident and test the capacity of the system, a model test on a full-scale railway system is conducted, as shown in Figure 7. The first foreign body is a stone from the ballasted bed, which is likely to be on the track when a train passes fast or when the track rail is under construction. The size of this kind of stone is usually within 8cm yet hard enough to disturb the wheels. The other foreign body is a metal block of 30cm×15cm. The reason for such choice is because metals are usually the hardest object on the track rails and it is slightly larger than the first object, which provides a comparison for the identification capability of the system. There are larger objects possibly to be found on the track, yet they are easier to be recognized compared to the picked objects.

Figure 7. Ballasted track model

Figure 8. Railway foreign body intrusion recognition system

The railway foreign body intrusion recognition system is shown in Figure 8. The system is consisted of a camera with CMOS sensors and a computer. The camera has 5 megapixel and the maximum resolution of the camera is 1280×720. The computer has a Core i5-7200U CPU. The pixel and resolution of the camera will affect the effective range of the system dramatically while the property of the CPU will affect the real-time property of the system. The distance between the camera and the rail track is 3m. The two intrusion objects are placed on the track and the sleeper respectively to see the influence of the position and the size of the objects. The result is illustrated in Figure 9. This application is quite meaningful for the practical use, due to the booming development of railways in China. With the help of such system, alert of the foreign body intrusion events will be discovered timely with lower cost.

CONCRETE CRACK TRACKING AND EVALUATION SYSTEM

Crack assessment is one of the most important inspections for concrete. Traditional ways of inspection for concrete crack rely on manual work with naked eyes which is inaccurate and inefficient (Liu et al., 2014). Also, it is impossible to inspect the length, width or direction of the cracks in a real-time manner. To tackle the problems, the concrete crack tracking and evaluation system based on computer vision technology is developed (Yamaguchi et al., 2010). This technology monitors the crack in real time from

Figure 9. Recognition of intrusion stone on rail track

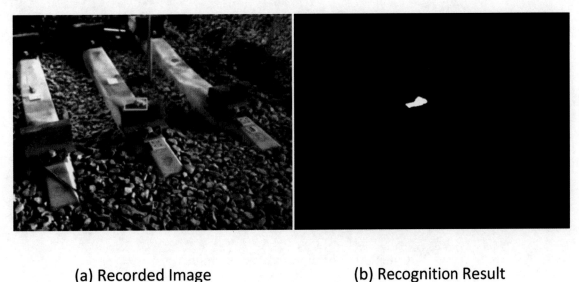

(a) Recorded Image   (b) Recognition Result

the beginning of the cracks to the expansion stage, and it can track the development of the cracks, analyze the damage state of the concrete, and provide support for the evaluation of the safety state.

This technique deploys several industrial digital cameras in different angles to obtain images of the cracks on the reinforced concrete. With the help of digital image processing technology and relevant crack identification and tracking algorithms, it analyzes images obtained by the cameras and recognizes characteristics of the cracks, such as the length, width, direction and distribution during the development. This technology monitors the whole process of the development of the cracks from their occurrence to the growth in real time. It analyzes the damage state of the reinforced concrete in real time and provides support for the evaluation of safety and life-time health state.

Brief Introduction of the System

The concrete crack tracking and evaluation system is consisted of hardware system featured by several digital cameras and relevant image processing software system. The cameras are deployed to monitor the surface of the concrete in target areas. The software is consisted of crack identification and tracking module, crack length and width calculation module, and crack distribution module. The main procedure of crack tracking and evaluation is showed in Figure 10. Among the procedure, the analysis of images is the most important step as illustrated in Figure 11.

Figure 10. Crack tracking and evaluation procedure

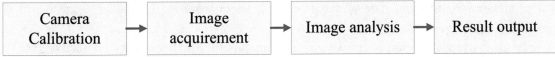

Figure 11. Image analysis procedure

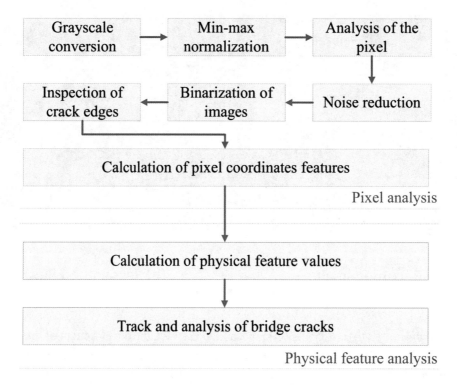

The graying of the images is processed according to the following equation:

$$Y = 0.3r + 0.59g + 0.11b \tag{5}$$

where Y stands for the value of the grayscale images, r, g and b are the component value of red, green and blue in the color image.

The grayscale value of the images will change at the border of the cracks which forms the edge of the crack image. This step aims at reducing the range of analysis for the images and enhances the difference between the cracks and their nearby areas. A linear convert equation is expressed by:

$$g(x) = \begin{cases} 0, x = T_{min} \\ \dfrac{(x - T_{min}) \times 255}{T_{max} - T_{min}}, T_{min} \langle x \langle T_{max} \\ 255, x = T_{max} \end{cases} \tag{6}$$

where $g(x)$ stands for the value of the grayscale images after conversion. T_{min} and T_{max} are the smallest and biggest value of grayscale.

Pixels are analyzed by means of checkerboard corner point method with the combined help of Harris algorithm and SV algorithm. The reduction of noise for the value of gray scale is based on the averaging method and median method. For example, if the gray scale value of the image is within $[\mu - 2\sigma, \ \mu + 2\sigma]$, Averaging method will be adopted and if the gray scale value of the image is beyond $[\mu - 2\sigma, \ \mu + 2\sigma]$, Median method will be adopted. The μ and σ is able to be obtained by the calculation of pixel gray scale value by:

$$\mu = \frac{1}{N} \sum_{i=1}^{N} x_i \tag{7}$$

The binarization of images aims at differentiating the crack district and its background to improve the accuracy of crack edge inspection and recognition of low resolution images. The inspection of crack edges is mainly implemented by means of Sobel operator and G Laplacian operator. The calculation of pixel coordinates features mainly refers to the calculation of crack width, perimeter and areas of cracks. Based on the relation between the pixel coordinates and the world coordinate system as well as the pixel coordinates features, actual feature values of the cracks will be able to be obtained. Taking pictures of a same area and analyzing the images to record the growth of the cracks.

System Test

Images of cracks are identified by the system to see the effect. Three kinds of crack images are identified for comparison study. They are image with narrow crack, image with narrow crack and noisy background, and image with wide crack. The results are shown in Figures 12-14. The concrete crack tracking and evaluation system is a promising application in the engineering practices. During the operational period, the most common damages for concrete structures like bridges and buildings is crack problems and the timely inspection will obtain considerable benefits such as the extension of structural service period and the reduction of maintenance cost.

Figure 12. Identification of image with narrow crack

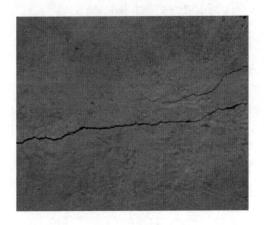 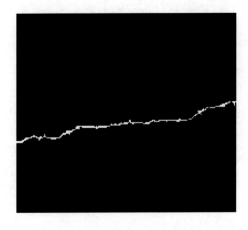

(a) Original Crack   (b) Binary crack image

Figure 13. Identification of crack image with noisy background

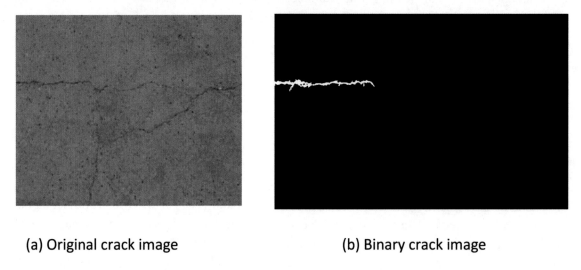

(a) Original crack image   (b) Binary crack image

Figure 14. Identification of crack image with wide crack

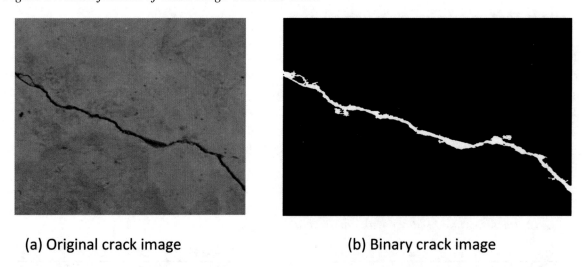

(a) Original crack image   (b) Binary crack image

CONCLUSION

Automatic bridge inspection and evaluation aircraft system is a promising application that will bring much convenience in the inspection of bridges. It is easily to get to some spots difficulty to reach and is usually faster than other methods. By reducing the need for human inspection on the spot, it will also reduce the cost and secure the inspection workers. Problems such as the flight stability and damage classification are still to be dealt. Crack tracking and evaluation system meets the need for the inspection work of the most commonly kind of damage for concrete structures. Similarly, it is able to reduce the cost and accelerate the inspection process. Due to the vast forms of concrete cracks and many other factors, crack identification rate and evaluation capacity of such application require more research work. Railway foreign body intrusion recognition system will benefit most the security of railways. The booming construction of railways brings great pressure on the inspection work of railway systems. By the timely discover and alert to related workers, this system will gain considerable commercial and social benefit. Problems like how to work at night efficiently and determine the accurate size of the foreign body are to be modified. Computer vision technology is now applied in many aspects among civil engineering, with the development of optical sensors and related algorithms, there will be more advanced technology developed and applied to the structural health monitoring of engineering structures.

ACKNOWLEDGMENT

The work described in this paper was jointly supported by the National Science Foundation of China (Grant No. 51778574), and the Fundamental Research Funds for the Central Universities of China (Grant No. 2017QNA4024).

REFERENCES

Chong, K. P., Carino, N. J., & Washer, G. (2003). Health monitoring of civil infrastructures. *Structural Health Monitoring*, *12*(3), 483–493.

Ellenberg, A., Kontsos, A., Moon, F., & Bartoli, I. (2016). Bridge related damage quantification using unmanned aerial vehicle imagery. *Structural Control and Health Monitoring*, *23*(9), 1168–1179. doi:10.1002tc.1831

Fukuda, Y., Feng, M. Q., & Shinozuka, M. (2010). Cost-effective vision-based system for monitoring dynamic response of civil engineering structures. *Structural Control and Health Monitoring*, *17*(8), 918–936. doi:10.1002tc.360

Hu, X., Wang, B., & Ji, H. (2013). A wireless sensor network-based structural health monitoring system for highway bridges. *Computer-Aided Civil and Infrastructure Engineering*, *28*(3), 193–209. doi:10.1111/j.1467-8667.2012.00781.x

Huang, H. W., Sun, Y., Xue, Y. D., & Wang, F. (2017). Inspection equipment study for subway tunnel defects by grey-scale image processing. *Advanced Engineering Informatics*, *32*, 188–201. doi:10.1016/j.aei.2017.03.003

Liu, Y. F., Cho, S. J., Spencer, B. F., & Fan, J. S. (2014). Automated assessment of cracks on concrete surfaces using adaptive digital image processing. *Smart Structures and Systems*, *14*(10), 719–741. doi:10.12989ss.2014.14.4.719

Morgenthal, G., & Hallermann, N. (2014). Quality assessment of unmanned aerial vehicle (UAV) based visual inspection of structures. *Advances in Structural Engineering*, *17*(3), 289–302. doi:10.1260/1369-4332.17.3.289

Schwitzke, M., Bachmann, V., Noack, M., Freyer, T., & Launert, B. (2010). Concrete-crack monitoring on structural elements by automatic digital image analysis. *Bauingenieur*, *85*, 455–459.

Yamaguchi, T., & Hashimoto, S. (2010). Fast crack detection method for large-size concrete surface images using percolation-based image processing. *Machine Vision and Applications*, *21*(5), 797–809. doi:10.100700138-009-0189-8

Ye, X. W., Ni, Y. Q., Wong, K. Y., & Ko, J. M. (2012). Statistical analysis of stress spectra for fatigue life assessment of steel bridges with structural health monitoring data. *Engineering Structures*, *45*(15), 166–176. doi:10.1016/j.engstruct.2012.06.016

Zhang, C. S., & Elaksher, A. (2012). An unmanned aerial vehicle-based imaging system for 3D measurement of unpaved road surface distresses. *Computer-Aided Civil and Infrastructure Engineering*, *27*(2), 118–129. doi:10.1111/j.1467-8667.2011.00727.x

Chapter 10
Machine Vision–Based Application to Structural Analysis in Seismic Testing by Shaking Table

Ivan Roselli
ENEA, Italy

Vincenzo Fioriti
ENEA, Italy

Marialuisa Mongelli
ENEA, Italy

Alessandro Colucci
ENEA, Italy

Gerardo De Canio
ENEA, Italy

ABSTRACT

In the present chapter the most recent and successful experiences in the application of machine vision-based techniques to structural analysis with main focus on seismic testing by shaking table are described and discussed. In particular, the potentialities provided by 3D motion capture methodologies and, more recently, by motion magnification analysis (MMA) emerged as interesting integrations, if not alternatives, to more conventional and consolidated measurement systems in this field. Some examples of laboratory applications are illustrated with the aim of providing evidence and details on the practical potentialities and limits of these methodologies for vibration motion acquisition, as well as on data processing and analysis.

DOI: 10.4018/978-1-5225-5751-7.ch010

INTRODUCTION

The use of machine vision-based data for application to structural dynamics is both promising and challenging. With respect to the state-of-the-art, further developments for on-the-field applications are still needed, but, at the present stage, such techniques are already feasible for laboratory dynamic tests. In fact, the growing interest in the use of machine vision-based technologies in this field of investigation, as well as in others, is mainly due to rapid advances in instruments performances available at falling costs. On the one hand, some issues related to the optimization of data processing and filtering algorithms still make these methods more complicated and less attractive than the use of more simple consolidated measurement techniques. On the other hand, many advantages in the vision-based instrumentation setup and its higher potentialities are hardly deniable. In particular, the potentialities of integration of such instrumentation output with computer graphics applications to visualize real-time data and results make these methods extremely effective in terms of science communication, with tremendous impact for technology transfer and diffusion purposes.

Several vision-based techniques have been developed in recent years and are quickly consolidating in their application to structural engineering. Some of them are derived from traditional techniques, but their impact and potentialities improved dramatically by latest hardware technology advances and new data processing algorithms. This is the case of the revival of stereo-photogrammetry applications due to the innovative Structure-from-Motion (SfM) data processing method (Micheletti, 2015) and the diffusion of relatively low-cost Unmanned Aerial Vehicles (UAV) equipped with vision-based systems for structural inspection and surveys (Remondino, 2011; Morgenthal, 2014; Mongelli, 2017). This kind of methods is particularly useful to provide an objective 3D documentation of the crack pattern, which is otherwise detected and documented by subjective visual inspection of experts. Application to shaking table testing is limited to spot surveys in static conditions to document the evolution of the crack pattern after each shaking trial.

Another practical and effective methodology called Digital Image Correlation (DIC) is now widely accepted and its use is already consolidated in many fields of experimental mechanics for quantitative deformation measurements of planar object surfaces. By comparing the digital images of a test object surface acquired before and after deformation the DIC directly provides full-field displacements to sub-pixel accuracy and full-field strains (Pan, 2009). This technique requires the preparation of the specimen surface with well-contrasted dots, for example by painted brush or spray, which makes it suitable for the study of small-scale objects subjected to slow pseudo-static deformations in lab tests. It was later successfully utilized for tensile and bond tests on composite reinforcements (Tekieli, 2017). Similar methodologies based on taking periodic photographs of tags or targets positioned on both sides of a crack were developed to monitor cracks aperture in damaged structures (Mangini 2017).

The application of computer vision and image processing techniques to shaking table tests in dynamic conditions, i.e. to detect motion parameters during seismic trials, has been investigated and developed in the last 20 years to remedy some drawbacks of conventional instrumentation, such as encumbrance, limitations in the number of sensors, range restrictions, and risk of damage of the devices (Berladin, 2004; Lunghi, 2012; Doerr, 2005; De Canio, 2013; Hutchinson, 2004). Moreover, a typical limitation of displacement devices, such as the Linear Variable Differential Transformers (LVDTs) and laser sensors, is due to their capacity of providing accurate data in only one direction. Consequently, the need of locating three sensors for tri-axial acquisition at each measurement point causes more difficult and time-spending set-up, also taking into account the usual range and encumbrance of conventional sensors. Among the several systems and techniques of this kind, particularly successful experiences were recently conducted in shaking table testing with the use of passive 3D motion capture systems. This kind of systems is capable of acquiring the positions of more than a hundred passive markers reaching interesting accuracy levels (0.01-0.1 mm) with a constellation of a proper number of high-resolution cameras. A crucial role for such accuracy levels is played by appropriate sub-pixel location estimation algorithms (Fillard, 1992).

For application to structural dynamics, the particular focus must be put on processing and filtering the markers data of 3D motion capture systems. Theoretically, the complete motion of a point can be calculated starting from the related marker trajectory, which means, in practical, to measure its absolute displacement. Subsequently, velocity and acceleration can be estimated through successive operations of numerical derivation. The accelerations estimated from markers can provide valuable indications for studying structures subjected to shaking table testing, even if accuracy cannot be as accurate as conventional accelerometers. Particularly interesting is the possibility of measuring even more than a hundred 3D points through passive cheap markers to obtain information also on the acceleration, as an alternative to installing numerous conventional expensive accelerometers that can be damaged in destructive shaking table tests. Data processing and digital filtering techniques are needed to reduce the measurement noise of acquired data in order to optimize the estimation of velocity and, especially, acceleration. Roselli et al. (2017) developed a methodology for optimizing the markers data processing to obtain an estimation of accelerations and of seismic damage/performance indices based on accelerations. In this optic, it is relevant to premise how displacement noise propagation to velocity and acceleration was modeled. Moreover, as well known in digital signal processing, even the numerical operation of derivation itself adds a further source of noise (Lyness, 1967). With the application of such processing and filtering steps, the acceleration can be obtained with an accuracy in the order of 0.01-0.02 g, which is still appropriate for providing interesting information for seismic tests with input peak accelerations that typically fall in the range 0.1-1.0 g (De Canio, 2016).

Markers displacements data and related estimated accelerations can be combined to obtain indications on the hysteretic behavior of the structure (or of any portions of the structure that is described by the positions of the markers), which constitutes one of the most relevant mechanical properties of construction materials and structural members in civil engineering (Sengupta, 2014).

Markers data proved to be able to provide very accurate Experimental and Operational Modal Analysis (EMA/OMA) in order to extract the modal parameters (De Canio, 2011) and to calibrate/validate the Finite Element Models (FEM) of the structure (Mongelli, in press). In particular, the combined used of OMA by a large number of markers and numerical modal analysis by FEM permits to compare the resulting modal shapes for a more refined and precise dynamic identification of the structures (De Canio, 2016).

Even more recently, also the application of the Magnified Motion (MM) technique provided remarkable results with the use of high-speed cameras. In the effort of extracting structural indications from video-based techniques, the feasibility of MM with the use of low-cost commercial cameras was explored and encouraging results were obtained in a laboratory environment. For many years attempts to produce qualitative (visual) and even quantitative analysis using videos of large structures have been conducted, but with poor results. This was because of the resolution in terms of pixels, of the noise, of the camera frame rate, computer time and finally because of the lack of appropriate algorithms able to deal with the extremely small motions related to a building displacement. These and others limitations have restricted in the past the applications of digital vision methodologies to just a few cases. Nevertheless, lately, important advances have been obtained by Wadhwa et al. (2017b) at the Massachusetts

Institute of Technology, namely the Motion Magnification Analysis (MM). In this Chapter, we explore the potentialities of vibration monitoring by means of the MM. The use of MM evaluate motion in video sequences, is similar to use a microscope to see the details of any sample, but used for observing the details on some groups of pixels of the motion in video sequences. The MM extract the physical properties of the object from the images, through the spatial resolution of the video camera, to analyze the dynamical behavior of the object under observation, e.g. to visualize at least the first mode shape, no matter its dimensions, since any point on the surface of the object can be considered a virtual sensor. A big number of experiments recently conducted on simple geometries on small and big objects have presented the reliability of this methodology compared to accelerometers and lasers

The motion magnification seems able to act like a microscope for motion and, more importantly, in a reasonably short computer time. The latter point is crucial, as it is well known that video processing takes a lot of time and resources. Therefore, any viable approach must consider the reduction of the calculation time as an absolute priority. The basic Eulerian MM version looks at intensity variations of each pixel and amplifies them, revealing small motions which are linearly related to intensity changes through a first order Taylor series for small pixel motions. Researchers have also explored the method's feasibility. A relevant interest by the scientific community was shown, since conventional devices are surely more precise, but more expensive and much less practical, as it requires a certain degree of specific expertise. Therefore, in this Chapter, we provide an introduction to MM and describe its application to the analysis of two full-scale historic masonry walls tested on shaking tables. This is an interesting point because the size of tested walls is larger than usually small experimental set-ups implemented in MM test-bed. Results prove that MMA is efficient for a visual identification of fractures in advance state. Conventional calculation for modal analysis of the walls is illustrated, such as FRF and PSD, on MM output data. Low-quality equipment (camera, tripod, and lighting) was used to test the methodology in an unfavorable environment with very high data noise. The first modal frequency was estimated, obtaining a good agreement with modal analysis by an optical system used as a reference.

In addition, the possible integration with computer graphics to visualize real-time data and results make these methods particularly effective in terms of communication for technology transfer and diffusion.

3D OPTICAL MOTION CAPTURE APPLIED TO SEISMIC VIBRATION

The 3D optical motion capture systems can be defined as motion measurement systems based on a constellation of cameras able to track the position of objects placed inside a certain measurement volume. In this sense they can be seen as local positioning systems, in analogy with the widely known Global Positioning System (GPS), as they track the positions of points in a local environment, typically a laboratory. In general, they extract three-dimensional information from the geometry and the texture of the visible surfaces in a scene through inherently non-contact procedures.

Some systems are laser-based (i.e. based on structured light emitted by the utilized instrumentation) can compute 3D coordinates of points on most surfaces. Other systems are stereo or photogrammetry-based and can operate with ambient light, even if most commonly these systems project additional external lighting toward the measurement volume, in order to ease the processing tasks. This is particularly important for passive systems, which are the ones that utilize retro-reflective markers located at the measurement points. In these cases, the additional lighting is optimized with markers coating reflective properties in order to improve their visibility and contrast with the surrounding environment. Consequently, the algorithm for markers detection is helped, as the scene rarely contains objects as brilliant as the markers, apart from undesired reflections, which can be masked either via hardware (avoiding the presence, or covering, reflecting objects in the measurement volume) or via software (through masking algorithms that exclude the unwanted areas in the camera view from the processing). To improve markers detectability their shape is usually studied so as to make them less ambiguous in a 3D environment, for example spherical, this makes them unchanged and so recognizable from any point of view. Consequently, passive markers are typically simple plastic balls coated with a retro-reflective material.

Other systems adopt active markers generally based on light-emitting diode (LED) technology. Active systems usually provide higher signal-to-noise ratio, resulting in lower marker jitter and potentially higher measurement accuracy. The markers identification is more unambiguous than in passive systems, especially if techniques are used for strobing one marker on at a time, or tracking multiple markers over time and modulating the amplitude or pulse width.

Moreover, since the active markers themselves are powered to emit their own light rather than reflecting light back, as in the passive systems, according to the inverse square law, they provide one quarter the power at two times the distance, which permit to increase the cameras distances and/or the measurement volume with the same power supply of a corresponding passive system. For the above reasons, active systems are able to provide better results in an outdoor environment, where lighting conditions and reflections are less controllable. On the other hand, passive systems do not require any cabling of measurement points and passive markers are much cheaper and replaceable than active ones, which makes them more suitable for laboratory seismic tests that can also lead to specimen collapse. Substantially, passive systems provide the advantage of locating the valuable instrumentation (acquisition units, additional lighting, cameras, cabling etc.) in the safety zone around the shaking table at a distance of 5-10 m from the measurement volume, so that the risk of costly instrumentation damage in case of specimen collapse is very limited.

System Preparation and Data Processing

In this section, some elements of system preparation and data processing are provided with reference to typical passive 3D motion capture systems applied to shaking table testing.

The first step for preparing the system for data acquisition is obviously the camera's setup, which essentially includes a proper cameras positioning around the measurement volume, an appropriate setup of each camera optics and acquisition parameters, executing a valid system calibration.

Cameras should be positioned at a safe distance from the shaking table and, in particular, outside its reaction mass area, which is the area corresponding to a "floating mass" that reacts to its hydraulic actuators (Figure 1). In fact, any vibration or floating motion should be avoided, since cameras are supposed to be perfectly still in their position. Moreover, the number and positioning of the cameras should be adequate to cover the desired measurement volume, i.e. the volume of the tested specimen plus the shaking table stroke in all directions. At this step, markers should already be located on the specimen to guide cameras positioning.

As a following step, each camera optic should be regulated to optimize focus. Also, the regulation of the acquisition parameters for each camera is particularly important in passive systems. They generally comprise gain and threshold settings on the acquisition software. After having verified that all markers are well visible and in focus, the system calibration is required in order to give the system a local geometric orientation and reference. In many passive systems, the calibration process is obtained

Figure 1. Shaking table cross section (left) and example of 3D optical motion capture cameras positioning (right, top view)

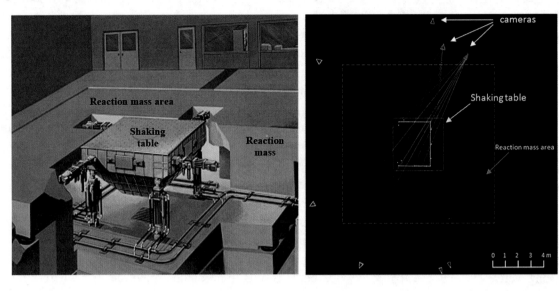

by manually waving a calibration device (it might be a passive or active wand, provided with a limited number of markers located in predefined positions) in the capture volume. Then, the x, y, and z coordinate origin are defined, e.g. by acquiring the calibration device on the ground at the axes origin location. The calibration process is particularly relevant because its efficiency substantially determines the final measurement accuracy, so most systems provide accuracy quality indicators to assess whether operators should repeat the process in case accuracy level is not as good as required. Present state-of-the-art systems are able to reach 0.1-mm accuracy in terms of Root-Mean-Square-Error (RMSE) in most acquisition configurations, but in some cases, the accuracy of about 0.01 mm was reached. After calibration, the system is ready for data acquisition.

Most advanced systems perform the first step of data processing onboard of each camera that detects each viewed marker and calculates its centroid and radius (Figure 2). This permits to send a limited amount of data to the acquisition units that perform only the triangulation and stereo-photogrammetric calculation required for the 3D reconstruction of the position of the marker.

Centroid calculation is usually based on circle fitting algorithms applied to the grayscale blobs produced by each spherical marker. Through such fitting algorithms, sub-pixel accuracy of markers centroids is obtained.

Marker centroids and radii are triangulated by 3D reconstruction algorithms for each acquired time in order to obtain the markers trajectories. Theoretically, the complete motion of each measurement point can be computed starting from such trajectories. The absolute displacements can be easily computed and processed with successive operations of numerical derivation to get an estimate of velocity and acceleration. In practical, each operation of numerical derivation produces data noise amplification, so that acceleration estimates are very poor. Nevertheless, numerical derivation can be implemented with data filtering algorithms that allow obtaining estimates of accelerations that can provide valuable indications for studying structures subjected to shaking table testing, even if they cannot be as accurate as conventional accelerometers. In fact, accuracy in the order of 0.01-0.02 g is still appropriate for providing interesting information for seismic tests that typically falls reach peak accelerations in the range 0.1-1.0 g.

Figure 2. Spherical marker grayscale blob (left) and its centroid and radius detection (right)

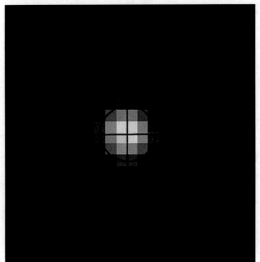

Data Filtering Techniques

Once all possible hardware sources of system error are eliminated or limited as much as possible, the markers trajectories are still affected by a residual error. To attenuate such error a variety of digital filters can be used depending on the objective that operators want to obtain from acquired data. For example, in most dynamic tests the main objective is typically to reduce the measurement noise in order to optimize the estimation of velocity and, especially, acceleration, since many seismic damage or performance indices are based on acceleration values. Consequently, how displacement noise propagates to velocity and acceleration must be characterized and modelled in both the time and the frequency domain. For example, if the error in the markers trajectory is a white noise, then the acceleration error amplitude propagates by quadratic growth with the frequency as shown in Figure 3. In addition, the numerical operation of derivation inherently adds a further source of noise. Therefore, a proper filtering strategy

Figure 3. Theoretical error propagation effect on the acceleration amplitude ($\Delta A\omega^2$) vs. signal frequency f (Hz) for maker's displacement error of 0.1 and 0.01 mm (left). FFT of real marker acceleration noise (right).

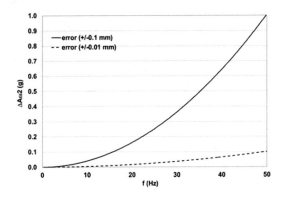
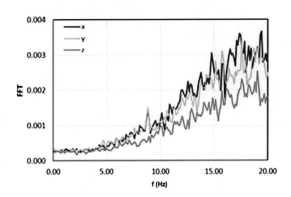

should provide good performance also in terms of attenuation of the error amplification generated by numerical derivative algorithms.

Among the several filtering approaches that can be considered a very common one is represented by the band-pass strategy, which is particularly effective when the frequency content of signals is clearly different from that of the noise. In seismic applications pass-band filters typically have a lower cut-off frequency at 0.1-0.5 Hz and an upper cut-off limit at 25-30 Hz, since natural earthquake spectra generally are within this range (Berardi, 1991).

A different strategy is based on the implementation of numerical differentiation algorithms. Classical methods based on higher-order central finite differences provide a somewhat low-pass effect. According to such methods a general expression of the first derivative f' can be written as:

$$f' = \frac{1}{\Delta t} \sum_{k=1}^{M} c_k \left(f_k - f_{-k} \right) \tag{1}$$

Where $M = (N-1)/2$ with N number of data within the moving window, Δt is the time step and c_k are called convolution coefficients. Several methods descending from such approach, also known as convolution methods, combine the derivative estimation with the filtering effect by choosing the appropriate c_k coefficients in Equation (1). A specific case of this general method can be implemented with convolution coefficients chosen in accordance with the Savitzky-Golay filtering algorithm (Savitzky, 1964).

The Savitzky–Golay smoothing filter, which is considered one of the most effective algorithms in removing random noise from experimental data without excluding completely the contribution of frequencies higher than an upper cut-off limit (De Canio, 2016). It essentially performs a local polynomial regression on a series of equally spaced data points (windowing), to determine smoothed values. This filter is characterized by two parameters: the fitting polynomial degree (n) and the windowing width (m points). The value of n is an even number, while m is an odd number.

The relevance of this approach in the present application is due to the notable possibility of implementing the Savitzky–Golay filter within the differentiation operations, which is particularly useful when successive derivatives of raw data must be calculated, like in the present case for obtaining velocity and acceleration from displacement data.

The above approach can be optimized by choosing the values of the filter parameters n and m that minimize the difference ΔF between the results obtained with filtered markers and with reference measurements at the same points on the structure (objective function). From a mathematical point of view the optimization problem can be formulated as follows (Boyd, 2004):

$$\begin{cases} min\Delta F\left(n,m \right) \\ n = 2k : k \in Z \\ m = 2k+1 : k \in Z \end{cases} \tag{2}$$

Where F can be whatever seismic parameters, such as the peak acceleration, the Arias intensity, the Housner intensity and others, we want to extract from the optimized markers data. In practice, this can be achieved by locating a limited number of reference instruments (e.g. conventional accelerometers) and comparing the recorded values of the chosen parameter with the ones calculated from the corresponding markers data. Some accelerometers can be used as checkpoints for evaluating the achieved accuracy. In this sense, a minimum of only one reference position and only one checkpoint could be use if accuracy is adequate. Obviously, increasing the number of reference positions and checkpoints would improve filtering parameters optimization, but such strategy remains convenient if very few costly instruments on the specimen are installed.

Recent Experiences of Application

A compendium of the most significant recent experiences is here illustrated in order to stress the potentialities in the application to seismic testing. In particular, the pros and cons of the use of a passive 3D motion capture technique at ENEA Casaccia research center in comparison with the conventional displacement sensors are emphasized and discussed. The examples exposed in the following are not meant to be exhaustive nor definitive, nonetheless, they can effectively provide evidence of practical situations in which the unconventional methodologies used at shaking table laboratory gave substantial advantages with respect to some typical limitations of other more traditional techniques.

Operational Modal Analysis (OMA) of a Tuff-Masonry Building

In this example, a 2/3-scaled two-story tuff-masonry specimen representing a typical Italian historic building was subjected to shaking table testing. The specimen is shown in Figure 4. Its dimensions were 3.50 m × 3.00 m and the inter-story height was 2.30 m.

The 3D motion data were acquired by a passive system installed at ENEA Casaccia Research Center. It was made up of a constellation of 10 high-frequency Near Infra-Red (NIR) cameras, based on VICON technology. The NIR cameras are equipped with a 4-Mpixel high-frequency CMOS digital sensor. Data were acquired at 100 fps (which as saying the sampling frequency in Hz).

Figure 4. Markers and accelerometers positioned on the specimen. Retro-reflecting markers appear lighted in photos taken by the common camera using a flash (right).

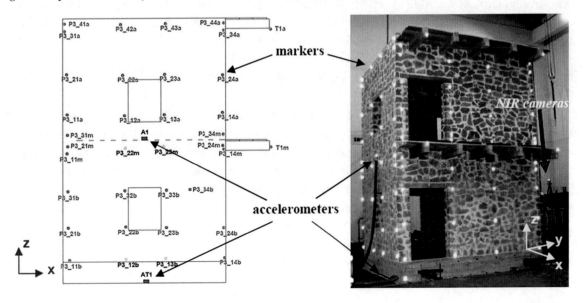

A total amount of 141 markers were used to monitor a large number of the measurement point of the specimen. Also, two triaxial accelerometers were located at the positions depicted in Figure 4, accelerometer AT1 was used as a reference for filtering parameters optimization and accelerometer A1 was used as a checkpoint of the acceleration estimate accuracy. With a measurement error in the range of 0.014-0.035 mm in terms of RMSE of markers displacement, the acceleration could be estimated with an accuracy of around 0.02 g.

According to a typical procedure, the experimental program comprised a sequence of scaled natural inputs, in this case based on the time-history recorded at Colfiorito station during Umbria-Marche earthquake in 1997 (Cinti, 2008) for the shaking table. In the testing sequence, the input Peak Ground Acceleration (PGA) was increased gradually at each trial by 0.05 g. Each trail of the sequence was alternated by an identification test, which, as usual, is a random test (white noise input) at 0.05 g. The random tests were processed with modal analysis techniques for the dynamic identification of the structure. In particular, the availability of a large number of measurement points allowed the application of the Frequency Response Function Multi-Input Multi-Output (FRF-MIMO) approach. The FRF-MIMO can be used when several signals record the base input (Multi-Input) and others the structural response (Multi-Output) so that the obtained modal parameters provide a refined and comprehensive description of the overall behavior of the studied structure. The FRF was computed in terms of the transmissibility function $H_1(\omega)$ given by:

$$H_1(\omega) = S_{io}(\omega) / S_{ii}(\omega) \tag{4}$$

where S_{io} is the power density spectrum of the input-output cross-correlation function, S_{ii} is the power density spectrum of the auto-correlation function of the input signal.

The markers at the specimen base were taken as input signals. In Figure 5 the FRF of markers at first story (MIMO 1) and at second story (MIMO 2) are compared with the FRF of the theoretical Single Degree Of Freedom (SDOF). Even the application of Operational Modal Analysis (OMA) approach can be applied in the case of white noise input. More specifically, in this study, a Frequency Domain Decomposition (FDD) with markers data was performed. Through OMA the numerous measurement points available made possible to provide a detail description of the modal shapes as shown in Figure 6 so that the comparison with the same mode extracted by numerical models, such as Finite Element Models (FEMs), indicating clearly a bending mode in the x-direction.

Structural Damage Detection of a 2-Storey R.C. Frame

This section illustrates the application of passive markers acquired to study a mock-up represented a typical 2-storey Reinforced Concrete (R.C.) frame with conventional R.C. slabs (Figure 7). The dynamic behavior of this structure was taken as a reference to evaluate the effectiveness of various anti-seismic techniques (such as an innovative lightened slab system and base isolation) applied to another geometrically identical mock-up. The position of the markers is shown in Figure 7.

Markers set-up was easy and fast with no problems of cabling and encumbrance typical of conventional displacement sensors. Besides, according to the max displacements that could be reached at the 2nd floor of the mock-up during this campaign the range of the displacement sensors should not be lower than 200 mm. At such high range, it is very challenging for conventional sensors to reach an accuracy of less than 0.1 mm, which would require a linearity of at least 0.05% of Full-Scale Output (FSO).

Figure 5. Frequency Response Function (FRF) of markers at first story (MIMO 1), at second story (MIMO 2), and at both stories (MIMO 1-2) with respect to markers at the specimen base compared with FRF of theoretical Single Degree Of Freedom (SDOF).

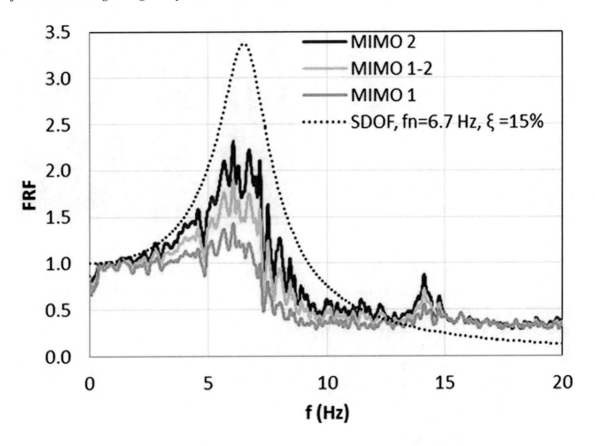

Figure 6. The modal shape and frequency f_n of the first mode by FEM (left) and by Frequency Domain Decomposition (FDD) with markers data (right).

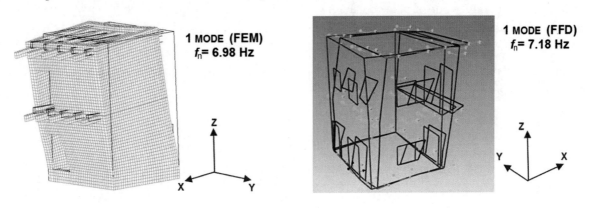

Figure 7. Reinforced-concrete frame mock-up dimensions (left) and positions of passive markers during shaking table tests (right)

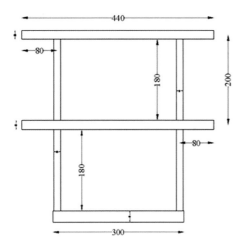 

The seismic input was mono-axial and its intensity was scaled and gradually increased from 0.05 g to 0.45 g of PGA along the test sequence with steps of 0.05 g. As in the previous case, the seismic tests were alternated random noise input at 0.05 g in order to obtain the structure fundamental frequencies. In particular, the evolution of the 1st mode frequency shift calculated through the FRF function of the markers data showed the gradual plasticization of the frame nodes (Figure 8).

Another crucial parameter for assessing the R. C. frame dynamic behavior is the drift at each level. In Figure 8 the measured drifts at the 1st floor with respect to the ground and at the 2nd floor with respect to the 1st floor are depicted versus the test PGA along the sequence.

Evolution of Masonry Wall Cracking

An experimental campaign on shaking table was executed on the dynamic resistance of masonry wall connections and, on the out-of-plane behavior of the façade. A mock-up, made up of a masonry façade connected to lateral walls, was built for shaking table tests (Figure 9). This case is interesting to com-

Figure 8. First mode frequency shift (left) and drifts at 1st and 2nd floor (right) versus test PGA along the shaking table sequence

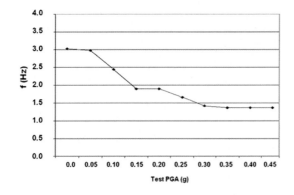 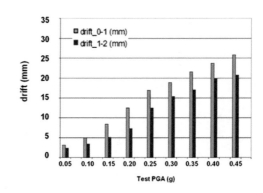

pare the 3D optical motion data with conventional sensors for absolute and relative displacement. For a refined description of the façade out-of-plane dynamic behavior, the displacement wire transducers and accelerometers were installed at the same position on the façade. The markers were positioned at façade connections, on the lateral walls, and also at the mock-up base.

The wire transducers housing were attached to a stiff frame in front of the façade as a reference for the out-of-plane absolute displacement (Figure 10). In order to protect the instrumentation and the reference frame from damage, a protection net was mounted to the mock-up to avoid the façade crashing down after the collapse.

The absolute displacement time-histories of a marker and a wire transducer at approximately the same position are compared in Figure 11. The wire sensors were affected by remarkable stretching and non-linear effects that caused a sensible underestimation of displacement in the negative x-direction (Figure 11). In fact, the error of wire transducers is typically in the range 0.1-1% of FSO and other conventional displacement sensors rarely reach higher levels of accuracy without dramatically increase instrumentation cost.

During the test, the cracks evolution has been monitoring by the use of markers to measure relative displacements at connections between lateral walls and the façade. The graph in Figure 12 shows that one of the connections failed at the early steps of the tests sequence and gradually opened, while the other one remained in place until the late tests before the façade collapse. This asymmetric behavior could be also investigated by analyzing the Root-Mean-Square (RMS) of the relative displacement between markers at both wall connections. Such RMS values indicate the oscillation of cracks openings during the seismic excitation. In Figure 12 it is evident that some damage appears in the early tests, but the crack propagation did not reach the top of the wall until the test before the final collapse. In similar cases in which relative displacement between points that can suddenly fall apart must be measured, the installation of traditional sensors requiring a physical connection (wire transducers, LVDTs, etc.) needs to be pondered very carefully to avoid the relevant risk of instrumentation damage.

Figure 9. Masonry mock-up on shaking table before tests (left) and after the collapse of the façade against the protection net (right).

Figure 10. Steel frame scaffolding in front of the façade for conventional displacement sensors set-up (left) and position of the 3D motion capture markers (right).

Figure 11. Displacement of a marker (E11) and a wire transducer (F2) during a shaking table test (left). Detail of the first 10 s of the same time-history (right).

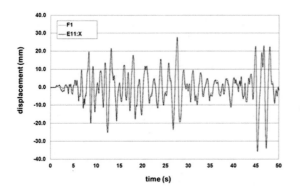 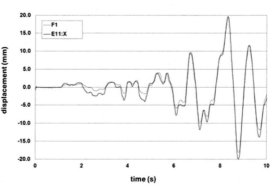

Figure 12. The behavior of lateral wall-façade connection (left): relative displacement after each shaking table tests at different height z; connection profiles (right) of relative displacement Root-Mean-Square (RMS) values.

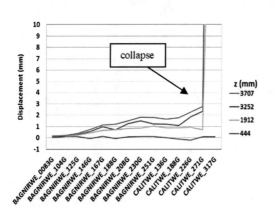 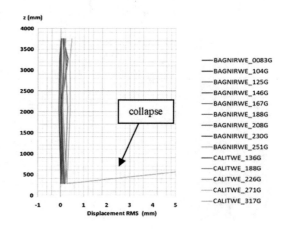

Description of the Dynamic Behavior of High Mock-Ups

As a noteworthy example of 3D motion capture acquisition of high mock-ups, a research project focused on a selection of most relevant architectural types of the historical heritage in the Mediterranean countries was conducted within the PERPETUATE project funded by the European Commission in the Seventh Framework Program (FP7/2007-2013).

Markers data proved to be very effective for the study of the dynamic behavior of a 1:6 scale model of the Laterano Obelisk. In detail, the 5-m-high mock-up, made up of three blocks, as well as the original obelisk located in Rome, was submitted to shaking table and pull-release tests.

Given the size and the typology of the model, it would be an extremely difficult task for conventional sensors to provide the required data for a complete description of the 3D absolute displacement of each block.

Besides, it would require a very engaging set-up preparation, with the use of high scaffoldings and the installation of sensors attached to 5-m-high stiff frames around the obelisk.

On the contrary, the markers could be located on each block on the floor before obelisk composition and test set-up (Figure 13). The 3D motion capture system offered the possibility of capturing the 3D rotational motion of each block (Figure 13) in a very complete way, allowing the analysis of the precession effects on rocking phenomenon under free oscillations Also, the real-time monitoring of the markers allowed to guide the operator in the correct timing for pulling and release steps according to the desired initial obelisk position.

An important added value consisted in the possibility of visualizing numerous windows with graphs of markers displacements and their relations. For example, the vertical distance between the markers at the base of each of the three blocks could be followed to detect the activation of rocking mechanisms at each level (Figure 14): an increase of such distance signaled a contact loss implying a rocking rota-

Figure 13. The composition of the 1:6 scale model of the Laterano Obelisk after markers installation (left); markers wireframe overlay during tests (center); markers rotation angles for rocking analysis (right).

  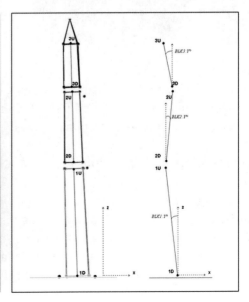

Figure 14. Monitoring of the Lateran Obelisk mock-up: a 3D reconstruction (top right); relative displacements in the vertical direction between blocks (right); top view (bottom left).

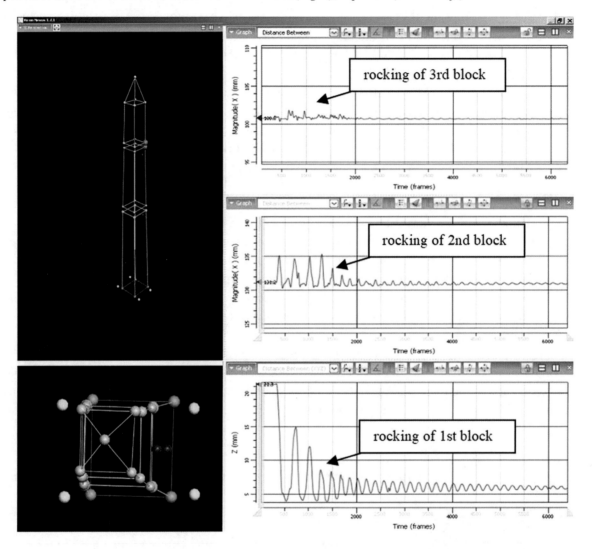

tion of the upper block. Another interesting contribution was given by the top view (Figure 14), which provided a unique point of view for understanding the real dynamic behavior to the mock-up, especially in terms of torsional and rotational effects.

Application to Small Multi-Block Mock-Ups

Another experimental campaign conducted within the PERPETUATE Project focused on the effect of different kinds of tie-rods to the seismic resistance of typical historic masonry arches. To this aim, a small-scale model of an arch was tested on shaking table (Figure 15). The arch was made up of 43 small discrete blocks of plastic material with the insertion of a thin membrane of Polyvinyl Alcohol (PVA) in order to obtain the desired friction angle between the blocks.

In consequence of the small size of the arch (Figure 15), it was very difficult to utilize conventional sensors to monitor the relative displacement between the blocks. In effect, common displacement sensors suffer from limits in housing encumbrance and weight that would affect the results of tests on small and light models.

In such circumstances, if miniaturized instrumentation can be used, some additional costs must be considered. Instead, a large number of markers were placed very easily and quickly with no further expense.

Also, a load cell was installed at the tie-rod and acquired along with the markers, giving the opportunity of comparing synchronized load data and displacement between the arch imposts (Figure 16).

Relative displacement between markers located on adjacent blocks was utilized to reveal temporarily contact loss (Figure 17) and sliding phenomena (Figure 18).

The collapse mechanism of the model could be observed in detail with no data loss, while trajectories of collapsing markers could be enhanced (Figure 19).

Estimate of Hysteretic Behavior

3D motion data provided an interesting added value in the characterization of anti-seismic devices and of other specimen for which is crucial the analysis of relative displacements and of the hysteretic behavior.

As a brief example, the characterization tests of seismic isolators is illustrated in the following. In Figure 20 a prototype of an EARLYPROT seismic isolator, a sliding-rolling (pendulum type) device provided with dissipating steel cables limiting the displacements is shown. Qualification tests on shaking table were set up with four isolators linked together and loaded with a vertical mass of 17 kN and 34 kN, reproducing a realistic configuration.

Markers were located on the table, on each isolator and on the loading plates for a complete 3D motion description of all mock-up pieces (Figure 21). A triaxial accelerometer and a laser sensor were also taken into account as a reference.

Figure 15. The tested arch with the markers overlays visualization on the shaking table (left) and model dimensions (right).

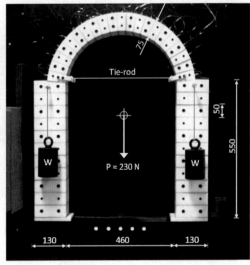

Figure 16. Monitoring of tie-rod failure during shaking table tests: load cell (top right) and the distance between markers at the imposts (bottom right).

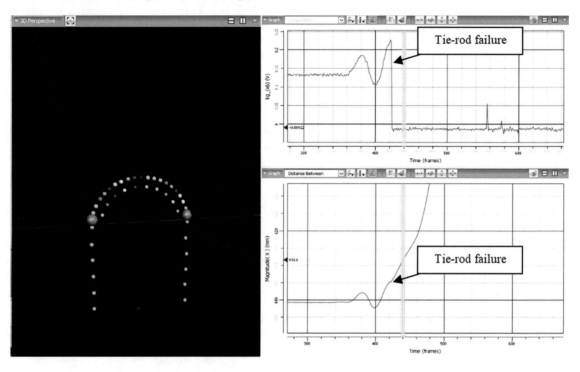

Figure 17. Dynamic behavior of adjacent blocks: temporarily contact loss between the two selected markers (larger blue points in left window).

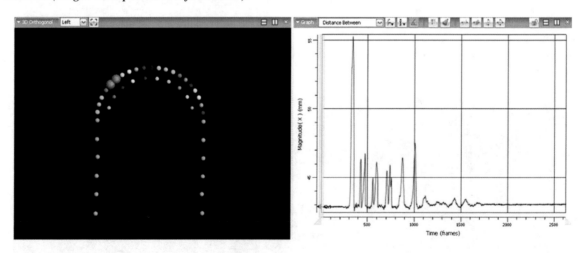

The device characterization was performed through a long sequence of pull-release, sine sweep and pure sine tests, varying the number of steel cables (Figure 21). The stiffness and the damping coefficients were calculated for each tested steel-cable configuration.

Markers data revealed interesting properties of the device hysteretic cycles: at low displacements, the sine sweep tests showed a constant friction coefficient of 3% in accordance with the theoretical pendulum

Figure 18. Permanent sliding between the two blocks (larger blue points in left window).

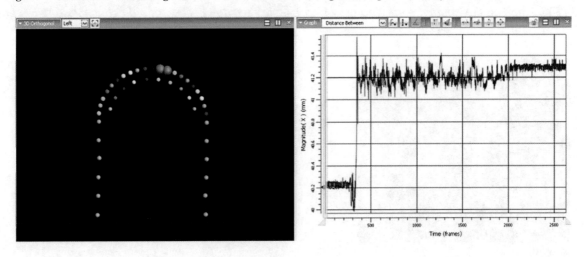

Figure 19. Following the trajectory of a selected marker (light blue line) during the arch collapse.

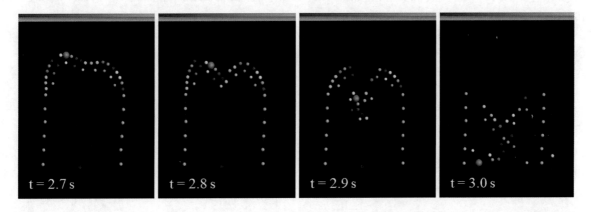

Figure 20. EARLYPROT seismic isolator (left). Pull-release test set-up with 2 cables per side (right).

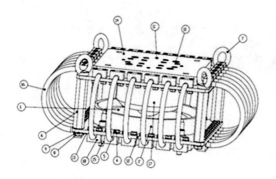

Figure 21. Sine test at device resonance frequency of 0.5 Hz: in the graph, the displacement of a marker (F1E) located on the isolated plate (orange wireframe, left window) is compared with a marker (T1) on the shake table (yellow wireframe, left window).

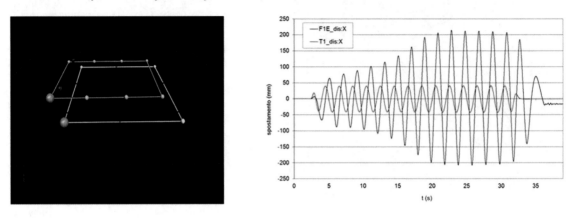

of same geometry, while the hardening effect caused by the steel cables appeared with the increase of the displacement over 100 mm (Figure 22).

Moreover, also the isolators torsional and recentering properties could be effectively characterized by markers data, giving indications for possible improvements in cables designing and distribution.

Figure 22. Hysteretic cycles of EARLYPROT isolators in several sine tests compared to the theoretical pendulum cycle with the same radius at 3% friction (pink cycle).

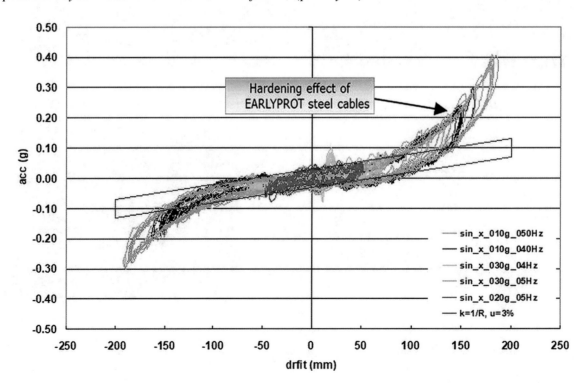

Integration Within a Virtual Lab

In order to keep research partners and students in contact for exchanging information on their scientific activities, research institutions, such as ENEA, are increasingly making use of remote technologies. A few years ago ENEA started several virtual labs, among which a new virtual lab dedicated to seismic and vibration experimentation and related numerical simulations. Therefore, it gave birth to the Structural Dynamics, Numerical Simulations, Qualification Tests and Vibration Control (DySCo) virtual lab, which is accessible through a dedicated web portal (Roselli, 2010). This web-based system is intended to allow the participation of authorized remote users to the experiments conducted at ENEA Casaccia R.C. by visualizing real-time videos and graphs of the tests underway while interacting with local researchers through a chat-based session. The ICT architecture of DySCo is shown in the flowchart in Figure 23. Measurement systems capable of displaying real-time signals along with computer graphic rendering of the test underway are very suitable and effective in such web-conference environment (Figure 24). From this point of view, 3D motion capture systems are provided with very friendly graphical interfaces for real-time monitoring, making use of computer graphic potentialities associated with the markers data and test movies that can be synchronized and calibrated together in order to obtain a 3D-overlay visualization.

Figure 23. Flowchart of DySCo virtual lab system architecture for real-time remote sharing of shaking table tests.

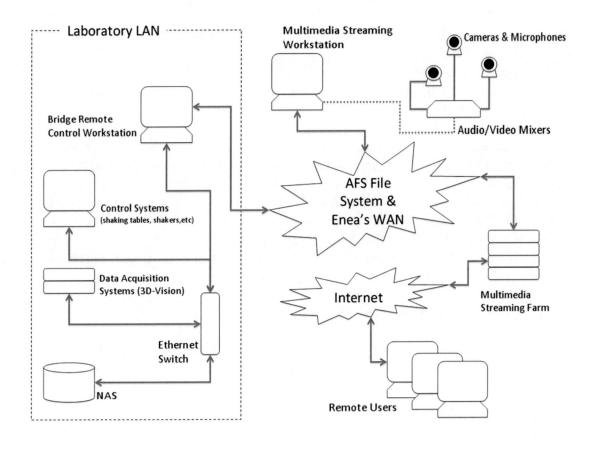

Figure 24. Graphic interface for web conference sessions through DySCo virtual lab.

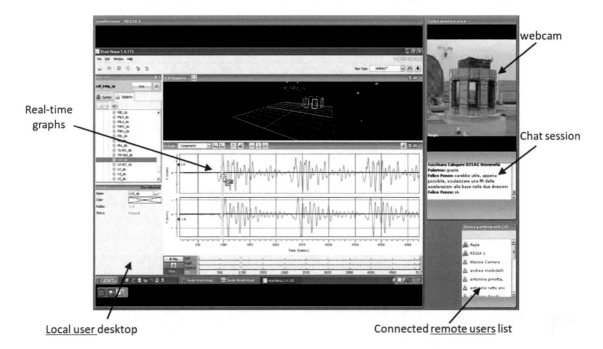

Furthermore, through the DySCo web portal research partners can be given access to software and hardware resources of the ENEA High-Performance Computing (HPC) system, called CRESCO (Computational Research Centre for Complex Systems), for advanced numerical simulations and computations (Ponti, 2014). This computing platform located in Portici, near Naples, constitutes the main node of the ENEA-GRID computing facility which links HPC platforms located in several ENEA Research Centers throughout the Italian territory.

Consequently, an outstanding contribution to the study of the dynamic behavior of specimen can be given by integrating and comparing experimental data with results from numerical models.

In particular, a good estimation of dynamic modal parameters through the modal analysis techniques, such as OMA and FRF, was obtained with markers data, showing relevant potentialities to implement the FEM calibration and validation (Figure 25). As well known, this is crucial to attenuate the uncertainties commonly affecting FEM boundary conditions, especially in terms of materials properties, model constraints and loads, so that the simulations significance and reliability for similar cases can be improved.

MOTION MAGNIFICATION ANALYSIS

The use of machine vision-based data for application to structural dynamics is both promising and challenging. With respect to the state-of-the-art, further developments for on-the-field applications are still needed, nevertheless at the present stage, such techniques are already feasible for laboratory dynamic tests. In fact, the growing interest in the use of machine vision-based technologies in this field of investigation, as well as in others, is mainly due to rapid advances in instruments performances available at falling costs. The most interesting among these methodologies is surely the motion magnification (MM).

Figure 25. Comparison of modal shapes by OMA with data from 3D motion data (left) and by FEM analysis (right).

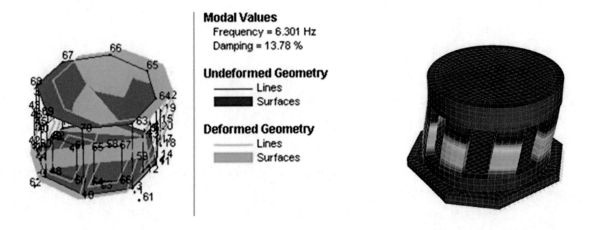

Motion magnification acts like a microscope for tiny motions in digital videos: small displacements, not visible to the naked eye become clear and evident after the magnification processing. The methodology does not rely on optic techniques but on algorithms capable to amplify the tiny changes in video frames, while the large ones remain unaltered. Currently, applications range from vibrometry to fluid dynamics, medicine, sound recovery, identification of materials, mechanics. In the past many attempts to produce qualitative and quantitative analysis using videos of large structures have been conducted, but with poor results. This is because of the pixel resolution, the noise, the frame rate, calculation time and finally because of the lack of proper algorithms to deal with the very small motions as such as a building. However, recently a number of experiments conducted on simple geometries and small objects (but with the notable exception of a bridge), showed the reliability of this methodology compared to conventional accelerometers (Wadhwa, 2017a), see some examples of magnified videos on the MIT site: http://people.csail.mit.edu/mrub/vidmag/.

Many advantages are to be gained: no wires, no physical contact, simplicity, low costs, and predictive capabilities. Most of all, due to each pixel provides an intensity variation time series; each pixel can be considered a contactless "virtual sensor". This way a very large number of such sensors are made available for an *ex-post* analysis, meaning that it is not necessary to decide in advance the sensor placements because every point of the surface recorded on the video will provide a signal. Of course, these signals may be used by the conventional quantitative analysis techniques such as the analysis in the frequency domain.

However, the extension of the MM to large structures as those of the seismic field is not straightforward: dimensions, illumination, distance from the recording device, vibration disturbances induced by the shaking tables on the video cameras, are all issues to be taken carefully into consideration. Here we report an example of the application of MM to seismic testing carried out by a large shaking table on a masonry in order to outline this novel methodology.

The basic MM algorithm looks at intensity variations of each pixel and amplifies them, revealing small motions which are linearly related to intensity changes through a first order Taylor series for small pixel motions (Wadhwa, 2017a). Videos are made of temporal sequences of 2D images, whose pixel intensity is $I(x, t)$. The 2D array of color intensity is the spatial domain, while the time domain is the temporal sequence. Here, in order to describe the Eulerian version of the magnification algorithm, only a 1D translating image with displacement $\delta(t)$ will be considered. At the image-position x and video-time $t = 0$ it is: $I(x, 0) = f(x)$. Translating for the quantity $\delta(t)$, we have, following a demonstration by (Whadwa, 2017b):

$$I(x, t) = f(x - \delta(t)) \tag{4}$$

The final expression of the motion magnified by the constant, α is defined as:

$$\Delta I = f(x - (1 + \alpha) \delta(t)) \tag{5}$$

If the displacement $\delta(t)$ is small enough, it is possible to expand the Equation (4) as Taylor's first order series around x at the time t:

$$I(x, t) = f(x) - \delta(t) (\partial f / \partial x) + \varepsilon \tag{6}$$

where ε is the error due to Taylor's approximation. The intensity change at each pixel can be expressed as:

$$\Delta(x, t) = I(x, t) - I(x, 0) \tag{7}$$

that takes into account Equation (6), becomes:

$$\Delta(x, t) = f(x) - \delta(t) (\partial f / \partial x) + \varepsilon - f(x) \tag{8}$$

and finally disregarding the error ε:

$$\Delta(x, t) \approx - \delta(t) (\partial f / \partial x) \tag{9}$$

meaning that the absolute pixel intensity variation Δ is proportional to the displacement and to the spatial gradient. Therefore, the pixel intensity can be written as follows:

$$I(x, t) \approx I(x, 0) + \Delta(x, t) \tag{10}$$

Magnifying the motion by a given constant α using of Equations (6) and (7), simply means that pixel intensity $I(x, t)$ is replaced by the magnified pixel intensity $I_{magn}(x, t)$ according to the following:

$$I_{magn}(x, t) \approx I(x, 0) + \alpha \Delta(x, t) \approx f(x) - \delta(t) (\partial f / \partial x) - \alpha \delta(t) (\partial f / \partial x) + O(\varepsilon, \delta) \tag{11}$$

where $O(\varepsilon, \delta)$ is the remainder of the Taylor series. Then, the magnified intensity can be calculated as:

$$I_{magn}(x, t) \approx f(x) - (1 + \alpha) \delta(t) (\partial f / \partial x) \tag{12}$$

but Equation (12) is immediately derived from the first order Taylor's expansion of the exact magnified motion:

$$\Delta I = f(x - (1 + \alpha) \, \delta(t)) \tag{13}$$

and therefore:

$$f(x - (1 + \alpha) \, \delta(t)) \approx f(x) - (1 + \alpha) \, \delta(t) \, (\partial f / \partial x) \tag{14}$$

Therefore, we can say that to magnify the motion displacement it suffices to add $\alpha \Delta(x, t)$ to $I(x, t)$, as long as Taylor's expansion of Equation (12) is valid, that is to say, if $O(\varepsilon, \alpha)$ is small. The limitation clearly depends on the linear approach of Taylor's expansion, an approach that does not hold either if the initial expansion of Equation (6) or the amplification α is too large. Essentially, in order to remain within the linearity range, we must consider slowly changing video sequences with sufficiently small amplifications. Moreover, the Nyquist sampling theorem imposes to the frame rate f_{fps} to be at least double than the highest frequency f_{max} we want to analyse, so that another condition is:

$$f_{sampling} \geq 2f_{max} \tag{15}$$

where f_{max} is the maximum frequency of the signal in the temporal domain, $f_{sampling}$ is the sampling frequency. Now Eq. 15 becomes:

$$f_{fps} \geq 2f_{max} \tag{16}$$

where f_{fps} is acting now as the sampling frequency.

Let us move to the example of MM applied to the seismic testing. Usually, the specimen, positioned on a shaking table, undergoes a series of tests, tuned to reveal particular dynamic behavior. In the beginning, the shakes are weak, then they increase in intensity and finally the specimen may result severely damaged. During the tests a number of accelerometers, or similar devices, record the vibrations of critical points on the specimen. This is the standard procedure. The first step of the MM analysis is to record a video of the structure, including the critical points we are interested in, taking good care to avoid large motions such as people passing by, swinging cables, oscillating mechanical parts. Indoor illumination is also a key aspect: shadows should be avoided as much as possible. Note that the large motions are the most significant source of noise for the MM algorithm. At the moment the issue is still open, however, several research groups are trying to fix the problem in Europe and in the USA.

Experimental Application of Motion Magnification to Shaking Table Tests

In this example (Fioriti, 2017), the utilized video camera was the one installed on a low-performance commercial tablet (720x1280 pixels, 28 fps, 8-bit greyscale). The test specimens were two masonry walls that were simultaneously subjected to shaking table tests. The two specimens represented real-scale inter-storey walls typical of historic building masonry typologies in Central Italy: one was made up of two-leaf limestone masonry and the other was a regular tuff-unit masonry. During shaking tests, videos were taken and subsequently processed by the above described MM algorithm (Whadwa, 2017b). The basic idea is to take advantage of a large number of pixels contained in each video frame. Theoretically, in the present experiment, the number of usable pixels is 921,600 that we can consider as "virtual sensors", meaning that each pixel has a time history of intensity variations (color or grey scale), from the first frame to the last one. These time series contain the information about the displacements of the physical points (although they are not the *real* sensors attached at each visualized point of the structure). Of course, it would be too cumbersome to analyze all the virtual sensors. Moreover, in general, not all the image contains points of the structure and therefore generates useful information. Consequently, an important step before the image processing is the identification of appropriate areas of interest of the frames, possibly represented by rather small portions of the specimen, preferably characterized by the high signal-to-noise ratio (SNR). In Figure 26 the setup of the studied masonry specimens on the shaking table is showed. The specimens are enclosed in a steel-frame structure that reproduces the constraints of the rest of the building and is protected by safety grids to prevent risks of catastrophic collapse. In Figure 27 the red box indicates the selected part of the image with high SNR. The selected area contains 19x78 pixels, providing, therefore, 1482 "virtual sensors", whose variations in the whole video were followed and analyzed by MM. Within the selected area there are also three markers of the 3D optical motion capture system motion capture system. The used optical motion capture system as described in the above section of this Chapter. The signals from the markers provide the reference for the comparison with the MMA, together with some conventional accelerometers. The presence of edges or texture is very helpful for the MM algorithm, but unfortunately, the area in the red boxes of Figure 27 is rather homogeneous. These circumstances, added to other disturbances, produce a certain amount of noise, responsible for the artefacts and blurred images in the processed videos.

As said before, signals from the magnified motion technique described does not provide directly the displacements, although they could be recovered (Whadwa, 2017b), nevertheless, on the other hand,

Figure 26. Setup of the two tested masonry walls on the shaking table (left). Frontal view showing the walls during test setup (right). The height of the two walls is 3.45 m.

they may be used to calculate the power spectral density (PSD), allowing the modal analysis and the calculation of the frequency response function (FRF). Note that to determine a modal frequency, it is necessary a number of frames compatible with the frequency resolution, since the frequency resolution is inversely proportional to the frame number. Thus with high resonant frequencies (rigid structures), we need a high frame rate but a short video length, and vice-versa with low frequencies (non-rigid structures). The intensity variation signals of the 19x79-pixel area are averaged to produce only one signal to reduce noise since the average is just a low pass filtering. The benchmark signal is generated by the 3D optical motion capture system tracking system. Ten high resolution (4M pixel) cameras are able to measure accurately the 3D axial absolute displacements of the markers during the seismic tests.

Now it is possible to calculate the FRF for both the 3D optical motion capture system and for the MM signals. It is well-known that the modal parameters allow evaluating the seismic vulnerability of a structure and that a significant variation after a period of time may indicate a major damage.

Figure 27. A frame from the magnified motion video (left): note how the image is blurred because of the noise. The red circle indicates a marker of the 3D optical motion capture system. I is the position of input signal point for the FRF calculation, while O is the position of the output signal.

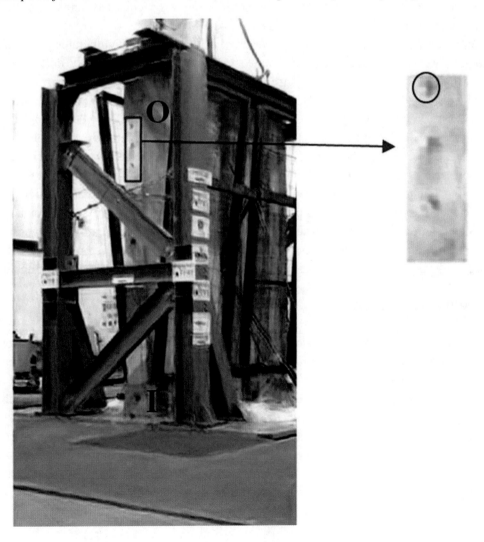

The frequency response function is a mathematical representation of the relation between two points on a structure (called the input and the output) in the frequency domain. The output signal is indicated with O and the input signal of markers positioned at the basis of the wall is indicated with *I*, as in Figure 27; deriving the input and the output accelerations we describe a functional relationship between these points, namely the FRF. This experimental methodology was applied making use of the transmissibility function $H_I(\omega)$ given of Equation (3).

In Figure 28, the first resonance peak identified by MM is at 7.9 Hz. The corresponding peak by 3D optical motion data was found at 7.8 Hz (up to the frequency resolution 0.1 Hz). The second peak MM at 11.1 Hz is not aligned with the peak by optical motion data, which is at 10.7 Hz. It should be considered that the 3D optical motion data were captured at a sampling rate of 200 Hz, while the video camera was able of a maximum frame rate of 28 Hz. Moreover, the two signals are not well synchronized in time, meaning the start recording instants do not coincide exactly. In particular, the different sampling rate has introduced in the FRF by MM some spurious peaks below 8 Hz and above 12 Hz. Actually, to face this problem it suffices a better recording device. To confirm further the 7.9 Hz (3D optical motion data) and the 7.8 Hz (MM) results, we compare them with an independent estimation of the FRF carried out by means of a standard accelerometer equipment, that is 7.89 Hz for the first FRF peak. Considering reliable the accelerometer value, the 3D optical motion capture system yields an error of -0.127% and the MM an error of -1.141% (Fioriti, 2017; Chen 2014).

Another interesting feature of the MMA is the direct video detection of vulnerable points of the structure in advance with respect to the sequence of the strongest tests. In Figure 26 the dotted red line in the frontal view indicates a crack produced by strong shakes at the end of the tests. It occurred exactly where the MM video showed the largest displacements during the low-intensity test, therefore, the magnified motion possesses predictive capabilities. Note that the original (meaning not MM processed) videos do not allow to see any displacement (videos of the masonry magnification are downloadable from the link: https://drive.google.com/drive/folders/0Bz540aXsdKTnZjZxU1FYZndfczQ?usp=sharing).

The above results on the identification of the fundamental frequencies of masonry specimens in laboratory dynamic tests can be extended even to larger structures for in-situ applications if a careful experimental setup is provided. Authors point out that noise is a pervasive obstacle, especially when the recording devices are of low quality and laboratory parameters (illumination, camera vibration insulation, shadows, unwanted large motions) are not optimal. Nevertheless, these developments in digital vision technologies are very promising, with an exceptional benefit/cost ratio and a variety of applications.

In-the-Field Applications

The practical potentialities of vision-based methods for in-the-field applications to real structures have been explored since the early 2000s, with interesting results for the structural health monitoring of large infrastructures, notably long-span bridges (Wahbeh, 2003). Apart from the specific interest about this kind of constructions, the decks of large bridges are usually characterized by very ample vibrations at low frequencies, which make them the most suitable structures to be monitored by means of reasonably cheap vision-based technologies, with limited fps speed and number of pixels. In fact, commercial vision-based instruments can be positioned at a long distance from the bridge deck (e.g. at a stable point of a spandrel, abutment or pier) and still be able to record displacements with sufficient accuracy.

A few examples of application in practical structures are already present in Hong Kong, such as the dynamic displacement measurements of the Tsing Ma and the Stonecutters Bridges (Ye, 2013).

Figure 28. FRF amplitude calculated from the 3D optical motion capture data (blue line) and from the magnified motion technique (red line). The first peak is recovered within an error of 0.1 Hz and the second within 0.4 Hz.

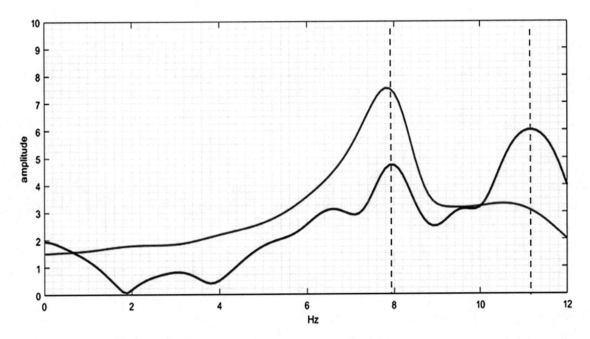

A major issue for outdoor applications is the instability of lighting conditions, which affects the effectiveness of image processing algorithms. In this context, a recent study (Lee, 2017) faced the processing problems related to the light-induced image degradation for in-service bridges. Other relevant issues associated with the environmental conditions are heat haze and thermic expansion effects.

An updated and comprehensive review of computer vision methods and their pros and cons for the damage detection in civil infrastructure was recently published (Feng, 2018). According to this review, in-the-field applications are promising, but still need significant advances in terms of accuracy, resolution, and robustness, especially for small and medium-size structures under operating excitations. From a technological point of view, future advances should also focus on the development of vision sensors with improved real-time on-site image processing performances in order to make long-term continuous monitoring practically feasible.

REFERENCES

Beraldin, J. A., Latouche, C., El-Hakim, S. F., & Filiatrault, A. (2004). Applications of photogrammetric and computer vision techniques in shake table testing. *Proceedings of the 13th World Conference on Earthquake Engineering.*

Berardi, R., Longhi, G., & Rinaldis, D. (1991). Qualification of the European strong motion data bank: Influence of the accelerometric station response and pre-processing techniques. *European Earthquake Engineering, 2*, 38–53.

Boyd, S. P., & Vandenberghe, L. (2004). *Convex Optimization.* Cambridge University Press. doi:10.1017/CBO9780511804441

Chen, J. G., Wadhwa, N., Cha, Y. J., Durand, F., Freeman, W. T., & Büyüköztürk, O. (2014). Structural Modal Identification through High Speed Camera Video: Motion Magnification. *Proceedings of the 32nd International Modal Analysis Conference, 7,* 191-197. 10.1007/978-3-319-04753-9_19

Cinti, F. R. (2008). The 1997-1998 Umbria-Marche post-earthquakes investigation: Perspective from a decade of analyses and debates on the surface fractures. *Annals of Geophysics, 51*(2-3), 361–381.

De Canio, G., Andersen, P., Roselli, I., Mongelli, M., & Esposito, E. (2011). Displacement Based approach for a robust operational modal analysis. In Sensors, Instrumentation and Special Topics, 6, 187-195. doi:10.1007/978-1-4419-9507-0_19

De Canio, G., de Felice, G., De Santis, S., Giocoli, A., Mongelli, M., Paolacci, F., & Roselli, I. (2016). Passive 3D motion optical data in shaking table tests of a SRG-reinforced masonry wall. *Earthquakes and Structures, 40*(1), 53–71. doi:10.12989/eas.2016.10.1.053

De Canio, G., Mongelli, M., & Roselli, I. (2013). 3D Motion capture application to seismic tests at ENEA Casaccia Research Center: 3D optical motion capture system and DySCo virtual lab. *WIT Transactions on the Built Environment, 134,* 803–814. doi:10.2495/SAFE130711

Doerr, K. U., Hutchinson, T. C., & Kuester, F. (2005). A methodology for image-based tracking of seismic-induced motions. *Proceedings of the Society for Photo-Instrumentation Engineers, 5758,* 321–332. doi:10.1117/12.597679

Feng, D., & Feng, M. Q. (2018). Computer vision for SHM of civil infrastructure: From dynamic response measurement to damage detection – A review. *Engineering Structures, 156,* 105–117. doi:10.1016/j.engstruct.2017.11.018

Fillard, J. P. (1992). Sub-pixel accuracy location estimation from digital signals. *Optical Engineering (Redondo Beach, Calif.), 31*(11), 2465. doi:10.1117/12.59956

Fioriti, V., Roselli, I., Tatì, A., & De Canio, G. (2017). Historic masonry monitoring by motion magnification analysis. *WIT Transactions on Ecology and the Environment, 223,* 367–375. doi:10.2495/SC170321

Hutchinson, T. C., & Kuester, F. (2004). Monitoring global earthquake-induced demands using vision-based sensors. *IEEE Transactions on Instrumentation and Measurement, 53*(1), 31–36. doi:10.1109/TIM.2003.821481

Lee, J., Lee, K. C., Cho, S., & Sim, S. H. (2017). Computer Vision-Based Structural Displacement Measurement Robust to Light-Induced Image Degradation for In-Service Bridges. *Sensors (Basel), 17*(10), 2317. doi:10.339017102317 PMID:29019950

Lunghi, F., Pavese, A., Peloso, S., Lanese, I., & Silvestri, D. (2012). Computer Vision System for Monitoring in Dynamic Structural Testing. *Geotechnical, Geological, and Earthquake Engineering, 22,* 159–176. doi:10.1007/978-94-007-1977-4_9

Lyness, J. N., & Moler, C. B. (1967). Numerical Differentiation of Analytic Functions. *SIAM Journal on Numerical Analysis, 4*(2), 202–210. doi:10.1137/0704019

Mangini, F., Dinia, L., Frezza, F., Beccarini, A., Del Muto, M., Federici, E., . . . Segneri, A. (2017). New crack measurement methodology: tag recognition. In *Proceedings of IMEKO International Conference on Metrology for Archaeology and Cultural Heritage* (pp. 433-436). Lecce, Italy: Academic Press.

Micheletti, N., Chandler, J. H., & Lane, S. N. (2015). *Structure from motion (SFM) photogrammetry. In Geomorphological Techniques*. London, UK: British Society for Geomorphology.

Mongelli, M., De Canio, G., Roselli, I., Malena, M., Nacuzi, A., & de Felice, G. (2017). 3D photogrammetric reconstruction by drone scanning for FE analysis and crack pattern mapping of the "Bridge of the Towers", Spoleto. *Key Engineering Materials, 747*, 423–430. doi:10.4028/www.scientific.net/KEM.747.423

Mongelli, M., Roselli, I., De Canio, G., & Ambrosino, F. (in press). Quasi real-time FEM calibration by 3D displacement measurements of large shaking table tests using HPC resources. *Advances in Engineering Software*.

Morgenthal, G., Hallermann, N., & Hallerman, N. (2014). Quality Assessment of Unmanned Aerial Vehicle (UAV) Based Visual Inspection of Structures. *Advances in Structural Engineering, 17*(3), 289–302. doi:10.1260/1369-4332.17.3.289

Pan, B., Qian, K., Xie, H., & Asundi, A. (2009). Two-dimensional digital image correlation for in-plane displacement and strain measurement: A review. *Measurement Science & Technology, 20*(6), 1–17. doi:10.1088/0957-0233/20/6/062001 PMID:20463843

Ponti, G., Palombi, F., Abate, D., Ambrosino, F., Aprea, G., Bastianelli, T., . . . Vita, A. (2014). The role of medium size facilities in the HPC ecosystem: the case of the new CRESCO4 cluster integrated in the ENEAGRID infrastructure. In *Proceedings of the 2014 International Conference on High Performance Computing and Simulation, HPCS 2014*, (pp. 1030-1033). Bologna, Italy: Academic Press. 10.1109/HPCSim.2014.6903807

Remondino, F., Barazzetti, L., Nex, F., Scaioni, M., & Sarazzi, D. (2011). UAV photogrammetry for mapping and 3D modeling - Current status and future perspectives. *Int. Archives of Photogrammetry, Remote Sensing and Spatial Information Sciences, 38*(1/C22).

Roselli, I., Mencuccini, G., Mongelli, M., Beone, F., De Canio, G., Di Biagio, F., & Rocchi, A. (2010). The DySCo virtual lab for seismic and vibration tests at the ENEA Casaccia Research Center. *Proceedings of the 14th European Conference On Earthquake Engineering (14ECEE)*.

Roselli, I., Paolini, D., Mongelli, M., De Canio, G., & de Felice, G. (2017). Processing of 3D optical motion data of shaking table tests: filtering optimization and modal analysis. In *Proceedings of the 6th International Conference on Computational Methods in Structural Dynamics and Earthquake Engineering (COMPDYN)* (*vol. 2*, pp. 4174-4183). Rhodes, Greece: Academic Press. 10.7712/120117.5714.18115

Savitzky, A., & Golay, M. J. E. (1964). Smoothing and Differentiation of Data by Simplified Least Squares Procedures. *Analytical Chemistry, 36*(8), 1627–1639. doi:10.1021/ac60214a047

Sengupta, P., & Li, B. (2014). Hysteresis behavior of reinforced concrete walls. *Journal of Structural Engineering, 140*(7), 1–18. doi:10.1061/(ASCE)ST.1943-541X.0000927

Tekieli, M., De Santis, S., de Felice, G., Kwiecień, A., & Roscini, F. (2017). Application of Digital Image Correlation to composite reinforcements testing. *Composite Structures*, *160*, 670–688. doi:10.1016/j.compstruct.2016.10.096

Wadhwa, N., Chen, J. G., Sellon, J. B., Wei, D., Rubinstein, M., Ghaffari, R., ... Freeman, W. T. (2017a). Motion microscopy for visualizing and quantifying small motions. *Proceedings of the National Academy of Sciences of the United States of America*, *114*(44), 11639–11644. doi:10.1073/pnas.1703715114 PMID:29078275

Wadhwa, N., Wu, H., Davis, A., Rubinstein, M., Shih, E., Mysore, G. J., ... Durand, F. (2017b). Eulerian video magnification and analysis. *Communications of the ACM*, *60*(1), 87–95. doi:10.1145/3015573

Wahbeh, A. M., Caffrey, J. P., & Masri, S. F. (2003). A vision-based approach for the direct measurement of displacements in vibrating systems. *Smart Materials and Structures*, *12*(5), 785–794. doi:10.1088/0964-1726/12/5/016

Ye, X. W., Ni, Y. Q., Wai, T. T., Wong, K. Y., Zhang, X. M., & Xu, F. (2013). A vision-based system for dynamic displacement measurement of long-span bridges: Algorithm and verification. *Smart Structures and Systems*, *12*(3-4), 363–379. doi:10.12989ss.2013.12.3_4.363

Chapter 11
Methods to Reduce the Optical Noise in a Real-World Environment of an Optical Scanning System for Structural Health Monitoring

Jesus E. Miranda-Vega
Universidad Autónoma de Baja California, Mexico

Oleg Sergiyenko
Universidad Autónoma de Baja California, Mexico

Moises Rivas-Lopez
Universidad Autónoma de Baja California, Mexico

Julio Cesar Rodríguez-Quiñonez
Universidad Autónoma de Baja California, Mexico

Wendy Flores-Fuentes
Universidad Autónoma de Baja California, Mexico

Lars Lindner
Universidad Autónoma de Baja California, Mexico

ABSTRACT

This chapter describes different methods and devices that can be used in optical scanning systems (OSS), especially applied to structural health and monitoring (SHM) in order to reduce the interference and losing of resolution in the measurements of the displacements and coordinates calculated by the OSS of a specific structure to be monitored. The principal parts of the OSS are a photo-detector, non-rotating emitter source of light, a DC electrical motor, lens, and mirror. All the measurements and experiments have been realized in a controlled environmental; the optical noise was simulated with a similar intensity than the intensity of the reference signal of the emitter source. Applying analogue filters has disadvantages because part of signal with important information for the performance of the system is removed, but particularly the components will often be too costly. However, there are digital filters and techniques of computational statistics that can solve these problems.

DOI: 10.4018/978-1-5225-5751-7.ch011

INTRODUCTION

Existing research in 3D & 2D machine vision technologies, medical scanning, and optoelectronic sensors represent a challenge, particularly in industrial applications of structural health monitoring (SHM) due to the environmental conditions. SHM methodology has received considerable attention in the technical literature, where there has been a concerted effort to develop a firm mathematical and physical foundation for this technology (Sohn, Hoon, Czarnecki, Jerry & Farrar, 2000). Optical scanning systems (OSS) allow monitoring and extracting patterns of structures. These systems consist principally of a set of elements as an optoelectronic sensor, lens, mirror and non-rotating incoherent light emitter; however, these systems and structures to be monitored are subject to the environmental conditions that can affect the measured signals. The basic premise of the OSS is to characterize the normal conditions of a structure, however, in occasional cases the system and structure are exposed to excessive bright sunlight that might lead to cause problems with measured data, because the optoelectronic sensor and incoherent light are in the same spectral ranges of environmental conditions like bright sunlight.

The principal environmental condition factor is the bright sunlight that affects the optical system due to the system could detect two or more patterns at the same time. One pattern belongs to the reference source of a light mounted on the structure and the second pattern is caused by other optical radiations (ultraviolet, visible light, infrared). In order to ensure accuracy and precision of the measurements, it is very important to filter the undesired signals and noise the best it could be. Likewise, the system should also include some filters to distinguish the noise of the reference source. These filters can be optical filters, digital and analog filters. Machine learning technics and statistical pattern recognition methodologies could also be applied to establish difference with a specific pattern from a structure to being monitored. Statistical pattern recognition methodologies have gained considerable attention for SHM applications to detect changes in a structure (Gul & Necati Catbas, 2009).

The bright of sunlight provoke an interference with the wavelength of the emitter source and the phenomena as reflection, diffraction, absorption, and refraction (Fischer, 2008). The signal generated by the optoelectronic sensor inside of Scanning Aperture (SA) is similar in shape to the Gaussian curve, when the SA system is exposed to the sunlight, the Gaussian curve changes dramatically affecting the signal energy center, which is the reference to measure the position of the incoherent light source mounted on the structure. If signal energy center of incoherent light is changing by environmental conditions factors, the coordinates measured using the SA system will also change, in other words the signal energy center is the reference to calculate the spatial coordinates, in this way it is detected a light emitter mounted on the infrastructure being monitored (Flores-Fuentes et al., 2016). In preliminary experimentations, it has been achieved satisfactory results with this method. It is possible to increase the resolution by decreasing the rate of scanning, and however, this creates sources of error that can be minimized with an appropriate method for the measurement *(Sergiyenko, Tyrsa, Hernandez, Starostenko, & Rivas, 2009)*. Despite good results in the laboratory, when the experiments were carried out in the exterior, the measurements were affected by environmental noise such as bright of sunlight, dust and high temperatures. The present chapter will provide a general overview of the methods and devices currently used in optical scanning system to minimize the effects of environmental conditions caused primarily by excessive exposure to the bright sunlight. Furthermore, it will be shown the results of a method to discriminate the classes generated by reference source and the environmental noise by using computational statistics and digital filters. These methods have been proved in optical scanning system with satisfactory results.

METHODS AND DEVICES TO REDUCE THE NOISE FROM ENVIRONMENTAL AND OPERATIONAL CONDITIONS

Nowadays, there are several devices and methods to reduce the noise from environmental and operational conditions such as fiber optic sensors, optical filters, computational statistics, digital and analog filters. These methods and devices improve the performance of the optical monitoring systems. However, applying these methods implies a cost, in this context, the correct method or device should be chosen by the designer. Each method or device applied has its advantages and disadvantages. Moreover, the need to reach a specific objective could be a determinant to select it. When it would be necessary to establish a connection between two points to send and receive information, the best device to reduce the optical noise is the optical fiber. In the case where the system only requires receiving the same pattern, optical scanning system could be chosen to execute this task. The best performance of the system to reduce de optical noise can be provided by a combination between a device and method selected.

Optic Fiber Sensors

The fiber optic is a material widely used in the area of communications, nowadays, is used as a sensor or medium to transmit information from different sources through one channel. This information is often carried by an electromagnetic carrier wave whose frequency can vary from a few megahertz to several hundred terahertz. A general optical fiber communication system consists of following:

1. **Transmitter:** The signal should be converted from a non-electrical into an electrical signal, and afterward this signal is modulated by a laser diode (coherent light system) or light emitting diode (incoherent light system).
2. **Information Channel:** The path between the transmitter and receiver, in other words, the channel is a glass or plastic fiber, and finally.
3. **The Information Receiver:** The signal or information is detected by an optical detector, in this step the optic wave is converted into an electric current by a photodetector and subsequent signal processing, such as filtering and amplification.

However, the fiber optic needs a protection to resist moisture and sunlight exposure. In order to protect optical fibers from the environmental conditions, it is necessary to incorporate them in some form of cable structure; the structure of the cable will largely depend on the type of installation (aerial, underground duct, mini-trench, buried underground or submarine, etc.) and other environmental conditions. The environmental conditions can be of two types, e.g., natural external factors such as temperature variations, wind pressure, water contaminations, and man-made factors such as smoke, air pollution, fire, etc. (Chakrabarti, 1998).

The fiber optic is fully immune to all types of external electromagnetic interference. Systems with fiber optic sensors FBG (Fiber Bragg Grating) are consolidated in the Structural Health Monitoring. In McCoy, Thomsen, Ibsen, and Richardson (2004) is presented an experimental investigation about the effects of filtering the noise in a spectrum-sliced incoherent light system, incorporating semiconductor optical amplifier (SOA) based noise reduction. SOA is a device formed by materials semiconductors that are able to amplify the light in a coherent way by stimulated emission.

Advantages of Optical Fiber

- Size and weight
- Large capacity to transmit large amounts of information.
- Is not affected by electronic interference.

Disadvantages of Optical Fiber

- Electrical to optical conversion.
- Needs special installation.
- It is difficult to repair a fiber optic cable damaged.
- Chromatic dispersion.

Optical Filters

The applications of optical filters are widely used in the areas like astronomy, spectroscopy, machine vision, optical communications and any other dealing with light. An optical filter is a medium that allows light passing through it which reject a specific wavelength (called color filters) or range wavelengths, these filters can attenuate the light by absorption or reflection, one example of an optical filter is the solar-blind that reject a specific wavelength, this kind of filter rejects visible and near-infrared (IR) light while transmitting only a passband of Ultraviolet (UV) light. The optical filter can be categorized either according to their construction (Interference Filters) or according to their spectral transmission range (Heat Filters, Neutral density filters (NDF)) (Kuzmany, 2009). The typical optical filters are the neutral density filters, polarized filters.

Neutral Density Filter

Reduce the intensity of the incident beam of light evenly over a broad spectral range (Burke, 2012), usually, this performance is defined in terms of the optical density D, shown in Equation (1):

$$D = \log_{10} \left(I_0 \middle/ I_T \right) \tag{1}$$

Where, I_0 .is the incident intensity, and I_T .is the transmitted intensity.

Advantages of Neutral Density Filters

- The decrease in the amount of light of a source.

Disadvantages of Neutral Density Filters

- High cost.
- Are fragile and can easily be damaged during use.

Polarized Filters

This kind of optical filters accept only polarized light and is constituted by a pair of orthogonal vector $\vec{E}, \vec{H}$ .oscillating in a direction perpendicular to propagation. The most common source of polarized light is a laser, when polarized filter is used a filter absorbs a percentage of the incoming light. When a perfect polarizer is placed in a polarized beam of light, the intensity, I. of the light that passes through is given by Malu's law, see Equation (2).

$$I = I_0 \cos^2 \theta \tag{2}$$

Where, I_0 .is the initial intensity, and θ .is the angle between the light's initial plane of polarization and the axis of polarizer.

Advantages of Polarizing Filters

- Reduce the effect of reflections off.
- Polarizer filters are characterized for accepting only polarized light.

Disadvantages of Polarizing Filters

- The polarizing filters are not cheap.
- Reduce the amount of light passing through the optical system (lens).

Optoelectronic Devices

The optoelectronic devices can provide a natural filter due that only operate with a range or specific wavelength of the light. Photodiodes and photoresistors such as a light-dependent resistor (LDR) are an example of photodetectors (Sergiyenko & Rodriguez-Quiñonez, 2016). The sensitivity of an LDR depends on a wavelength of the light (Regtien, van der Heijden, Korsten, & Otthius, 2004). LDR's can be used for detecting light of a source that is present for the scanning system, for example in Figure 1, the relative spectral sensitivity for LDR is shown, it is clear that the typical LDR detects with very low sensitivity the infrared waves.

The relative response of LDR 1 is shown in Table 1, in both cases, the LDR responds to the wavelength in a similar way to the human eye, below of 450nm and above 800nm the sensitivity of the LDR is very low. According to the Table 1, the LDR typical can discriminate infrared.

Digital Filters

Digital filters are based on analog filters and are the main part of digital signal processing (DSP), these filters are used in signals separation and restoration applications. This kind of filters have been used for optical communication system to mitigate the chromatic dispersion (Savory, 2008), where the signal transmitted on the optical fiber presents this problem, for example, when an optical pulse is transmitted through the fiber, the propagation velocity of every wave is different according to wavelength, there-

Figure 1. Relative spectral sensitivity for two types of CdS resistors (image from Measurement Science for Engineers)

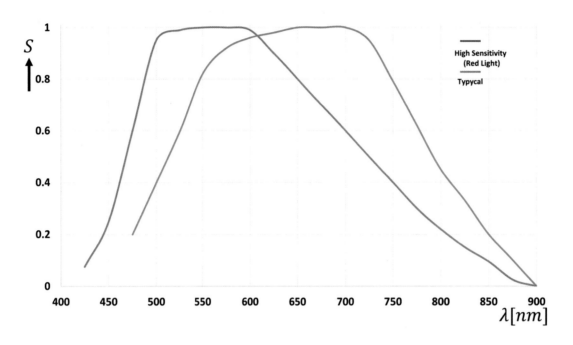

Table 1. Relative spectral sensitivity for two types of CdS resistors

Wavelength (in nm)	LDR Typical Relative Response of LDR (%)	LDR With High Sensitivity From Red Light Relative Response of LDR (%)
480	40	20
500	90	40
550	100	80
600	99	95
650	80	100
700	60	99
750	40	80
800	20	40

fore this will cause delay of the signal due to the different refraction for each wavelength on the signal. There are two types of digital filters: the finite impulse response digital filters (FIR) and infinite impulse response digital filters (IIR). However, the form to implement these filters can raise questions such as: What type of filter it should be used? the answer will depend on the specifications of the system. The principal characteristic of the OSS of this chapter is that measurements of the angles are taken according to the time where the peak of the signal is sensed, in other words, the response of the filter should be linear phase. The principal parts of digital filter FIR and IIR such as filter specification and methods to design them are described in detail below.

Filter Specification

The frequency response of an ideal low-pass filter with linear phase and a cutoff frequency ω_c .is shown in Equation (3).

$$h_d\left(e^{j\omega}, \omega_p\right) = \begin{cases} e^{-j\omega}, & |\omega| \le \omega_C \\ 0, & \omega_C < |\omega| \le \pi \end{cases} \tag{3}$$

$$h[n] = \frac{1}{2\pi} \int_{-\pi}^{\pi} H\left(e^{jw}\right) e^{jwn} dw \tag{4}$$

Considering an ideal low pass filter showed in Figure 2, the Equation (4) can be represented by Equation (5) in the following way, the passband cutoff frequency ω_p . the stopband cutoff frequency ω_s . the passband deviation δ_p . and the stopband deviation δ_s .(Hayes, 1999).

$$h[n] = \frac{1}{2\pi} \int_{-w_c}^{w_c} e^{jwn} dw \tag{5}$$

Its impulse response sequence is expressed in Equation (6).

$$h[n] = \frac{sin\left(nw_c\right)}{n\pi} \tag{6}$$

Figure 2. Filter specification

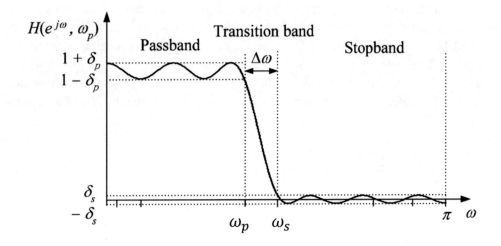

Where *h[0]* is computed by Equation (7).

$$h\left[0\right] = \frac{\dfrac{d}{dn}\,sin\left(nw_c\right)}{\dfrac{d}{dn}\,n\pi}\Bigg|_{n=0} = \frac{w_c}{\pi} \tag{7}$$

The Filter FIR

The filter FIR is characterized by finite-duration impulse response and has a linear-phase response. This kind of filter can be expressed of the form.

$$H\left(z\right) = b_0 + b_1 z^{-1} + ... + b_{M-1} z^{1-M} = \sum_{n=0}^{M-1} b_n z^{-n} \tag{8}$$

Hence the impulse response $h\left(n\right)$ .is

$$h\left(n\right) = \begin{cases} b_{n,} & 0 \leq n \leq M-1 \\ 0, & else \end{cases} \tag{9}$$

The difference equation representation is given by

$$y\left(n\right) = b_0 x\left(n\right) + b_1 x\left(n-1\right) + ... + b_{M-1} x\left(n-M+1\right) \tag{10}$$

Equation (10) corresponds to a linear convolution of finite support. There are different methods at a time to design FIR filter such as windowing method that is a widely used by designers in DSP, however, there is another method for the design digital FIR filters called frequency sampling method. The last popular methods mentioned before for the design of the digital FIR filters are windowing method and frequency sampling method (Sharma, Katiyal, & Arya, 2015). As a matter of fact, there are many window techniques available for designing the FIR filter and they are: Rectangular window, Hamming window, Blackman window, Triangular window or Bartlett window, Hanning window, among other windows (Tan & Jiang, 2007), see Table 2 (Yu, Ching Lim, & Shi, 2009). When it is applied a hamming window *w[n]* the ideal low pass FIR filter coefficients can be calculated and represented by *Hd[n]* in Equation (11):

$$h_d\left[n\right] = w\left[n\right]h\left[n\right] = \left[0.54 - 0.46\,cos\left(\frac{2\pi n}{N-1}\right)\right]\frac{sin\left[w_c\left(n-M\right)\right]}{\pi\left(n-M\right)} \tag{11}$$

Table 2. Type of window and windows functions

Type of Window	Window Function $w[n]$
Rectangular	1
Hamming	$0.54 - 0.46\cos\left(\dfrac{2\pi n}{N-1}\right)$
Blackman	$0.42 - 0.5\cos\left(\dfrac{2\pi n}{N-1}\right) + 0.08\cos\left(\dfrac{4\pi n}{N-1}\right)$
Hanning	$0.5 + 05\cos\left(\dfrac{2\pi n}{N-1}\right)$

The Filter IIR

The filter IIR is characterized by infinite-duration impulse responses. The system function of an IIR filter is given by Equation (12).

$$H(z) = \frac{B(z)}{A(z)} = \frac{\sum_{n=0}^{M} b_n z^{-1}}{\sum_{n=0}^{N} a_n z^{-1}} = \frac{b_0 + b_1 z^{-1} + \ldots + b_M z^{-M}}{1 + a_1 z^{-1} + \ldots + a_N z^{-N}} ; a_0 = 1 \qquad (12)$$

Where, b_n .and a_n .are the coefficients of the filter. It has assumed without loss of generality that $a_0 = 1$.

The difference equation representation of an IIR filter is expressed as

$$y(n) = \sum_{n=0}^{M} b_m x(n-m) - \sum_{m=1}^{N} a_m y(n-m) \qquad (13)$$

There are several methods available for the design of IIR type digital filters, such as the impulse invariance method, bilinear transform method, analog approximation method, and so on (Ibrahim, 2013). Table 3 shows the descriptions of these methods which can be designed in MATLAB (The MathWorks, 2018). In recent years, optimization methods have been intensively investigated for digital IIR filter design problem (Lu & Antoniou, 2000).

For example which converts an analog filter to a digital filter is the bilinear transformation and the sequence of steps to design it are as follows:

Step 1: Determine the Butterworth polynomial for the required filter from the following Table 4.

Table 3. Type of window and windows functions.

Filter Method	Description
Analog Prototyping	Using the poles and zeros of a classical lowpass prototype filter in the continuous (Laplace) domain, obtain a digital filter through frequency transformation and filter discretization.
Direct Design	Design digital filter directly in the discrete time-domain by approximating a piecewise linear magnitude response.
Generalized Butterworth Design	Design low pass Butterworth filters with more zeros than poles.
Parametric Modeling	Find a digital filter that approximates a prescribed time or frequency domain response.

Table 4. Type of window and windows functions

n	Butterworth Polynomial in Factored Form
1	$\left(s+1\right)$
2	$\left(s^2 + \sqrt{2}s + 1\right)$
3	$\left(s+1\right)\left(s^2 + s + 1\right)$
4	$\left(s^2 + 0.7654s + 1\right)\left(s^2 + 1.8478s + 1\right)$
5	$\left(s+1\right)\left(s^2 + 0.6180s + 1\right)\left(s^2 + 1.6180s + 1\right)$
6	$\left(s^2 + 0.5176s + 1\right)\left(s^2 + \sqrt{2}s + 1\right)\left(s^2 + 1.9319s + 1\right)$
7	$\left(s+1\right)\left(s^2 + 0.4450s + 1\right)\left(s^2 + 1.2470s + 1\right)\left(s^2 + 1.8019s + 1\right)$
8	$\left(s^2 + 0.3902s + 1\right)\left(s^2 + 1.1111s + 1\right)\left(s^2 + 1.6629s + 1\right)\left(s^2 + 1.9616s + 1\right)$

Step 2: Calculate the normalized cut-off frequency of the digital filter.

$$\omega_{dc} = \frac{2\pi f_c}{f_s} \tag{14}$$

Step 3: Calculate the normalized cut-off frequency of the digital filter.

$$\omega_{ac} = tan\left(\frac{\omega_{dc}}{2}\right) \tag{15}$$

Step 4: De-normalizing the analog filter transfer function by replacing s with $\dfrac{s}{\omega_{ac}}$.

Step 5: Convert the analog filter into a digital filter using the bilinear transformation.

$$s = \frac{z-1}{z+1} \tag{16}$$

Step 6: Convert z^k into z^{-k} in the polynomial.
Step 7: Expand the polynomial to obtain the filter coefficients by comparing with the next Equation (17).

$$H\left(z\right) = \frac{b\left(0\right) + b\left(1\right)z^{-1} + b\left(2\right)z^{-2} + \ldots + b\left(M\right)z^{-M}}{a\left(0\right) + a\left(1\right)z^{-1} + a\left(2\right)z^{-2} + \ldots + a\left(N\right)z^{-N}} \tag{17}$$

Digital Filter Structures

The reason to apply a digital filter is that signal captured by OSS system is susceptible to electrical noise and this causes errors during the measurements of the systems. However, in practice, there are several FIR structures and the most commonly used ones are the direct form and cascade form. The structure of the filter is the most important part in order to implement the digital filter and this is an important concern because different filter structures dictate different design techniques (Ingle & Proakis, 2011). The Figure 3, shows the direct form structure and the difference equation is represented Equation (18).

$$\left(n\right) = b_{0}x\left(n\right) + b_{1}x\left(n-1\right) + b_{2}x\left(n-2\right) + b_{3}x\left(n-3\right) + b_{4}x\left(n-4\right) + b_{5}x\left(n-5\right) \tag{18}$$

Moving Average

In the context of digitals filters for smooth viewing, the average filter is widely used in DSP, because it is the easiest digital filter to understand, it is used to reduce a random noise in the signal (Smith, 1997), analyze data, and identify the direction of the trend. This kind of filter operates by averaging a number of points from the input signal to produce each point in the output signal. The Equation (19) expresses the moving average filter:

Figure 3. Direct form FIR structure

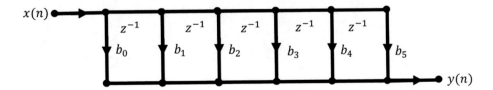

$$y[n] = \frac{1}{M} \sum_{j=0}^{M-1} x[i+j]$$ (19)

Where $x[\ ]$ is the input signal, $y[\ ]$ is the output signal, M denotes the number of points or samples in the average, in other words is, the size of window. The moving average filter applied to a signal with noise is illustrated in Figure 4, where blue line represents the original signal and red line the signal filtered.

The moving average filter has been used in an optical communication system (Luo, Zhang, Liu, Han, & Li, 2012) where approach the short-range of ultraviolet and strong absorption of high altitude ozone that reduces the background radiation. The results improved the system performance. Another digital filter that is used in optical scanning system is the median filter in Ho Ling and Ming Kai (2011) this work is based on laser technology in order to get information like coordinates of 3D measurement where the radiation of the laser is projected on the detection of objects, and use the CCD of sensitive, different positions and then the analysis image algorithms.

The next function can be implemented in MATLAB to filter and smooth the signal with noise, this code returns in *Xvect* variable a signal filtered, only the column vector *signal X* is needed, *w* denotes the size of the window and method is for choosing whether it will calculate with 'mean' or 'median'.

Figure 4. Signal smoothing using moving average

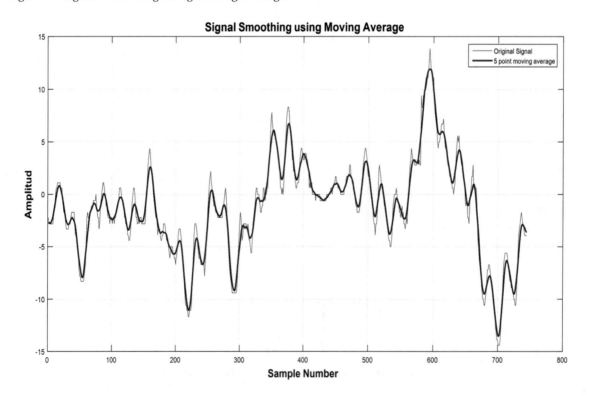

Computer Code

```
function [Xvect]= filtermu_median(signalX,w,method)
clc
w1=[];%extrem left of the window
w2=[];%extrem right of the window
Xvect=[];
X= signalX;
% w = corresponds the size of window
% method = 'mean'; calculates the mean
% method = 'median'; calculates the median
for i=1:length(X)
    w1=i-floor(w/2);
    w2=w1+w-1;
    if w1<1
        w1=1;
    end
    if w2>length(X)
        w2=length(X);
    end
    switch method
    case 'mean'
        temp=mean(X(w1:w2));
        Xvect=horzcat(Xvect,temp);
    case 'median'
        temp=median(X(w1:w2));
        Xvect=horzcat(Xvect,temp);
    otherwise
        disp('only, the value: mean or median');
    end
end
end
```

In some contexts the filters are called spectrum analyzers, this kind of filters are designed for discrete systems in both the time and frequency domains. Structurally, a filter can be classified into two categories – recursive or non-recursive type (Lewis, Lakshmivarahan, & Dhall, 2006). These are often referred to as infinite impulse response (IIR) filters and finite impulse response (FIR) filters, respectively.

Advantages of Digital Filters

- Linear-phase response (FIR).
- Easy to implement, similar to IIR.
- Digital filter can make practical many designs.
- Specification of the filter can be modified by software.

Disadvantages of Digital Filters

- Digital filters are more expensive than analog filters.
- Have problematic latency.
- Requires advanced knowledge of programming.
- The number of coefficients requires more memory (FIR).
- High filter order (FIR).

Computational Statistics

Computational Statistics is a collection of techniques for statistical pattern recognition, probability density estimation, Monte Carlo method and others, in order to analyze data by using computer algorithms. Now and days, computational statistics allows to process massive amounts of data and deal with data sets with very large and high-dimensional. Now and days SHM has many techniques and methods to ascertain if the damage is present or not, and one of them is statistical pattern recognition, also used in the areas of statics, engineering, artificial intelligence, computer science, speech recognition, and in any other applications. The goal of pattern recognition is to clarify the decision-making process to discrimination and classification on a real-world problem, in other words, is for building classifiers (Fukunaga, 2013). According to the complex nature of the SHM, the properties of a system and measurements can be recorded to extract the features. The pattern is represented by p–dimensional data vector $x = (x_1, \ldots, x_p)^T$ obtained of the properties of the structure monitored. According to Sohn et al. (2003), the statistical pattern recognition paradigm for SHM is addressed to four topics:

1. Operational Evaluation.
2. Data Acquisition and Cleansing.
3. Feature Extraction.
4. Statistical Modeling for Feature Discrimination.

The present chapter is focused on topics 2, 3 and 4. Before to build a model to distinguish the noise of the reference source, it is necessary to follow the previously mentioned, following the same procedure from topic 2 in *Data Acquisition and Cleansing* the data should be transformed into forms which are appropriate for creating a model to classify.

In the next step in the *Feature Extraction,* the features will be selected giving a better description of the particular pattern. In some cases, raw data contains more features than necessary, feature construction and selections simplify the problem to analyze. For this reason, the pattern generated by the incoherent light emitter and environmental conditions has been transformed to be presented in a convenient form, these data had to be re-expressed to produce effective visualization or, more informative analysis.

Statistical Modeling for Feature Discrimination is the final step of the procedure, at this point the data have been transformed, and features were extracted, and it can already create the model, one of the methods used was Linear Discriminant Classifier (LDC), this method was proposed as a discriminant functions, although Gaussian Mixture Model is also used to make it possible to make reliable results comparisons with each model.

A discriminant function which relates numerically an instance to class C_i in particular was used. This function uses the criterion of maximization of the distance between two groups. There are as many possible discriminant functions as there are classes in a problem analyzed. The classifier based on functions assigns to object X, to class i if d_i if the largest distance. A Gaussian Mixture Model (GMM) has been proposed in SHM to model the vibration signal (Nair & Kiremidjian, 2007). GMMs cluster by assigning query data points to the multivariate normal components that maximize the component posterior probability given the data. That is, given a fitted GMM, the constructed cluster from Gaussian mixture distribution assigns query data to the component yielding the highest posterior probability (The MathWorks, 2017). In this chapter is show novel methods in vision-based theories and applications, 3D & 2D technologies and practical applications for structural health monitoring.

Active and Passive Methods

The optical scanning systems can work with a source that emits energy radiant to a specific object and then the system OSS has the task to collect the data and process them, this method is called *active method* due to the system employs its own source of radiation in order to send and receive the light reflected by the object monitored. There is another method that considers to the sun as a source of energy, this method is used to detect the sun's energy reflected from the specific object. These methods have been used mainly in remote sensing (Sharkov, 2003; Sergiyenko et al., 2009).

Advantages of Active Methods

- The ability to obtain measurements anytime.
- The system can control its own source of radiation.

Disadvantages of Active Method

- The distance between the system and the object increment the cost of the source light.
- Susceptible to environmental conditions.

Advantages of Passive Method

- The cost of the system is cheaper than active method.
- It is possible to use a source of radiation situated in a structural to monitor.

Disadvantages of Passive Method

- The intensity of the sunlight reflected from the object is low.
- The distance between OSS and the radiation source mounted on the object may increment the cost.
- Environmental conditions can cause conflict with the sensors used of OSS.

DATA ACQUISITION AND CLEANSING

The acquisition of the signal is the main process that ensures the accuracy of the SA system because with this information, the SHM system determines if there is a damage or not, each SA system has two sensors to capture the electrical signals from the reference source that is mounted on a specific point of the structure under monitoring. The first sensor of the scanning apertures consist of a sensor as a photodiode to convert the optical signal reference in the electric signal, this signal is acquired in one digital channel of Data acquisition system (DAQ). The second sensor measures the revolution of the (Direct Current) DC motor or the angular frequency, where each revolution of this motor is counted. With this kind of sensors, two pulses per revolution are acquired, processed by a DAQ as a digital input. The sensor mentioned before consist of a packaging system as an infrared emitting diode coupled with a silicon phototransistor in a plastic housing.

In this chapter, the signals were acquired at 5000 sampling rate per channel by a DAQ. The data processing has been worked with DAQ USB-6003, this device provides eight single-ended analog input (AI) channels with resolution of 16-Bit and allows 100 kS/s, two analog output (AO) channels at 5 kS/s/ch, 13 DIO channels, and a 32-bit counter with a full-speed USB interface. It was used only two analog inputs, the first was used to measure the angular frequency of motor DC, and the second was used to acquire the signal of radiation of the incoherent light. The operation of the OSS can be appreciated in Figure 5, in a) the mirror has already reflected the reference source of incoherent light to the photodiode where gains in voltage is changing in time, when the mirror is situated in front of source a maximum voltage is detected by the photodiode, this can be showed in b), finally in c) the gain in voltage by the photodiode has decremented because mirror has moved away from the reference source of light. With this approach, the light is converted to an electrical signal, note that the shape is similar to a Gaussian signal, for more details see Rivas Lopez et al., (2010). Depending on the number of light sources the shape of pattern could change, for example in Figure 11 is shown a pattern for two sources. The infrastructure of industrial and public sector need to be properly monitored over time to predict damage or deteriorations in the structure that was subjected to excessive loads. The safety of the people depend on structural health all time, that is why it is so important the data acquisition in real time to study the behavior of the structure, besides is the medium of real operation with the algorithms and methods to determinate if a damage is present.

The OSS system always is exposed to environmental conditions, this causes that could appear some problems with data acquisition such as electrical and optic noise, however, sometimes this cannot be visualized in the time domain. For this reason, all observation have to be transformed in a proper way to visualize better the data set by applying mathematical functions.

All datasets acquired from OSS and presented for this chapter have been used the transformation of data. The first step to transform the data is choosing the range of the variables and apply the Equation (20), the ranges a, b are bounded by zero and one, x represents the observations scanned by photodiode and z is for data transformed.

$$z = a + \frac{\left(x - \min\left(x\right)\right)\left(b - a\right)}{\max\left(x\right) - \min\left(x\right)} \tag{20}$$

Figure 5. Optical scanning system with rotatory mirror

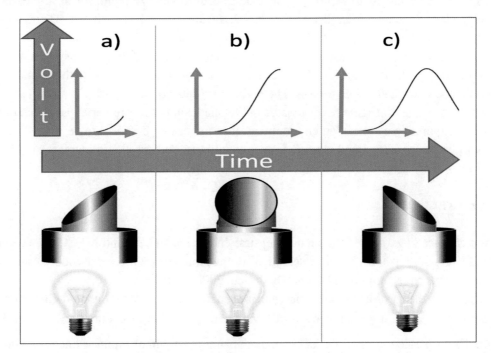

The MATLAB function that expresses equation mentioned before is shown in the following code.

Computer Code

```
function [Normdata] = normtrans(y,a,b)
%y =Signal; typical range a=0; b=1;

Normdata =[];
for i=1:length(y);
    miny=min(y(i,:));
    maxy=max(y(i,:));
    Ytrans=a+((y(i,:)-miny)*(b-a))/(maxy-miny);
    Normdata =vertcat(Normdata,Ytrans);
end

end
```

This function implemented in MATLAB returns in **Normdata** variable all the observations transformed between the two extremes *a* and *b*.

Analyze raw data is not a convenient form because, these data should be re-expressed to produce effective visualization or, more informative analysis, and select the principal features. In Figure 6, Linear predictive coding (LPC) are utilized to express the signal of pattern with one and two sources and have been calculated 3 coefficients to express a convenient form to analyze them, for example, red circles correspond to class 1 and blue triangles are for class 2, the abscise of Y is coefficient 3 and abscise X is for coefficient 2. We can appreciate in this figure that signals are linearly separable.

The features extraction can be used to determine the current state of the health system, for example when OSS is working with one source, LPC coefficients represent a transformation of this signal, moreover, the outliers can be visualized in a better way. The outliers are data points that are statistically inconsistent with the rest of the data. In Figure 6, the class 2 in blue has points that are very apart from others. All datasets were acquired in a controlled environment, for this reason, outliers are very few according to Figure 6 with patter for class 1 and class 2. In real conditions, outliers would increase significantly in the case of feature extraction from the pattern with one source and two sources.

Feature Extraction

Linear predictive coding and Mel-frequency cepstral coefficients (LPC and MFCC) were selected to extract the features of a pattern in this research. These techniques are the most widely used feature in state-of-the-art speech recognition systems (Diederich, 2008).

LPC is based on theory of linear prediction and is used for the representation, modeling compression, and computer generation of speech waveforms (Vaidyanathan, 2008), the output $\hat{x}(n)$ (see Equation (22)) is the linear prediction of $x(n)$, is given as the sum of the input minus a linear combination of

Figure 6. Signals of patterns transformed in LPC coefficients

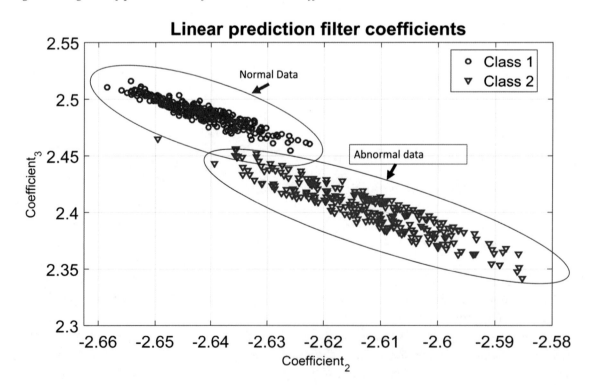

past outputs. The coefficients LPC are calculated of a *pth*- order linear predictor that predicts the current value based of the time series.

$$\hat{x}(n) = -a(2)x(n-1) - a(3)x(n-2) - \ldots - a(p+1)x(n-p)$$

(21)

$$\hat{x}(n) = -\sum_{i=1}^{N} a_i x(n-i)$$

(22)

$$e(n) = x(n) - \hat{x}(n) = x(n) + \sum_{i=1}^{N} a_i x(n-i)$$

(23)

The goal of the LPC analysis is to find the best prediction coefficients a_i which minimize the quadratic error function:

$$E = \sum_{n=0}^{L-1} \left[e(n) \right]^2 = \sum_{n=0}^{L-1} \left[x(n) - \sum_{i=1}^{N} a_i x(n-i) \right]^2$$

(24)

Where, p is the order of the prediction polynomial filter. The Levinson-Durbin block solves the nth-order system of linear equations.

$$Ra = b$$

(25)

Where: R is a Hermitian, positive-definite, Toeplitz matrix. b is identical to the first column of R shifted by one element and with the opposite sign.

$$\begin{bmatrix} r(1) & r^*(2) & \cdots & r^*(n) \\ r(2) & r(1) & \cdots & r^*(n-1) \\ \vdots & \vdots & \ddots & \vdots \\ r(n) & r(n-1) & \cdots & r(1) \end{bmatrix} \begin{bmatrix} a(2) \\ a(3) \\ \vdots \\ a(n+1) \end{bmatrix} = \begin{bmatrix} -r(2) \\ -r(3) \\ \vdots \\ -r(n+1) \end{bmatrix}$$

(26)

There are at least three techniques for calculating the predictor coefficients (Harrington & Cassidy, 2012), from the autocorrelation method, from the covariance method, and from reflection coefficients using a lattice filter. In this work to find the filter coefficients, the autocorrelation method of autoregressive modeling has been used.

One way to compute the Linear Predictor Coefficients is by applying a native function of MATLAB as follows:

Computer Code

```
[a,e]=lpc(x,N)
%LPC   Linear Predictor Coefficients
```

Where x, is a signal and N is Nth order forward linear predictor.Mel-frequency cepstral coefficients (MFCCs) are features that try to simulate the human ear, this acoustic vector characterizes certain features which are unique to a specific individual. MFCC can represent a signal as a sequence of spectral vectors computing mel frequency cepstral coefficient features from a given speech signal, the Mel scale can be calculated by Equation (27):

$$Mel(f) = 2595 \log_{10}\left(1 + \frac{f}{700}\right) \tag{27}$$

Where, $Mel(f)$ is the logarithmic scale of the normal frequency scale f. Mel scale has a constant mel frequency interval, and cover the frequency range of 0 Hz -20050 Hz (Aizawa, Nakamura, & Satoh, 2004), the process of calculating MFCC is illustrated in Figure 7.

The MFCC features are calculated and processed as follows:

Step 1: Pre-emphasis process by using a first order FIR filter.
Step 2: The signals are divided into small frames of N samples.
Step 3: Windowing.
Step 4: Apply the mel bank filter.

Figure 7. Steps to compute the MFCC feature

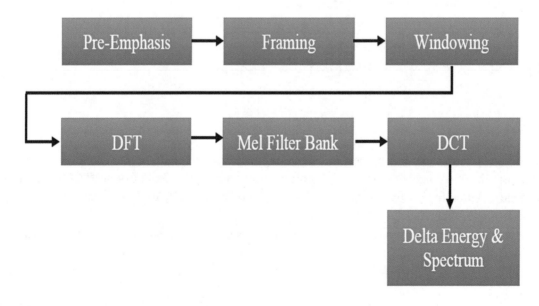

Step 5: Take the logarithm of all filterbank energies.
Step 6: Take the discrete cosine transform (DCT) of the log filterbank energies.
Step 7: Keep DCT coefficients 2-13, discard the rest.

Windowing is the process to window each frame to reduce discontinuities and leakage at the beginning and end of each frame. $W(n)$, $0 \leq n \leq N - 1$.

Where, $W(n) =$ Hamming Window, $N =$ number of samples in each frame, $Y[n] =$ Output signal, and $X(n) =$ Input Signal.

$$Y(n) = X(n)W(n) \tag{28}$$

$$W(n) = 0.54 - 0.46 \cos\left(2\pi \frac{n}{N-1}\right), \; 0 \leq n \leq N - 1 \tag{29}$$

Take the DCT of the log filterbank energies.

$$S_i(k) = \sum_{n=1}^{N} s_i(n) h(n) e^{-j2\pi kn/N} \quad 1 \leq k \leq K \tag{30}$$

In order to calculate the MFCC vectors of the signal a speech processing toolbox can be used and downloading from Brookes (1997), this useful toolbox is called VOICEBOX and consists of MATLAB routines.

STATISTICAL MODELING FOR FEATURE DISCRIMINATION

Statistical modeling allows us to know and predicts the behavior of a process or phenomenon in nature.

A Gaussian Mixture Model is characterized by the mixture component and its means, variances's/ covariance's and each case is represented by a GMM model (λ). The probability that $\vec{x}$ given λ density is a weighted sum of K mixture component weights, as depicted by the following Equation (31):

$$p(\vec{x} \mid \lambda) = \sum_{i=1}^{K} m_i b_i(\vec{x}) \tag{31}$$

Where $\vec{x}$ is a D-dimensional random vector (LPC, MFCC), and the mixture component weights are defined as: m_i and $i = 1, ..., K$ are the component densities, and, $i = 1, ..., K$, are the mixture weights. Each component density is a D-variate Gaussian function as follows:

$$b_i(x) = \frac{1}{\sqrt{(2\pi)^K |\Sigma_i|}} \, exp\left(-\frac{1}{2}(\vec{x} - \vec{\mu}_i)^T \Sigma_i^{-1}(\vec{x} - \vec{\mu}_i)\right) \tag{32}$$

Here, $\vec{\mu}$ is the mean vector and Σ_i is the covariance matrix. On the other hand, the mixture weights satisfy the constraint that $\sum_{i=1}^{K} m_i = 1$. The whole GMM model is parameterized by the mean vectors, covariance matrices and mixture weights for Gaussian densities.

$$\lambda = \left\{ \mathrm{m}_i, \vec{\mu}, \Sigma_i \right\}_i \; \forall i = 1, \dots, \mathrm{K} \tag{33}$$

Linear Discriminant Analysis is a discriminant function relates numerically an instance to class C_i in particular. There are as many possible discriminant functions as there are classes in a problem analyzed. The classifier based on functions assigns to object. X, to class i if:

$$d_j(X) > d_i(X) \; \forall_i = 1 \cdots l; \neq j \tag{34}$$

The algorithm computes the values of every discriminant functions and then select it the class C_j as true prediction to which the highest score $d_j(X)$, that is the probability of C_i given an object X is expressed by next equation (35), where $P(X|C_i)$ is the class-conditional probability density function for X given class C_i and $P(C_i)$ is the prior probability for class $P(C_i)$ and $P(X)$, is the probability of object X.

$$P(C_i|X) = \frac{P(X|C_i)P(C_i)}{P(X)} \tag{35}$$

Let $d_i(X) = P(C_i|X)$, where $P(C_i|X)$ is the posterior probability that the true class is C_i given $X \in \mathbb{R}^n$, is clear that all probabilities are divided in $P(X)$ the equation also can be expressed as follows in Equation (36):

$$d_i(X) = P(X|C_i)P(C_i) \tag{36}$$

Considering the following:

$$P(X|C_i) = \frac{1}{(2\pi)^{n/2} |\Sigma_i|^{1/2}} e^{\left[-\frac{1}{2}(X-\mu_i)\sum_i^{-1}(X-\mu_i)\right]} \tag{37}$$

Assuming that all class-covariance matrices are the same, $\Sigma_i = \Sigma$:

Finally, it is obtained the discriminant function to select the class that has the highest score shown in Equation (38).

$$d_i(X) = \mu_i \Sigma^{-1} X^t - \frac{1}{2} \mu_i \Sigma^{-1} \mu_i^t + ln\left[P\left(C_i\right)\right] \qquad (38)$$

The next code can be entered at the command window of MATLAB and running, it would only be needed the signals *Xtest* a *Xtrain*, where every row is a signal. The next step is to transform and to extract the features (LPC or MFCC).

Computer Code

```
%data transformation
[Xtest] = normtrans(Xtest,0,1);%Xtest 500x193
[Xtrain] = normtrans(Xtrain,0,1);%Xtrain 500x193
%Features extration of LPC's
a=[];
b=[];
co=4;%number of coefficients

for i=1:length(Xtrain);
temp= lpc(Xtrain(i,:),co);
a=vertcat(temp,a);
temp1= lpc(Xtest(i,:),co);
b=vertcat(temp1,b);
end

%train data set with 500 signals
PL = a(1:500,3);
PW = a(1:500,5);
%test data set with 500 signals
PM = b(1:500,3);
PN = b(1:500,5);

species = celldata(1:end);%label of each class

h1 = gscatter(PL,PW,species,'rb','.',[],'off');

h1(1).LineWidth = 2;
h1(2).LineWidth = 2;
legend('Class_1','Class_2','Location','best')
hold on
```

```
X = [PL,PW];
Y = [PM,PN];

MdlLinear = fitcdiscr(X,species);
MdlLinear.ClassNames([1 2])
K = MdlLinear.Coeffs(1,2).Const;
L = MdlLinear.Coeffs(1,2).Linear;
f = @(x1,x2) K + L(1)*x1 + L(2)*x2;
h2 = ezplot(f);
xlabel('\fontsize{20}Coefficient 5');
ylabel('\fontsize{20}Coefficient 3');
title(['\fontsize{22}LDC with ',num2str(co),' LPC coefficients'])
h2.Color = 'g';
h2.LineWidth = 2;
grid
[class,err,POSTERIOR] = classify(Y,X,species,'linear');
err
```

PROCESSING SIGNAL WITH OPTOELECTRONIC SCANNING SYSTEM

Nowadays, optical scanning systems play an important role to measure structural deformation and displacements of buildings to monitor and prevent undesirable damage. Additionally, these systems have several applications and its necessary accuracy and resolution in each activity in order to measure minimal changes to assess structural conditions (Lopez et al., 2010). However, one of the principal targets to increment accuracy and resolution in optical scanning systems is reducing noise due these systems are subject to changing environmental and operational conditions that affect measured signals during data acquisition.

The environmental conditions affect the statistical model for damage classification, the principal source of noise and loss of resolution from the optical scanning system is the excessive bright of sunlight as previously mentioned. Recently, in the optoelectronic and measurements laboratory of Engineering Institute of Universidad Autónoma de Baja California has been developed a method for the task of SHM in which case the principle is based on 45°-slopping mirror and embedded into scanning aperture. This scanning aperture is constituted by elements such as a non-rotating incoherent light emitter that is mounted on the structure, a 45°-slopping mirror inside of the SA system, a double convex lens and an optoelectronic sensor as a photodiode.

The basic function of the system depends on the two passive rotating optical apertures sensors SA#1, SA#2, which are shown in Figure 8, they are used for dynamic triangulation, the system first captures a Gaussian curve generated by the scanning of the incoherent light source mounted on the infrastructure. Each peak of the Gaussian curve is related to the angle of position of the source, in this way are obtained the angle B_i for the SA #1 and angle C_i for the second SA #2, considering that distance between aperture called a is known, the coordinates of the object or structure to analyze can be calculated by using the theorem of sines and the correlation between the sides in the triangle.

Figure 8. Dynamic triangulation with scanning aperture

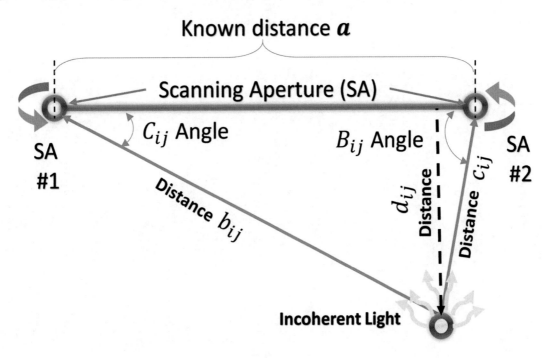

In this work the peak of the Gaussian curve represents the geometric centroid energy center (Flores-Fuentes et al., 2016), the voltage encoder of DC motor supplies two pulses per revolution $T_{2\pi}$, this is known as the period or a complete revolution. (see Figure 9).

$$N_{2\pi} = T_{2\pi} f \tag{39}$$

$$N_{\alpha} = T_{\alpha} \tag{40}$$

Where f is the scanning frequency, and $N_{2\pi}$ is the number of samples per revolution of motor dc, N_{α} represents the number of the samples taken from first pulse of the encoder to peak of a Gaussian curve. The Equation (41) calculates the angle of the source.

$$\alpha = 2\pi \left(\frac{N_{\alpha}}{N_{2\pi}} \right) \tag{41}$$

In Figure 10 is shown the OSS system and principal elements such as opto-interruptor to generate the pulses of the every revolution of the motor dc (yellow color in oscilloscope), the photodiode to detect the presence of light of the source (turquoise color in oscilloscope), and finally a 45°-slopping mirror inside of the OSS.

Figure 9. Data acquisition of a Gaussian curve and two pulses that encoder sends per revolution.

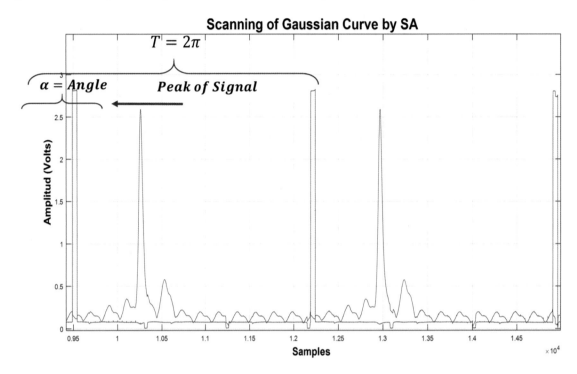

Figure 10. Data acquisition of the OSS system.

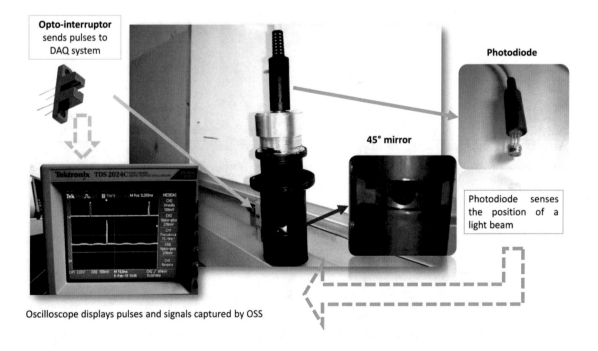

EXPERIMENTAL RESULTS

In this section are presented the results and measurements of the experiments in the research, the first method used was the moving average that belongs to digital filters. As a computational statistics method, it was applied LDC and GMM to detect when scanning aperture detects interference or abnormal patterns. The measurements from an abnormal pattern and the reference source pattern were taken in an environmental controlled, this means that noise was caused to conduct the experiment. The two sources of light, where one source was used in order to be the reference of the position from SA and another source was employed to cause optical noise (abnormal pattern) to the SA system. In the first instance, the signal was captured with optical noise, considering that optical noise as the little peak of the signal, this caused by the second source. In Figure 10 the Gaussian red represents the signal with noise as previously mentioned, also we can appreciate in this figure that appears two peaks for every revolution of SA system, this characterizes the Gaussian curve captured with optical noise in an environmental controlled. The maximum and minimum peak of the signal in red color results of the two sources that the SA detects in every revolution. However, the difference between these sources is the intensity, the little peak is undesired signal and the maximum peak is the reference of the source.

The procedure for the first experiment of the moving average was as follows: firstly, the data of the signal was acquired with noise in an environmental controlled with a scanning frequency of 43Hz, and were taken 5000 samples per seconds using two analog inputs from DAQ 6003, and after that it was applied the method of moving average to the signal with optical noise that was captured from the scanning aperture. In Figure 11 are illustrated the results of moving average where the pulses are in a blue color that corresponds to each revolution of the dc motor, and the Gaussian green results of the signal smoothed by the application of a moving average.

The real angle between the SA and the source was established arbitrarily at 120 degrees (according to the position of OSS) for this experiment, in Table 5 it is shown a more detailed information about the experiment applying this digital filter (moving average) to reduce the effects of the noise in an environ-

Figure 11. Signals of a Gaussian curve captured with two peaks

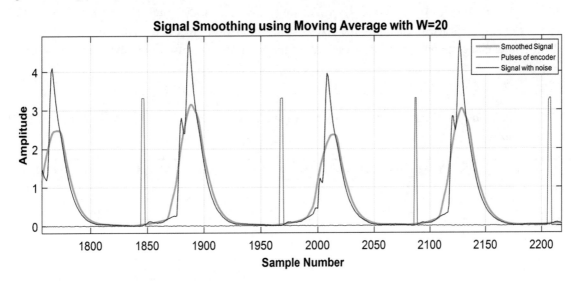

mental controlled. It was also calculated the statistical properties of the signals as the mean, standard deviation, mean square error (MSE), minimum and maximum measurements for each experiment.

The data in the Table 5 was calculated by developing an algorithm in Mat Lab software and comparing the real angle of the source (120 degrees) with respect of the angle detected by the algorithm of SA by returning two column vectors, where the first column vector has the measurements of the angle calculated with two peaks and it considers a signal with optical noise, it is evident that in this measurements (x_{exp1}) the number of the elements of the column vector is greater than the second measurements (x_{exp2}) due that in the second the signal was smoothed and the minimum peak was filtered by moving average method with of window size of $W = 20$, for example, in the first experiment the column vector x_{exp1} has $m = 53$ elements, and the second column vector x_{exp2} has $n = 43$ elements.

$$x_{exp1} = \begin{bmatrix} 304.615 & 252 & 26.341 \ldots x_{exp1(m-2)} & x_{exp1(m-1)} & x_{exp1(m)} \end{bmatrix}^T \tag{42}$$

$$x_{exp2} = \begin{bmatrix} 127.241 & 131.478 & 133.8053 \ldots x_{exp2(n-2)} & x_{exp2(n-1)} & x_{exp2(n)} \end{bmatrix}^T \tag{43}$$

Where, x_{exp1} = Column vector with optical noise, x_{exp2} = Column vector applying moving average, m = number of elements of column vector for x_{exp1} , n = number of elements of column vector for x_{exp2} .

The measurements of the angles from x_{exp1} , are much further way from real position (120° degrees), in the case of x_{exp2} the data are more near to the real angle according to Table 5. The mean calculated 179.938 of the vector x_{exp1} relates the signal with noise, and the mean resulted of the vector x_{exp2} 131.002 are significantly better applying moving average filter. The measurements of the MSE are significantly different for each experiment, in the first experiment is not used a moving average and the result is MSE=13558, for the second experiment in which was applied the method of moving average, it was calculated a MSE= 149.8513. It is important to mention, that the discussed in this section are compared the MSE calculated with $W = 20$ and $W = 7$ using moving average filter, however, in Figure 12 is presented the results of MSE changing the size of windows from 1 to 20. In order to get better results of MSE, and can see the behavior by incrementing the window size of moving average. According to

Table 5. Results with the filter moving average with the size of windows (= 20 and W = 7) applied to signal with noise

Statics	Results Without Filter	W=20 of Moving Average	W=7 of Moving Average
Mean	179.938	131.002	125.45
Standard deviation	100.7797007	5.431920377	1.636
Minimum	26.34146341	123.0769231	121.034
Maximum	306	144.6428571	129.2307692
MSE	13558	149.8513	32.32

Figure 9, we need to work with size of windows of $W = 7$ to obtain a MSE of 32.32, on the other hand, the computational cost will be reduced.

As previously mentioned, it was applied two methods of computational statistics to discriminate the interference. In Figure 13, it is showing the two classes of patterns captured in domain time, where Gaussian in blue color represents a reference source with optical noise and another pattern is acquired only with one source (without optical noise), and the main idea is knowing when there is an abnormal pattern between the scanning aperture and the reference source using LDC.

The computational statistics method applied was LDC and LPC coefficients as feature extractor, for this experiment the class 1 are correctly classified as C_1, the percentage of classification depend on LPC used, the classifier can perfectly identify C_1 as C_1 with minimal classification error, for example taking the third and fifth feature of the LPC coefficient it was achieved an error rate of 0.8%.

In Figure 14 it is showed results of applying LDC and LPC coefficients expressed in the error rate based on the percentage of the observation that is misclassified, the blue line represents the error rate. The values from horizontal axis are the numbers of coefficients used and the vertical axis corresponds to the error rate. According to the results based on the error rate of classification, using $(4 - 9)$ LPC coefficients is enough to determine is exist an abnormal pattern or interference. For the experiments the prior probabilities for every class were 0.5, this was calculated as follows, the number of signals of C_1 and C_2 divided between the prior probabilities for class C_i. When extracted 4 coefficients of LPC and the features taken are 3 and 5, the error rate is 1.4% that means that 7 points of 500 were incorrectly classified, the lower the increment of the coefficients LPC, better classification of LDA. If 10 LPC coefficients are chosen, the error rate increments to 8.4%, this means that 42 points are incorrectly classified out of 500 total points.

Figure 12. MSE of size of window for moving average (Line blue is the MSE calculated for each Window the filter)

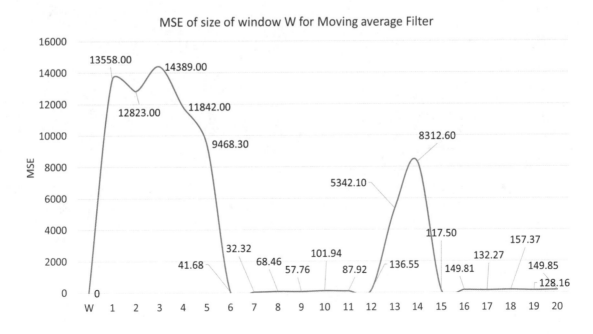

Figure 13. Pattern generated by OSS with optical noise (figure with blue) and without interference (figure with color)

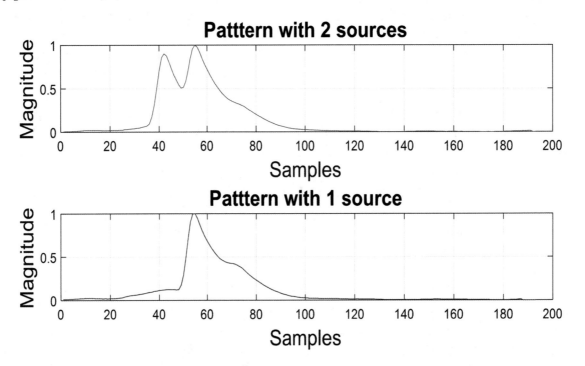

Figure 14. Error rate based on the percentage of the observation that is misclassified

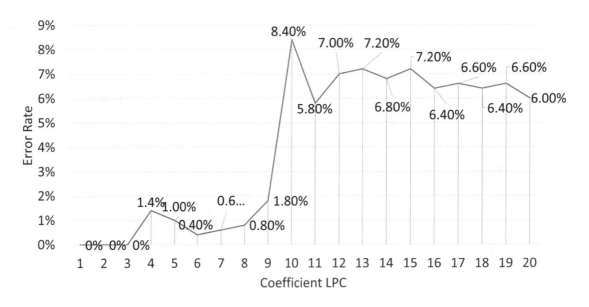

The algorithm applied to SA system could detect the two patterns with very good results that is why the classes are perfectly separable with minimum error rate. Figures 15 and 16, show the LDC with 4 and 10 LPC coefficients respectively, and also can be appreciated that the classes are mixed while the number of LPC coefficients are incremented.

It is clear that the vector of LPC coefficients represents an accurate method for characterizing the pattern of one source and two sources to discriminate with discriminant analysis. In Salehi, Burgueño, Das, Biswas, and Chakrabartty (2016) LDA method has good performance in terms of recognizing potential damage since this method deals directly with the discrimination between classes.

Other methods of computational statistics were applying GMM and MFCC, the models GMM's were created using 250 signals for each class, and 250 signals were used as test data set in each case, that is mean that 500 signals were in total to train the models and another 500 were to test the methodology. In Table 6 are presented the results from the evaluation of the test dataset of class 1 and class 2, the algorithm used to create the model was ***gmdistribution.fit*** from Mat Lab (The MathWorks, 2017).

Figure 15. LDC with 4 LPC coefficients, the vertical axis corresponds to the third coefficient. The horizontal axis corresponds to fifth LPC coefficient

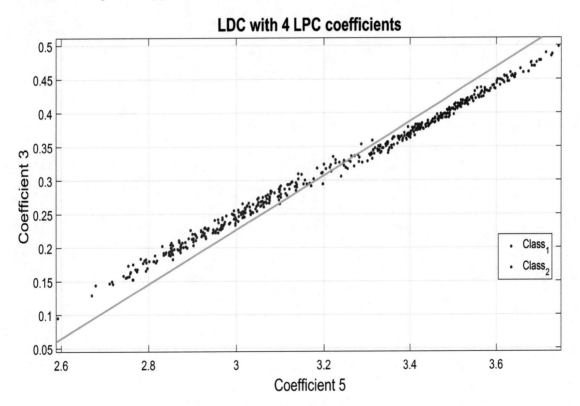

Figure 16. LDC with 10 LPC coefficients, the vertical axis corresponds to the third coefficient. The horizontal axis corresponds to fifth LPC coefficient

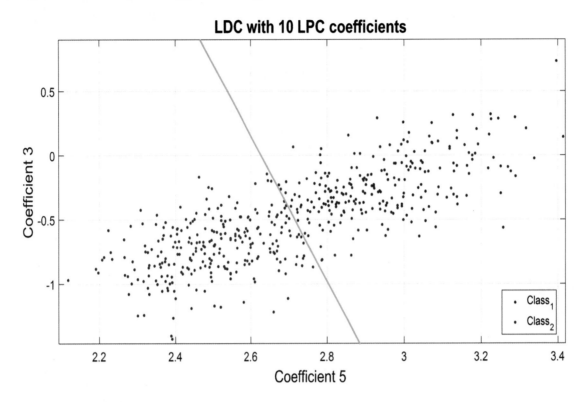

Table 6. Results applying GMM with 20 components and MFCC

Class Model	Evaluation With Class 1	Evaluation With Class 2
	Numbers of Successes/ Failures	Numbers of Successes
Class 1	250 / 0	0 / 0
Class 2	0 / 0	250 / 0

Computer Code

```
modClass1=gmdistribution.fit(Class_1,20,'SharedCov',true);
```

Where Class_1 is a matrix with MFCC features, 20 is the k components for data in the n-by-d matrix Class_1, and 'SharedCov' corresponds to the covariance matrices are restricted to be the same.

Table 7 shows the performance of the algorithm expressed in the numbers of successes and failures for each class. The measures of classification accuracy were calculated from the training dataset and evaluated with test dataset. For example, a model GMM with $k = 5$ components, evaluated with test data set of class 1 the error rate $(\%)$ is $1 / 250$, that is means that the algorithm recognize 1 signals from

Table 7. Results applying GMM with 20 components and MFCC for class 1

K Component	Evaluation With Class 1	Evaluation With Class 2
	Numbers of Successes/ Failures	Numbers of Successes/ Failures
5	249 / 1	247 / 3
10	245 / 5	250 / 0
15	241 / 9	250 / 0
20	250 / 0	250 / 0

250 test data set as class 2 when the real label is class 1. The error rate $(\%)$ evaluated with test dataset of class 2 is $3 / 247$, this means that 3 signals of test dataset of class 2 was classified incorrectly. The potential of this model was demonstrated in Mayorga-Ortiz, Druzgalski, Miranda Vega, and Zeljkovic (2016) where the efficiency in normal and traditionally noisy environments in light of very low intensities of these auscultation signals used as diagnostic indicators.

In Figure 17 each block gray correspond to a particular MATLAB code used in this chapter, the orange blocks represent the statistical pattern recognition paradigm for SHM. Furthermore, it should be emphasized that this chapter was focused in data acquisition and cleansing, feature extraction and statistical modeling for feature discrimination.

Figure 17. Block diagram of MATLAB code used

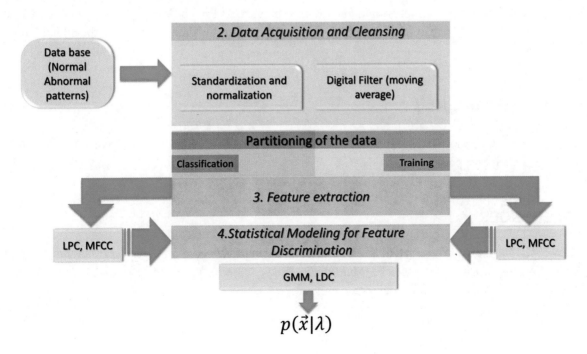

CONCLUSION

Digital filters and computational statistics showed them potentiality to smooth and discriminate the signal respectively, however, all the measurements are captured in a controlled environment. The photodetectors of the OSS can be affected in a hostile environmental due to the wavelength of other sources of light. Different types of filters and methods were analyzed for the implementation, the most appropriate were digital filters and computational statistics. Furthermore, according to the literature as previously mentioned this kind of digital (FIR) filters/mathematical methods are easy to implement and require only a computer to develop. Analogue filters were not implemented in this work due to that part of the signal with important information for the performance of the system is removed, but particularly due to the price of the components will often be too costly. However, there are a number of techniques where can be applied the digital filters and computational statistics in order to solve the problem of the cost and accuracy. The techniques of computational statistics such as LDC with LPC had an error rate of 1.4% with 4 coefficients it means that 7 points of 500 were incorrectly classified by LDC. For the best results, it is recommended 6 coefficients of LPC, this minimizes the error rate to 0.4% due to that will better adapt to the data generated by the Gaussian curve of the OSS. The results applying GMM with 20 components and MFCC as features extractor, were applied achieving satisfactory results. It is clear that 20 components to build the model GMM make the system more robust because the alignment is better, but the complexity and computational cost iteration increases, this considering that we are using 12 MFCC coefficients. Digital filter, particularly the moving average showed robustness at the moment of filtering the signal with optical noise. The MSE resulted in 7 window applying moving average, revealed an excellent compromise between the computational cost and complexity. The moving average can be implemented in an OSS by employing a microcontroller such as PIC or Arduino. This implementation would reduce the cost instead of using a DAQ system to capture the data and process the signal with a software and a computer.

ACKNOWLEDGMENT

This research was supported by the engineering institute at the Universidad Autónoma de Baja California (UABC) and support provided by CONACYT.

REFERENCES

Aizawa, K., Nakamura, Y., & Satoh, S. (2004). *Advances in Multimedia Information Processing - PCM 2004: 5th Pacific Rim Conference on Multimedia, Tokyo, Japan, November 30 - December 3, 2004, Proceedings*. Springer Berlin Heidelberg.

Brookes, M. (1997). *VOICEBOX: Speech Processing Toolbox for MATLAB*. Retrieved from http://www.ee.ic.ac.uk/hp/staff/dmb/voicebox/voicebox.html

Burke, M. W. (2012). *Image Acquisition: Handbook of machine vision engineering*. Springer Netherlands.

Chakrabarti, P. (1998). *Optical Fiber Communication*. Mc Graw Hill.

Diederich, J. (2008). *Rule Extraction from Support Vector Machines*. Springer Berlin Heidelberg. doi:10.1007/978-3-540-75390-2

Fischer, R. (2008). *Optical System Design* (2nd ed.). McGraw-Hill Education.

Flores-Fuentes, W., Sergiyenko, O., Rodriguez-Quiñonez, J. C., Rivas-López, M., Hernández-Balbuena, D., Básaca-Preciado, L. C., ... González-Navarro, F. F. (2016). Optoelectronic scanning system upgrade by energy center localization methods. *Optoelectronics, Instrumentation and Data Processing*, *52*(6), 592–600. doi:10.3103/S8756699016060108

Fukunaga, K. (2013). *Introduction to Statistical Pattern Recognition*. Elsevier Science.

Gul, M., & Catbas, N. (2009). Statistical pattern recognition for Structural Health Monitoring using time series modeling: Theory and experimental verifications. *Mechanical Systems and Signal Processing*, *23*(7), 2192–2204. doi:10.1016/j.ymssp.2009.02.013

Harrington, J., & Cassidy, S. (2012). *Techniques in Speech Acoustics*. Springer Netherlands.

Hayes, M. H. (1999). *Schaum's Outline of Digital Signal Processing*. McGraw-Hill Companies, Incorporated.

Ibrahim, D. (2013). *Practical Digital Signal Processing Using Microcontrollers*. Elektor International Media.

Ingle, V. K., & Proakis, J. G. (2011). *Digital Signal Processing Using MATLAB*. Cengage Learning.

Kuzmany, H. (2009). *Solid-State Spectroscopy: An Introduction*. Springer Berlin Heidelberg. doi:10.1007/978-3-642-01479-6

Lewis, J. M., Lakshmivarahan, S., & Dhall, S. (2006). *Dynamic Data Assimilation: A Least Squares Approach*. Cambridge University Press. doi:10.1017/CBO9780511526480

Ling, Ho, Fu, & Ming Kai. (2011). *Home-Made 3-D image measuring instrument data process and analysis*. Paper presented at the 2011 International Conference on Multimedia Technology.

Lopez, M. (2010). Optoelectronic Method for Structural Health Monitoring. *Structural Health Monitoring*, *9*(2), 105–120. doi:10.1177/1475921709340975

Lu, W. S., & Antoniou, A. (2000). *Design of digital filters and filter banks by optimization: A state of the art review*. Paper presented at the 2000 10th European Signal Processing Conference.

Luo, P., Zhang, M., Liu, Y., Han, D., & Li, Q. (2012). *A Moving Average Filter Based Method of Performance Improvement for Ultraviolet Communication System*. Academic Press.

Mayorga-Ortiz, P., Druzgalski, C., Miranda Vega, J. E., & Zeljkovic, V. (2016). Determinación del Tamaño Óptimo de Modelos HMM-GMM para Clasificación de las Señales Bioacústicas. *Revista Mexicana de Ingeniería Biomédica*, *37*, 63–79.

McCoy, A. D., Thomsen, B. C., Ibsen, M., & Richardson, D. J. (2004). Filtering effects in a spectrum-sliced WDM system using SOA-based noise reduction. *IEEE Photonics Technology Letters*, *16*(2), 680–682. doi:10.1109/LPT.2003.821077

Nair, K. (2007). Time series based structural damage detection algorithm using Gaussian mixtures modeling. *Journal of Dynamic Systems, Measurement, and Control, 129*(3), 285–293. doi:10.1115/1.2718241

Regtien, P., van der Heijden, F., Korsten, M. J., & Otthius, W. (2004). *Measurement Science for Engineers*. Elsevier Science.

Rivas Lopez, M., Sergiyenko, O., Tyrsa, V. V., Hernandez Perdomo, W., Devia Cruz, L. F., Hernandez Balbuena, D., ... Nieto Hipolito, J. I. (2010). Optoelectronic Method for Structural Health Monitoring. *Structural Health Monitoring, 9*(2), 105–120. doi:10.1177/1475921709340975

Salehi, H., Burgueño, R., Das, S., Biswas, S., & Chakrabartty, S. (2016). Structural Health Monitoring from Discrete Binary Data through. *Pattern Recognition*.

Savory, S. J. (2008). Digital filters for coherent optical receivers. *Optics Express, 16*(2), 804–817. doi:10.1364/OE.16.000804 PMID:18542155

Sergiyenko, O., Hernandez, W., Tyrsa, V., Cruz, L. F. D., Starostenko, O., & Peña-Cabrera, M. (2009). Remote Sensor for Spatial Measurements by Using Optical Scanning. *Sensors (Basel), 9*(7), 5477–5492. doi:10.339090705477 PMID:22346709

Sergiyenko, O., & Rodriguez-Quiñonez, J. C. (2016). *Developing and Applying Optoelectronics in Machine Vision*. IGI Global.

Sergiyenko, O. Y., Tyrsa, V. V., Hernandez, W., Starostenko, O., & Rivas, M. (2009). Dynamic laser scanning method for mobile robot navigation. 제어로봇시스템학회 국제학술대회 논문집, 4884-4889.

Sharkov, E. A. (2003). *Passive Microwave Remote Sensing of the Earth: Physical Foundations*. Springer.

Sharma, S., Katiyal, S., & Arya, L. D. (2015). Performance Comparison of Teaching-Learning-Based Optimization and Differential Evolution Algorithms Applied to the Design of Linear Phase Digital FIR Filter. *IUP Journal of Telecommunications, 7*(1), 23-38.

Smith, S. W. (1997). *The scientist and engineer's guide to digital signal processing*. California Technical Publishing.

Sohn, H., Czarnecki, J. A., & Farrar, C. R. (2000). Structural health monitoring using statistical process control. *Journal of Structural Engineering, 126*(11), 1356–1363. doi:10.1061/(ASCE)0733-9445(2000)126:11(1356)

Sohn, H., Farrar, C. R., Hemez, F. M., Shunk, D. D., Stinemates, D. W., Nadler, B. R., & Czarnecki, J. J. (2003). *A review of structural health monitoring literature: 1996–2001*. Los Alamos National Laboratory.

Tan, L., & Jiang, J. (2007). *Fundamentals of Analog and Digital Signal Processing*. AuthorHouse.

The MathWorks, Inc. (2017). *gmdistribution*. Author.

The MathWorks, Inc. (2018). *IIR Filter Method Summary*. Author.

Vaidyanathan, P. P. (2008). *The Theory of Linear Prediction*. Morgan & Claypool.

Yu, Y. J., & Lim, C. (2009). Low-Complexity Design of Variable Bandedge Linear Phase FIR Filters With Sharp Transition Band. Academic Press.

Section 4

Technical Vision for Avoiding Obstacles and Autonomous Navigation

One of the main goals of technical vision systems is to detect and identify targets covered by obstacles. Section 4 also describes the application of technical vision systems to autonomous mobile robots (AMR), which represent mobile machines and which can move independently in their surroundings, carrying out specific tasks. Chapter 12 presents mobile robot path planning using continuous laser scanning, while Chapter 13 describes a technical vision based on radar technology.

Chapter 12
Mobile Robot Path Planning Using Continuous Laser Scanning

Mykhailo Ivanov
Universidad Autónoma de Baja California, Mexico

Lars Lindner
Universidad Autónoma de Baja California, Mexico

Oleg Sergiyenko
Universidad Autónoma de Baja California, Mexico

Julio Cesar Rodríguez-Quiñonez
Universidad Autónoma de Baja California, Mexico

Wendy Flores-Fuentes
Universidad Autónoma de Baja California, Mexico

Moises Rivas-Lopez
Universidad Autonoma de Baja California, Mexico

ABSTRACT

The main object of this book chapter is an introduction and presentation of mobile robot path planning using continuous laser scanning, which has significant advantages compared with discrete laser scanning. A general introduction to laser scanning systems is given, whereby a novel technical vision system (TVS) using the dynamic triangulation measurement method for 3D coordinate determination is found suitable for accomplishing this task of mobile robot path planning. Furthermore, methods and algorithms for mobile robot road maps and path planning are presented and compared.

INTRODUCTION

Robots are used to carry out mechanical work for humans and are designed as a stationary or mobile vehicle. *Autonomous Mobile Robot* (AMR) thereby represents mobile machines, which can move independently in their surroundings and can carry out specific tasks. *Robotics* represents a scientific

DOI: 10.4018/978-1-5225-5751-7.ch012

branch that deals with the construction and design of robots. Furthermore, robotics is strongly related to the scientific branches of electrical engineering, computer science, and mechanics. Mechatronics has developed from these three disciplines. There exist different versions of mobile robots, each using the required actuators for different terrains. For example, in the case of flat terrain, mostly wheels are used for locomotion, while for uneven terrains usually chains or legs are used.

The hardware used for mobile robots can be divided mainly into two groups: the *Sensors* and the *Actuators*. The sensors measure actual physical parameters of the environment, as well as the current axis positions and speeds of the robot. The sensors furthermore can be divided into internal and external sensors, whereby the internal sensors measure current status data about the mobile robot (A/D converter, odometer, etc.) and the external sensors record data from the environment (accelerometer, gyrocompass, etc.). The actuators represent the counterpart of the sensors and are used to manipulate the mobile robot position in space or the robot environment. The actuators can also be classified into internal and external actuators. Internal actuators are primarily used to alter the state of the robot, while external actuators drive the mobile robot or move external objects.

The environment of a mobile robot is typically measured with CCD cameras and/or laser scanning systems. In Ohnishi & Imiya (2013) for example, a mobile robot is navigated using a "visual potential", which is computed using a sequence-capturing of various images by a camera mounted on the robot. Work (Correal, Pajares, & Ruz, 2014) uses an automatic expert system for 3D terrain reconstruction, which captures the robot environment with two cameras in a stereoscopic way, similar to the human binocular vision. Laser scanning systems, as remote sensing technology, instead are known as Light Detection and Ranging (Lidar) systems, which are widely used in many areas, as well as in mobile robot navigation. Work (Kumar, McElhinney, Lewis, & McCarthy, 2013) for example uses an algorithm and terrestrial mobile Lidar data, to compute the left and right road edge of a route corridor. In Hiremath, van der Heijden, van Evert, Stein, and ter Braak (2014), a mobile robot is equipped with a Lidar-system, which navigates in a cornfield, using the time-of-flight principle.

However, other sensors and methods are also used to measure the mobile robots environment. Paper (Benet, Blanes, Simo, & Perez, 2002) for example uses infrared (IR) and ultrasonic sensors (US) for map building and object location of a mobile robot prototype. One ultrasonic rotary sensor is installed on the top and a ring of 16 infrared sensors are distributed in eight pairs around the perimeter of the robot. These IR sensors are based on the direct measurement of the IR light magnitude that is back-scattered from a surface placed in front of the sensor. The typical response time of these IR sensors for a distance measurement is about $2m$. Distance measurement with this sensor can be realized from a few centimeters to $1m$, which represents one limitation of this approach. The range of coordinate measurements by triangulation can be far over $1m$. The work (Volos, Kyprianidis, & Stouboulos, 2013) even experiment with a chaotic controlled mobile robot, which only uses an ultrasonic distance sensor for short-range measurement to avoid obstacle collision. The experimental results show the applicability of chaotic systems to real autonomous mobile robots.

An optical 3D laser scanning system for navigation of autonomous mobile robots, called *Technical Vision System* (TVS), was developed at the Laboratory of Optoelectronics and Automated Measurement of the Universidad Autónoma de Baja California (Sergiyenko, 2010). This TVS consists mainly of a laser scanning system, which uses the Dynamic Triangulation Method, to obtain 3D coordinates of an object under observation. This developed autonomous robot navigation system has its main task in the prevention of obstacle collision in an unknown environment. More work on the TVS can be found in Sergiyenko et al. (2011), Rodríguez-Quiñonez, Sergiyenko, Gonzalez-Navarro, Basaca-Preciado, and

Tyrsa, (2013), Rodríguez-Quiñonez et al. (2014); Flores-Fuentes et al. (2016); Flores-Fuentes et al. (2016), Lindner, Laser Scanners (2016), Lindner et al. (2016), Murrieta-Rico et al. (2016), Lindner et al. (2017) and Flores-Fuentes et al. (2017).

Robot Motion Planning is described as "necessary since, by definition, a robot accomplishes tasks by moving in the real world" (Latombe, 1991). The problem of robot motion planning can be defined as follows. Given a start and desired final position of the robot, a geometric description of the robot and world, find a trajectory that moves the robot from the start to the final position, while preventing the collision with any obstacle. Two main goals of robot motion planning can be defined by the need of collision-free trajectories and that the robot should reach one final location as fast as possible. The Configuration Space (C-space) describes the space of all robot configurations, which specifies the positions of all robot points relative to a fixed coordinate system. The configuration space of a robot can be determined by moving the robot along all obstacle edges and increasing the obstacles dimensions using the robot radius. For mobile robot path planning, the continuous C-space must be discretized, whereby generally exist two approaches: The combinatorial planning and the Sampling-based planning.

The present chapter defines the basic principles for mobile robot path planning using continuous laser scanning. Thereby, the prior introduced TVS is used to perform a continuous laser scan of a mobile robot environment, to create roadmaps and to find an optimal trajectory of the robot in the C-space.

TECHNICAL VISION SYSTEM

For mobile robot path planning using continuous laser scanning, a new prototype of a *Technical Vision System* (TVS) is currently under construction, which implements the *Dynamic Triangulation* measurement principle, to obtain 3D coordinates of an object under observation. Figure 1 shows the TVS, containing the two principal parts of the TVS, the *Positioning Laser* (PL) and the *Scanning Aperture* (SA).

Dynamic Triangulation

The measurement principle of Dynamic Triangulation is shown in Figure 2 (top view) and Figure 3 (lateral view). It can be seen in these two figures, that the PL strikes a laser beam under the angle γ to the object to be measured, which gets reflected (total or diffuse) on the object surface and afterward captured by the SA under the angle β.

Thereby, both angles β and γ are measured instantaneous and using the fixed design parameter a, the striking distance d of the laser beam is determined by:

$$d = a \cdot \frac{\sin \beta \cdot \sin \gamma}{\sin(\beta + \gamma)} \tag{1}$$

This striking distance is used to determine the x-coordinate of the measured object point A:

$$x = \frac{d}{\tan \beta} = a \cdot \frac{\cos \beta \cdot \sin \gamma}{\sin(\beta + \gamma)} \tag{2}$$

Figure 1. Technical Vision System (Lindner et al., 2016)

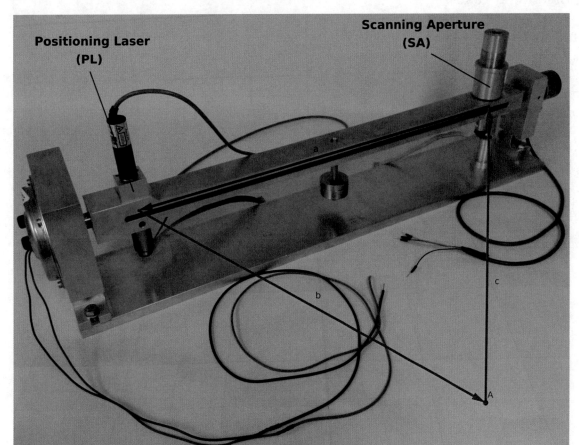

As shown in Figure 3, when the PL angle $\eta \neq 0$, the striking distance is not the same as the y-coordinate of the measured object point A, which thereby gets calculated by:

$$y = d \cdot \cos \eta = a \cdot \frac{\sin \beta \cdot \sin \gamma \cdot \cos \eta}{\sin(\beta + \gamma)} \tag{3}$$

Using the same PL angle η, the z-coordinate of the measured object point A is determined:

$$z = d \cdot \sin \eta = a \cdot \frac{\sin \beta \cdot \sin \gamma \cdot \sin \eta}{\sin(\beta + \gamma)} \tag{4}$$

The point $A(x; y; z)$ now represents a point in a three-dimensional Cartesian coordinate system.

Positioning Laser

The design of the positioning laser is depicted in Figure 4, which shows the optical path of the transmitted laser beam (I), a DC motor Maxon RE-max29 (II), a $45°$ staggered mirror (III), a pancake motor Printed Motor Works GPM9 (IV), a laser diode module Coherent StingRay-514 (V) and the TVS main rod (VI). Here, the DC motor II is used for horizontal and the pancake Motor IV for vertical positioning of the laser beam. The PL angle γ (Figure 2) corresponds directly to the actual angular position of the DC motor (II). Also, because the TVS is rotated around its horizontal axis without using a transmission, a motor with high torque and flat dimensions is needed. These two conditions are fulfilled by a pancake motor. Also here, the PL angle η (Figure 3) corresponds directly to the actual angular position φ_o of the pancake motor (IV).

Figure 2. Triangulation in top view

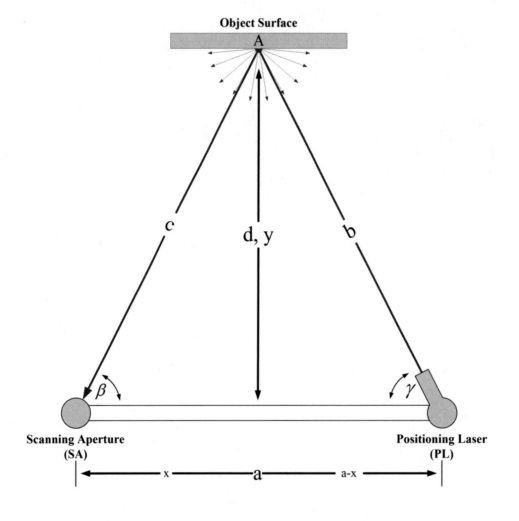

Figure 3. Triangulation in lateral view

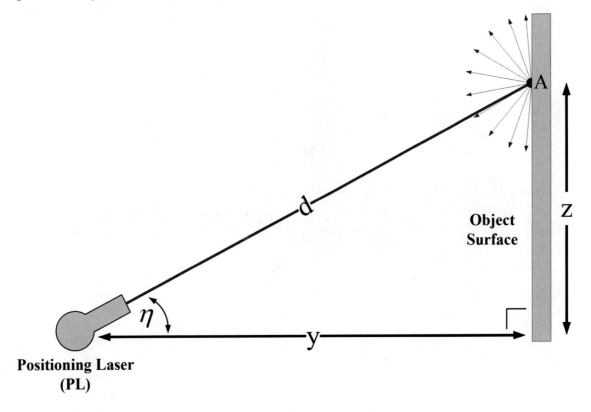

Scanning Aperture

The design of the scanning aperture is depicted in Figure 5, which shows the optical path of the received laser beam (I), a DC motor Maxon A-max16 (II), a $45°$ staggered mirror (III), two biconvex lenses (IV), an optical filter (V) and a high speed photodiode (VI).

The integrated quadrature encoder with index channel of the DC motor represents a zero sensor, which produces one pulse per revolution. The optical filter has a center wavelength of 515nm, the full width-half wavelength of 10nm and a transmission of 45%. The center wavelength of the filter corresponds to the laser diode module wavelength. The photodiode is used with a transimpedance amplifier, to convert the small photocurrent of the diode to a higher voltage, which thereby is measured using the ADC of the microcontroller. The Maxon A-max16 is digitally controlled by a microcontroller (Arduino Uno), which by use of a PI-algorithm controls the actual rotating speed of the $45°$ mirror.

The mirror is rotated by a DC motor that redirects the reflected laser beam from the observed object towards the lenses, which concentrate them on the stop sensor (high-speed photodiode). When the received laser beam is located in the orthogonal plane of the $45°$ staggered mirror, the stop sensor receives the maximum amplitude of the laser beam and converts it to an electrical pulse. It must be noted, that the maximum amplitude of the laser beam can be represented as one laser beam pulse, which is thereby received by the stop sensor during one revolution. The zero sensor produces a reference signal each full revolution and starts two counters to accumulate pulses. The first counter accumulates pulses from a standard reference f_p between two consecutive zero sensor pulses, which defines the pulse number for

Figure 4. Positioning Laser (PL)

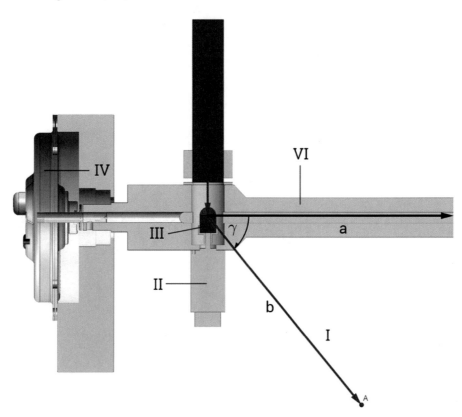

$N_{2\pi}$ in each full revolution. The standard reference signal represents a pulse train with fixed period $T_P = \dfrac{1}{f_P}$ and is generated by á microcontroller.

The second counter accumulates pulses of the same standard reference between the zero sensor pulse and the stop sensor pulse, which defines the pulse number for N_A in each revolution. Hence, N_A represents the received beam angle encoded in the pulse number of this variable. The received laser beam angle then is calculated using following equation:

$$\beta = 2\pi \cdot \frac{N_A}{N_{2\pi}} \tag{5}$$

Figure 6 depicts the standard reference signal f_P, the zero sensor signal $f_{2\pi}$, the period for one revolution $T_{2\pi}$, the full revolution pulse number $N_{2\pi}$, the time phase shift T_A and the SA angle pulse number N_A:

It must be noted, that the measurement speed of the scanning aperture is limited practically by the nominal speed of the used DC motor Maxon A-max16, since the high-speed photodiode saturates and desaturates in a range of some nanoseconds and the sampling rate of the used microcontroller is in the range of 9 kHz.

Figure 5. Scanning Aperture (SA)

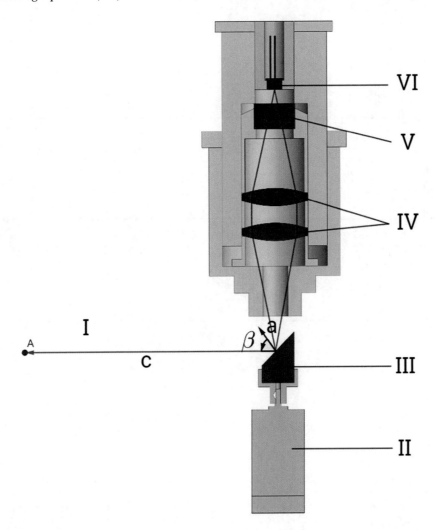

Discrete Field-of-View

The use of stepping motors for the PL leads to a discrete field-of-view (FOV) of the TVS, depicted in Figure 7. In the discrete FOV can exist scanning objects, that have dimensions smaller than the resolution of this field and which will not be detected by the TVS.

Figure 7 also shows, how the resolution of the discrete FOV is increased, the farther is the scanning plane distance and two mutual excluding conditions, when using a discrete FOV: A) Narrow FOV with high resolution or B) Wide FOV with low resolution. It must be noted, that by discretization of the scanner FOV, the continuous equations of triangulation get converted to equations using discrete values for the PL and SA angle. Distance a remains a constant value:

$$x_i = a \cdot \frac{\cos \beta_i \cdot \sin \gamma_i}{\sin(\beta_i + \gamma_i)} \tag{6}$$

Figure 6. Scanning Aperture Signals

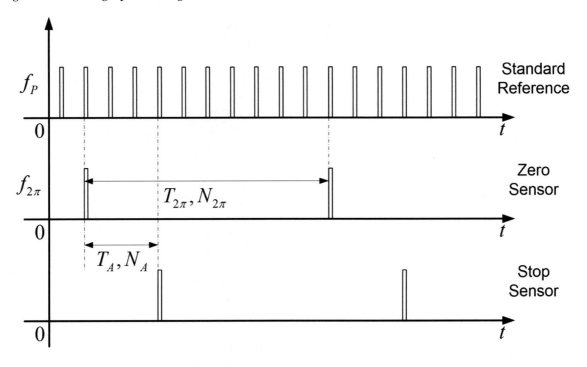

Figure 7. Discrete FOV

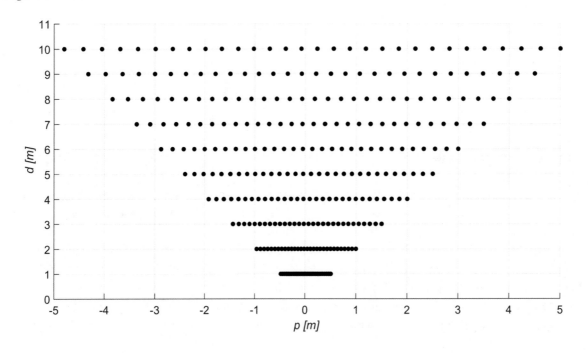

$$y_i = a \cdot \frac{\sin \beta_i \cdot \sin \gamma_i \cdot \cos \eta_i}{\sin(\beta_i + \gamma_i)} \tag{7}$$

$$z_i = a \cdot \frac{\sin \beta_i \cdot \sin \gamma_i \cdot \sin \eta_i}{\sin(\beta_i + \gamma_i)} \tag{8}$$

The point $A_i(x_i; y_i; z_i)$ now represents a discrete point in a three-dimensional Cartesian coordinate system.

Continuous Field-of-View

Using equation (2), the horizontal distance Δx between two adjacent points $A_1(x_1; y_1)$ and $A_2(x_2; y_2)$ can be defined by:

$$\Delta x = x_1 - x_2 \tag{9}$$

Using $a - x = \dfrac{d}{\tan \gamma}$ and a constant striking distance d for both points, they are distinguished by a different PL angle γ:

$$\Delta x = \frac{d}{\tan \gamma_2} - \frac{d}{\tan \gamma_1} \tag{10}$$

Since $\gamma_1 = \gamma_2 + \Delta \gamma$, equation (10) can be rewritten:

$$\Delta q = d \cdot \frac{\cos(\gamma_2)}{\sin(\gamma_2)} - d \cdot \frac{\cos(\gamma_2 + \Delta \gamma)}{\sin(\gamma_2 + \Delta \gamma)} \tag{11}$$

Using $\gamma_2 = \gamma$ and the trigonometric sum equations, equation (11) can be expressed as:

$$\Delta q = \frac{2d \cdot \sin(\Delta \gamma)}{\cos(\Delta \gamma) - \cos(2\gamma + \Delta \gamma)} \tag{12}$$

Hence, the distance Δq between two adjacent points is a function of the PL angle γ and the PL angle increment $\Delta \gamma$. In the case of $\gamma = \dfrac{\pi}{2}$, equation (12) gets simplified to:

$$\Delta q = d \cdot \tan(\Delta \gamma) \tag{13}$$

Figure 8. Continuous FOV

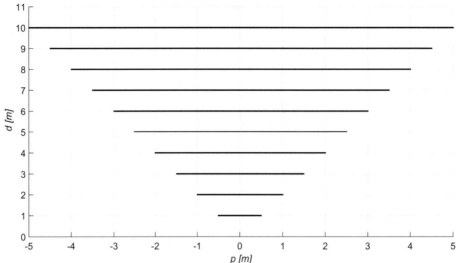

Since the $tan(x)$ for the interval $x\epsilon\ [0,\pi\ /\ 2]$ represents a monotone increasing function, equation (13) shows the resolution decreasing of the discrete FOV, when the PL angle increment $\Delta\gamma$ or the scanning distance d is increasing. Also, equation (13) shows the transition from the discrete to a continuous FOV when using $\Delta\gamma=0$, depicted in Figure 8:

$$\Delta q = d \cdot \tan\left(0\right) = 0 \tag{14}$$

The continuous FOV, proposed and researched in Lindner, Sergiyenko, Tyrsa, and Mercorelli (2014), Lindner et al. (2015) and Lindner et al. (2016) originates from the use of DC motors instead of stepping motors for the TVS positioning laser.

FIELD-OF-VIEW DISCRETIZATION

The information about an obstacle for the robot to avoid collision must contain general data (width, depth, height and large convexities of its surface which can obstruct the movement). In other words, obstacle should be represented for the robot as a simple 3D figure with low resolution.

However, in general, it is irrelevant what kind of TVS (with continuous or discrete FOV) is used for storing data it is necessary to perform a discretization of received information about obstacles. TVS with discrete FOV gives a fixed angle of scanning this means that the obtained points on obstacles storing frequency will be equal to this opening angle or can depend on the integer coefficient that determines the obstacle resolution (it is possible to take points in each second step of the motor), see Figure 9.

Discretization can be done by calculating the equivalents of opening angles that are needed on the different striking distances to maintain the same resolution of the received TVS image.

Figure 9. Example of over-detailed surface scanned by TVS

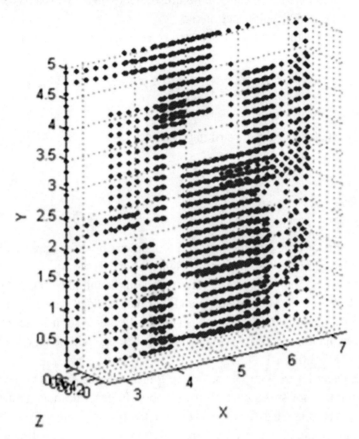

Garcia-Cruz et al. (2014) recommends three different angles 14.5636°, 5.5455°, 1.9091°. They are representing three types of resolution presented as linguistic variables: "Low", "Medium" and "High". Defining the amount of points (using the equation 14) per one meter of arc with 160° FOV next results are received: The resolution (ρ), in other words – density of image points, is 11 points per meter that is enough for the robots described in Básaca-Preciado et al. (2014).

$$\rho = \frac{\lambda}{\beta p_1} \tag{15}$$

where λ – opening angle of an arc (FOV), β – opening angle of the TVS, p1 – length of an arc of a size of 1 meter.

In general length of an arc can be calculated using:

$$p = \frac{\pi r \lambda}{180} \tag{16}$$

where r – radius of an arc (striking distance).

Results of the calculation are represented in Figure 10. Here there is a radius of one-meter arc equal to 0.358m with a resolution of 11 points per meter.

Opening angle, to prevent the changes in selected resolution, will be calculated using Equation 17:

$$\beta = \frac{180}{\pi r \lambda} \tag{17}$$

Average opening angle for each of the striking distance zone:

$$\beta_i = \frac{\sum_{j=0}^{n} \beta_{ij}}{n} \tag{18}$$

where β_i – opening angle for each striking zone, β_{ij} – opening angle for each striking distance in zone i. Results of the calculation are represented using Figure 11 and Table 1.

According to the calculation, the average angles based on the initial resolution of 14.5636° shall be 10.059°, 3.011°, 1.34°. As can be seen, average angle for the "critical" range in the end of the distance will give a small resolution equal to 5-6 points per meter. In this case, it is necessary to increase the resolution. That is why the limit value of angle for "critical" range will be taken. The set of angles will be changed to next 5.209°, 3.011°, 1.34°

Basing on the idea splitting striking distance into three scanning ranges zones (Figure 10) will be implementing this approach for speed control of the robot. In general, they can be described the next way: zone of effective scanning range the radius at which the system is able to determine with sufficient accuracy an obstacle to initiate rescheduling of the trajectory; zone of optimal scanning range is area in which the robot is able to determine the size and shape of the obstacles for its classification and more detailed adjustment of motion. In this area robot need to decrease his speed; zone of critical scanning range being at a given distance from an obstacle agent is already necessary to stop to avoid a collision.

Figure 10. FOV with opening angle for low resolution for one-meter arc length

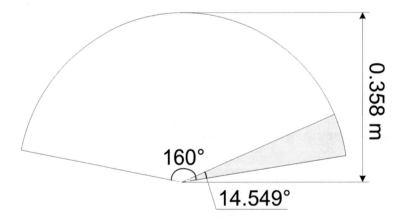

Figure 11. Dependencies of opening angle

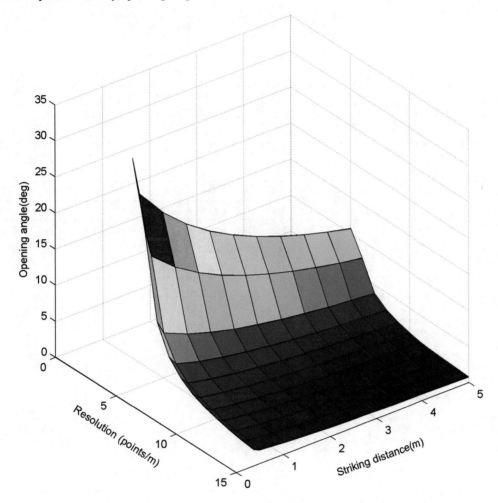

The speed control implementing can be managed on the use of linguistic variable "striking_distance". To determine the variable three membership functions (z-shaped for the "critical" distance, trapeze-like for "optimal" and s-shape – "effective") are used.

Using the striking distances zones and opening angles the data for fuzzy logic rules can be represented in Table 2.

For the conclusion let's make a review of the data amount reducing for transferring (Table 3). The calculation is based on the general opening angle (step) of the stepper motor (0.9°) comparing to low-resolution angles from Garcia-Cruz et al. (2014) and obtained the angle of 5.209°. Suggesting that each point is described like three floating numbers (X, Y, Z coordinates). The amount of data needed to store one floating number equal to 4 bytes. Using the extended algebra of algorithms for describing the passes of speed control determination will be:

Table 1. Comparison of scanning angles.

		Striking Distance	Length of Arc	Resolution (Points/m)									
				6	7	8	9	10	11	12	13	14	15
				Opening Angle (deg)									
Striking distance range	Critical	0,358	1,000	26,674	22,863	20,006	17,783	16,004	14,549	13,337	12,311	11,432	10,670
		0,500	1,396	19,099	16,370	14,324	12,732	11,459	10,417	9,549	8,815	8,185	7,639
		1,000	2,793	9,549	8,185	7,162	6,366	5,730	5,209	4,775	4,407	4,093	3,820
	Optimal	1,500	4,189	6,366	5,457	4,775	4,244	3,820	3,472	3,183	2,938	2,728	2,546
		2,000	5,585	4,775	4,093	3,581	3,183	2,865	2,604	2,387	2,204	2,046	1,910
		2,500	6,981	3,820	3,274	2,865	2,546	2,292	2,083	1,910	1,763	1,637	1,528
		3,000	8,378	3,183	2,728	2,387	2,122	1,910	1,736	1,592	1,469	1,364	1,273
	Effective	3,500	9,774	2,728	2,339	2,046	1,819	1,637	1,488	1,364	1,259	1,169	1,091
		4,000	11,170	2,387	2,046	1,790	1,592	1,432	1,302	1,194	1,102	1,023	0,955
		4,500	12,566	2,122	1,819	1,592	1,415	1,273	1,157	1,061	0,979	0,909	0,849
		5,000	13,963	1,910	1,637	1,432	1,273	1,146	1,042	0,955	0,881	0,819	0,764
		Average opening angle for width	Low	18,441	15,806	13,830	12,294	11,064	10,059	9,220	8,511	7,903	7,376
			Medium	5,520	4,731	4,140	3,680	3,312	3,011	2,760	2,548	2,366	2,208
			High	2,464	2,112	1,848	1,643	1,478	1,34	1,232	1,137	1,056	0,986

Table 2. Comparison of scanning angles.

Striking Distance(SD) (m)	Radius Type (Linguistic Variable)	Opening Angles (deg.)	Resolution (Linguistic Variable)
SD ≤ 1	Critical	5.209	Low
$1 < SD \leq 3$	Optimal	3.011	Medium
$3 < SD \leq 5$	Effective	1.34	High

Computer Code

```
SPEED=<INPUT(striking_distance)*[zone_state]
(dsn∨don∨den)*RULES>
RULES={[critical](SPEED_ZERO∧RESOLUTION_LOW) *
* [optimal](SPEED_SLOW∧RESOLUTION_MEDIUM) *
* [effective](SPEED_FAST∧RESOLUTION_HIGH)}
```

where "striking_distance" – receive the value "c" (critical) if obstacle in critical scanning range, "o" (optimal) if obstacle in optimal scanning range and "e" if ineffective scanning range; n – fixing statement.

In terms of fuzzy logic, it can be described using next IF-THEN rules type:

Computer Code

```
IF distance IS critical THEN speed EQUAL zero AND resolution EQUAL low
IF distance IS optimal THEN speed EQUAL slow AND resolution EQUAL medium
IF distance IS effective THEN speed EQUAL fast AND resolution EQUAL high
```

Implementing variable frequency of data storing will help to leave the conventionally equal level of obstacles resolution in the memory of robot. On this stage occur the need for data exchange between the individuals in group solving the task of obstacle avoidance and direct motion planning for reaching the goal by a group of robots.

After determining the amount of data needed to store $1m^2$ appears a question how to share this knowledge between the robots in the group. For this, it is necessary to determine the communication system for data transferring.

PATH PLANNING

Localization and Mapping

Simultaneous Localization and Mapping (SLAM) is a set of tasks and processes used by mobile robotic groups to digitalize terrain by using obstacles scans fusions of each robot to create a map while tracking their current locations according to obtained data (Gamini Dissanayake, 2001).

The reviewing approach uses a group of robot movement in densely cluttered terrain. The environments where they are moving is represented using the Cartesian coordinate system as a cloud point (Figure 12). This surrounding will be called World Coordinate System (WCS). Each of this robots works as an autonomous vehicle using TVS as a sensor for obstacle avoidance and mapping the terrain. Each robot uses its own Global Coordinate System (GCS). Moreover, TVS works in another Local Coordinate System (LCS). That is why to convert the obtained point on obstacle to the WCS, so each of the robotic group could use it. In this case for simultaneous mapping of obstacles by the group of robots, it is necessary to go through several steps: data obtained from TVS need to be converted to GCS of the individual robot and then global coordinates need to be converted to WCS.

For SLAM process some authors are using a method based on wireless sensors networks (Thierry Dumont & Sylvain Le Corff, 2014) other introducing R-SLAM method (Balcılar, Yavuz, & Amasy, 2017). Common for most of the proposed methods by other authors is to use filters for obtained data.

For the localization of the robot, it is proposed to use the Internal Navigation System based on IMU sensor path generation. Mentioned method is represented in Castro-Toscano (2017) and is using Kalman filter-based algorithm to reduce errors obtained from IMU. According to the IMU, data robot can solve the task of self-localization and construct its GCS.

Sectoring of the Terrain

Most of the researchers review motion planning in the frames of path building like Grymin, Neas, and Farhood (2014) where authors represented an approach that uses motion primitive libraries. Other examples are Kovács, Szayer, Tajti, Burdelis, and Péter Korondi (2016) representing attempt to implement

Figure 12. Surroundings representation with cloud point

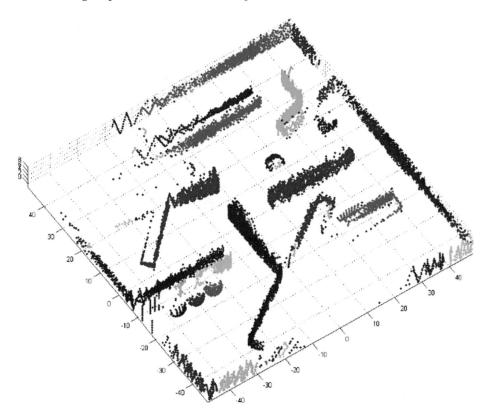

animal motion for robot behavior, or Ali,. Rashid, Frasca, and Fortuna (2016) suggested an algorithm of collision-free trajectory for robots. The articles cover some aspects of certain subtasks and widely describe special cases of behavior in a group of robots. However, no one considered the task from the point of view of the interconnected global approach. It must include the correlation of robots technical vision systems with communication and navigation rules.

As all real-time measurement made by TVS goes in its own Cartesian coordinates system it is possible to allocate two types of obstacles that are positive (exist above the local zero) (Figure 13, Figure 14) and negative (exist below the local zero) (Figure 15, Figure 16).

Laser scanning TVS, in a difference to camera vision, can give the exact coordinates of any selected point on the surface of an obstacle. However, they cannot process all surface at the same time and requires certain time to scan a 3D sector. Because of this, it is expedient to split terrain into sectors, sharing the task between robots in the group; it could be useful also due to a different position of each robot in the sector such distributed scanning will give more explicit information about obstacles position inside this sector.

According to the advantages and disadvantages of TVS, the group of robots will split terrain into sectors for their movement. To substantiate consider the example (Figure 17). While single robot moves (Figure 17a), it detects an obstacle A and the goal point still is in a "blind spot". Figure 17b represents another situation, similar to the group movement in Xidias and Azariadis (2016). Here the first robot still detects obstacle A, the second robot moves close to him. In the FOV of the second robot appears part of obstacles "A" and "B", but not complete both of them. In this case, robots have an over detailed

Figure 13. Positive obstacle (2D view)

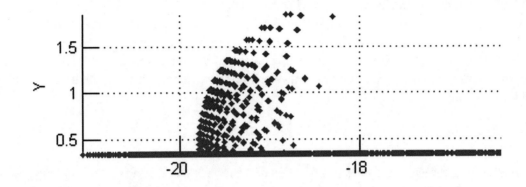

Figure 14. Positive obstacle (3D view)

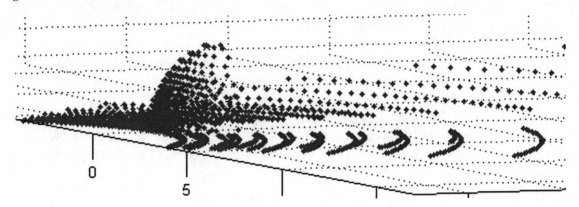

Figure 15. Negative obstacle (2D view)

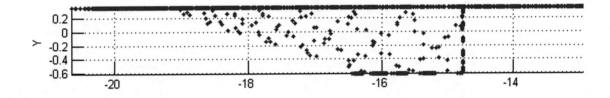

Figure 16. Negative obstacle (3D view)

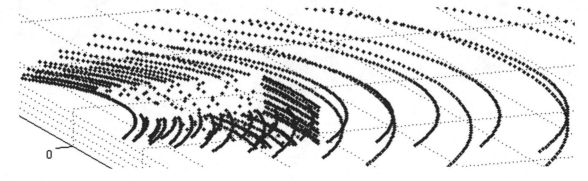

knowledge about one part of obstacle "A" and not so complete knowledge about obstacle "B". The goal is still invisible in this particular case Figure 17b.

In Figure 17c zone is separated by distancing of robots into two sectors ("sector A" for the first robot, "sector B" – for the second). As can be observed in case Figure 17c agents have enough resolution of an obstacle "A", information about the existence of obstacle "C" and moreover they have found the goal (object of interest). Using this information, trajectories of each robot can be recalculated for reaching a goal for scanning.

Motion Planning

Based on this logic the agent moves in the direction of the target as long as no obstacle to being detected. After the presence of obstacles is detected by TVS in real scene robot initialize clear A* path planning (Figure 18a). Points that are causing the necessity to change the vector of movement are separated from the full path and forming the graph of movement. Such points are called the breakpoints (Figure 18b). To improve the smoothness of the trajectory goes the approximation – the second step of the post-processing (Figure 18c). Thereafter, the agent starts to move to the first of the breakpoints, for which the heuristic evaluation is minimal. During its movement, the robot calculates new breakpoints after each new warning of TVS and moves to them until the heuristic evaluation of the distance to goal stops decreasing (reaches a local minimum).

For selecting the method of individual trajectory approximation of robots movement let us review the example presented in Figure 19. As can be observed, the robot often detects an obstacle and need to get from point A to point B. In current case to follow the rout robot, need to stop in point A and then turn

Figure 17. The need for separating the territory into sectors

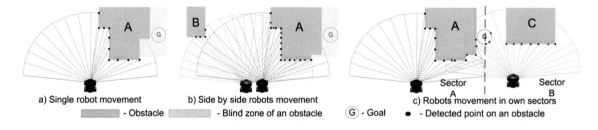

Figure 18. Dead reckoning with two-step post-processing

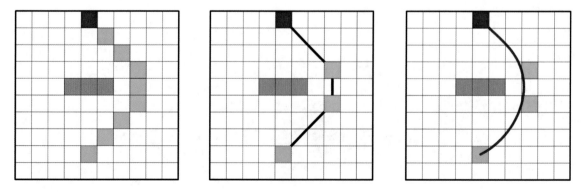

Figure 19. Obstacle avoidance

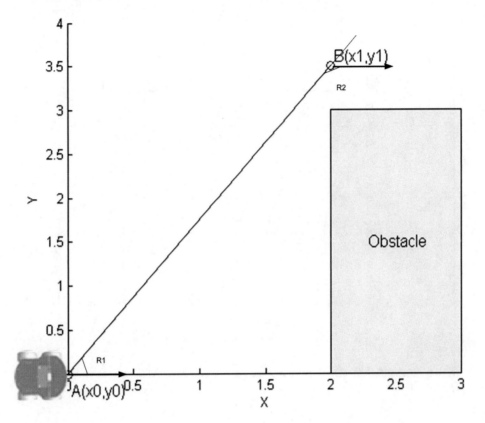

on an R1 angle. Reaching point B the same procedure must be done, but turning on angle – R1. This approach intensifies the mechanical load on the servos and the wear of the wheels.

In order to decrease this undesired influence the more smooth trajectories should be used. Trajectory smoothness describes coherence between the decisions interrelation of the navigation system actions and the ability to anticipate and make feedback to events with sufficient speed.

In Básaca-Preciado et al. (2014) was shown that using the ten points path is enough for building a smooth trajectory (example Figure 20).

Handling with this amount of data needs time and recalculation of the movement vector at every point. One of the brief solutions is to use less amount of point with approximation function like Bezier curve (Kawabata, Ma, Xue, Zhu, & Zheng, 2015).

Pierre Bezier proposed a method of any shape curves and surfaces creation. Beziers put the mathematical basis of his method of geometrical considerations (Bezier, 1971). But in Forrest (1972) and Gordon and Riesenfeld (1974) works was shown that the result is equivalent to Bernstein basis functions or a polynomial approximation. Mathematical parametric representation of a Bezier curve has the form:

$$P\left(t\right) = \sum_{i=1}^{n} B_i J_{n,j}\left(t\right), 0 \leq t \leq 1,$$ (19)

Figure 20. Obstacle avoidance

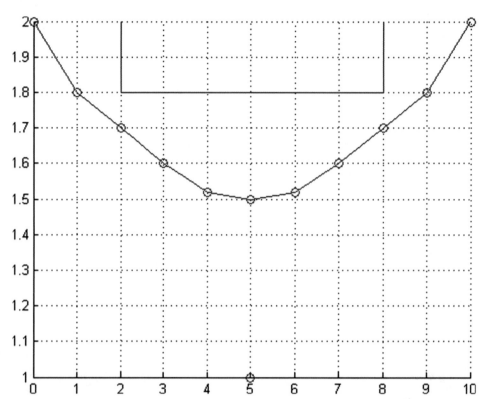

where t is a parameter, n is the degree of Bernstein's polynomial basis, i is the summation index and Bi represents the i-th vertex of the Bezier polygon.

The i-th Bezier or Bernstein function (approximation function) is given as:

$$J_{n,j}(t) = \binom{n}{i} t^i (1-t)^{n-i} \tag{20}$$

with

$$\binom{n}{i} = \frac{n!}{i!(n-i)!}, \tag{21}$$

Bezier curve equation can be written in matrix form, as well as the equation for cubic spline interpolation and parabolic:

$$P(t) = [T][N][G] = [F][G], \tag{22}$$

$$[F] = [J_{n,0}, J_{n,1}, ..., J_{n,n}], \tag{23}$$

$$[F]^T = [B_0, B_1, ..., B_n], \tag{24}$$

where T – vector of time, F - basis functions, F^T – vertices of the Bezier polygon, N – Bezier's basis matrix, G – matrix of vertices. In general case:

$$[T] = [t^n, t^{n-1}, ..., t^1], \tag{25}$$

$$[N] = \begin{bmatrix} \binom{n}{0}\binom{n}{n}(-1)^n & \cdots & \binom{n}{0}\binom{n-n}{n-n}(-1)^0 \\ \vdots & \ddots & \vdots \\ \binom{n}{0}\binom{n}{0}(-1)^0 & \cdots & 0 \end{bmatrix}, \tag{26}$$

Let us compare an example represented in Figure 21. Bezier curve is based on three points, where A and B are initial and final points. Point C is a virtual temporal point to make a curve. Matrix [N] for this case:

$$[N] = \begin{bmatrix} 1 & -2 & 1 \\ -2 & 2 & 0 \\ 1 & 0 & 0 \end{bmatrix}, \tag{27}$$

Figure 21. Obstacle avoidance with Bezier curve

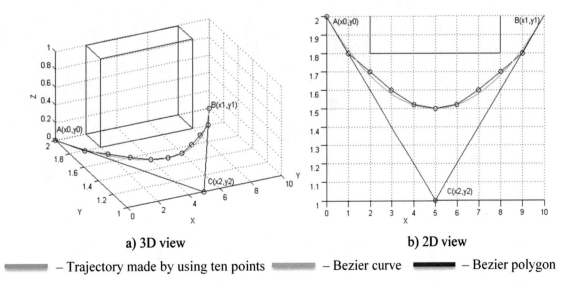

a) 3D view b) 2D view

▬ – Trajectory made by using ten points ▬ – Bezier curve ▬ – Bezier polygon

The results of the calculation are shown in Figure 21 (Figure 21a shows 3D view, Figure 21b – 2D view). It proves that in order to smooth the trajectory only three points need.

Let's return to example represented in Figure 22. Using the calculation based on three points (Figure 22a) helps to avoid movement in situ on point B but not on point A. That is why will be using Bezier matrix for four-point (Equation 28). The result of calculations is shown in Figure 22b. By adding temporal points C and D trajectory becomes smooth and it avoids the need for movement in situ on all parts of the trajectory.

$$[N] = \begin{bmatrix} -1 & 3 & -3 & 1 \\ 3 & -6 & 3 & 0 \\ -3 & 3 & 0 & 0 \\ 1 & 0 & 0 & 0 \end{bmatrix},$$ (28)

As an example, let us solve the same task using a Dubins path (Dubins, 1957), (Meyer, Isaiah, & Shima, 2015). Following equations describe Dubins path:

$$\begin{cases} x = V\cos(\theta) \\ y = V\sin(\theta), \\ \quad \theta = u \end{cases}$$ (29)

where (x, y) the position of the robot, θ to its heading, the robot is moving using the constant speed V. The optimal path type is based with robots making "right turn (R)", "left turn (L)" or driving "straight

Figure 22. Obstacle avoidance with Bezier curve

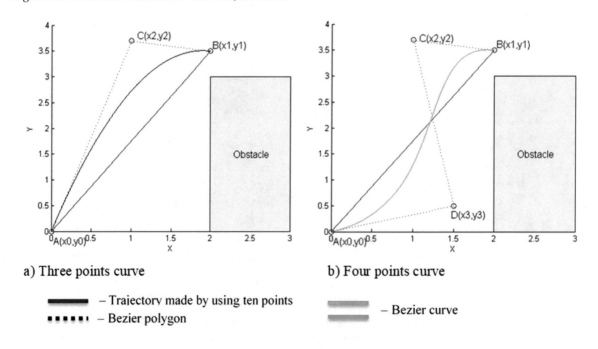

a) Three points curve

b) Four points curve

■■■■■ – Trajectory made by using ten points
▪▪▪▪▪▪ – Bezier polygon

▬▬▬ – Bezier curve

(S)". Every time optimal path will be one of the next six types: RSR, RSL, LSR, LSL, RLR, LRL (illustrated by Figure 23).

Using Dubins path (Figure 24) distance becomes longer and in initial point movement in situ happens. Making a detailed comparison of these two algorithm the result of the path made by Bezier approximation is 10.3% – 12.7% more effective than the solution of the same situations using Dubins path.

As a for a conclusion let's compare these methods more detail using parameters of the total length of a path and smoothness of trajectory where it is represented as an amount of bending energy of trajectory.

The bending energy is signifying a curvature function k. This function is used to estimate the smoothness of the robot's navigation trajectory. Bending energy calculated as the sum of the squares of the curvature at each point of the line, along its length. The bending energy of the trajectory of a robot is given by:

$$BE = \frac{1}{n}\sum_{i=1}^{n}k^2\left(x_i, f\left(x_i\right)\right) \tag{30}$$

Figure 23. Obstacle avoidance using Dubins path

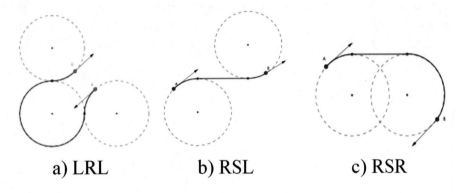

a) LRL   b) RSL   c) RSR

Figure 24. Obstacle avoidance with Dubins path and Bezier curve

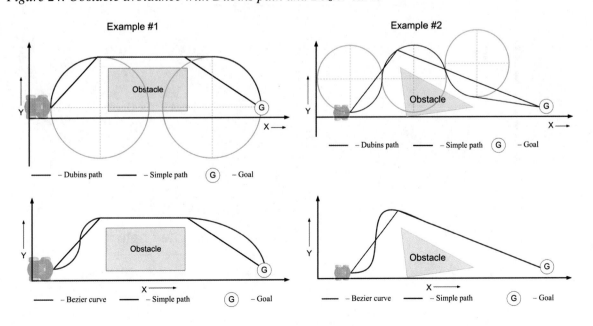

where: $k(x_i, y_i)$ is the curvature at each point of the robot's trajectory and n is the number of points in the trajectory. Here the curvature of the trajectory at the point is the reverse value of the circle radius, along with the arc of which the point moves at a given time instant.

On Figure 25, Figure 26, Figure 27 and Figure 28 is represented a simple example of modeling of four different motion planning methods. Figure 25 shows the 10 point curve which is taken as a basis for comparing. Other is a Dubins path (Figure 26), Bezier using three (Figure 34) and four (Figure 35) points polygons.

The results of modeling are compiled in Tab. 4 and for more detail comparing as a histogram (Figure 29). Here as main indications are taken the data achieved from 10 point curves (all represented as 100%). Summarizing result are represented as an average value between the total length and bending energy characteristic. Speaking other words each of them has the same weight for calculation.

Comparing three of motion planning methods (Dubins path, Bezier approximation using 3 and 4 points) the extremums can be found in Dubins path where received the highest length but the best Banding energy saving. Ultimately the satisfying average result gives Bezier approximation using 4 point polygon.

These methods, in total, solve the task of motion planning for the independent robot in-group. It is obvious that data transferring exchange between the robots in the group is a good tool for additional information obtaining with the aim of a complete implementation of all mentioned above methods.

Describing all of the parts of behavior logic of the robotic group steadily can go to the modeling of the real processes. It will be considered a simplified scenario of modeling for better understanding.

Figure 25. Motion planning using 10-points curve

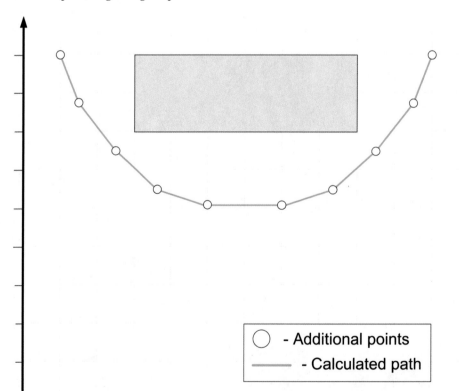

Figure 26. Motion planning using Dubins path

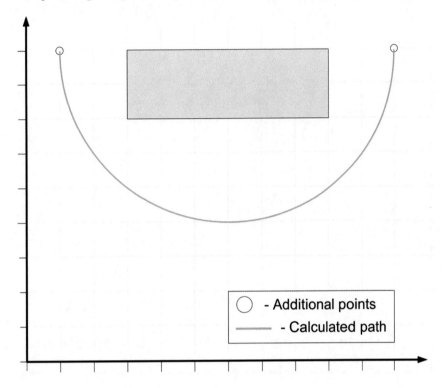

Figure 27. 3-points Bezier approximation

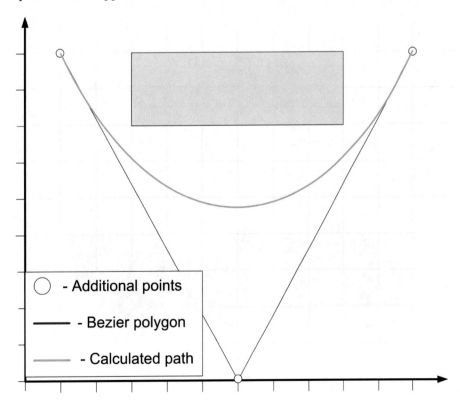

Figure 28. 4-points Bezier approximation

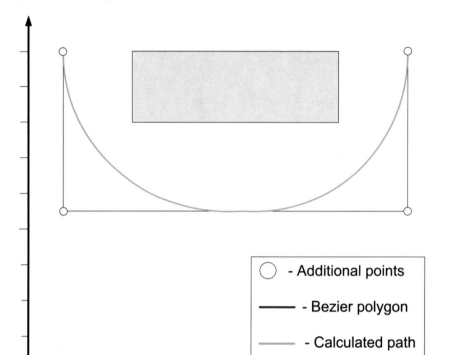

Table 3. Motion planning comparing results

	10 points curve [12]	Dubins path	Bezier with 3 points	Bezier with 4 points
Total length	100%	141%	113%	105%
Bending energy	100%	17%	75%	29%
Total	100%	79%	94%	67%

Figure 29. Motion planning comparing results

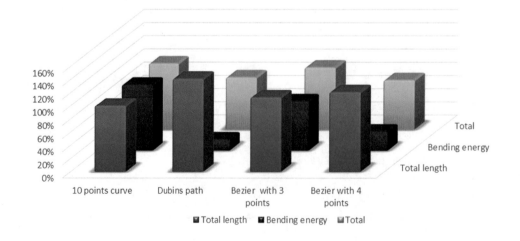

The scene on Figure 30, Figure 31, Figure 32 and Figure 33 represents the modeling of behavior during the motion planning while reaching the goal under a specific angle. Figure 30 is an initial state where "Robot 1" moving in its sector can observe the area in general and fully see the "goal". In FOV of "Robot 3" just a part of a goal is visible, most part of it is covered by an obstacle. "Robot 2" in his turn can only see an obstacle.

In this case "Robot 2" becomes a hotspot for data exchange within the group. "Robot 1" and "Robot 3" are providing additional information that they know about the environment. The question how to choose a hotspot can be solved basing on the voting process described in a previous research paper (Sergiyenko et al., 2016;Sergiyenko et al., 2016).

On Figure 31 represented preplanned trajectories for "Robot 1" and "Robot 3" based on Bezier approximation and trajectory patterns. In here trajectory pattern is a part of a not approximated trajectory that helps to reach the goal under the specific angle. For "Robot 2" it is harder to reach the goal because of an obstacle. In this case, he needs to avoid it (Figure 31 shows the first part of the "Robot 2" trajectory B1).

Figure 30. Initial state

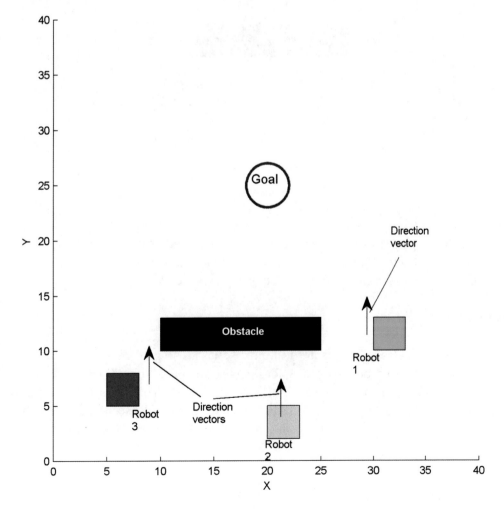

Figure 31. Basic dead reckoning

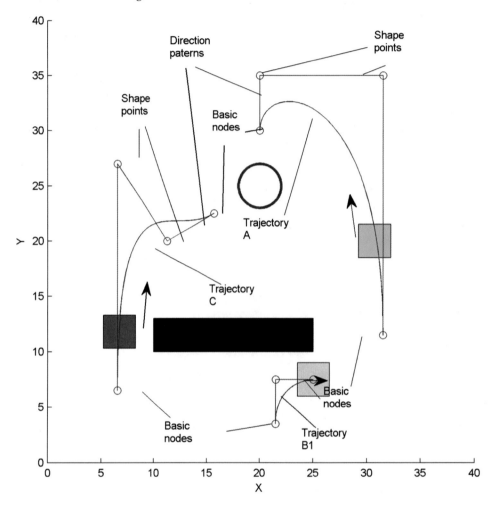

Using the data exchange process according to knowledge obtained by "Robot 1" the second part of "Robot 2" trajectory (Figure 16, trajectory B2) can be built. Figure 33 represents the reaching of goal by all of three robots where each of them directed on a preplanned angle.

Simulations Results

More complicated examples are represented in three scenarios (Figure 34 – Figure 36). Each of the scenarios represent four cases: dead reckoning for Robot#1, Robot#2, Robot#3 (Figure 34 a, b, c, Figure 35 a, b, c, Figure 36 a, b, c) and movement with common knowledge base of environment after data transferring process (Figure 34 d, Figure 35 d, Figure 36 d). Each of the scenes has the same purpose, where all of three robots need to reach the goal starting the movement from the same position.

Figure 32. All trajectories built

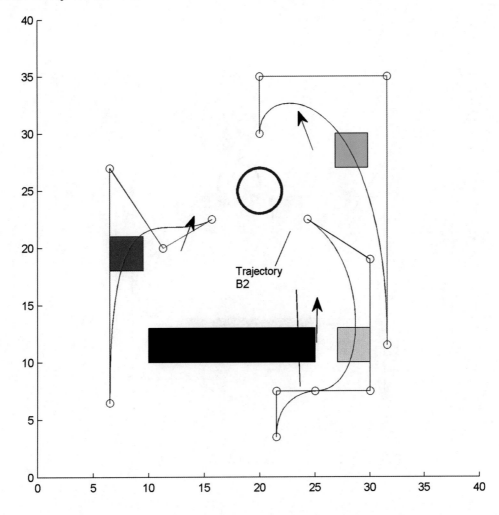

Reviewing the "Scenario 1" it is obvious to see that path of Robot#1 changed and became shorter for 9.56% after data exchange method implementing. The same happens in "Scenario 2" where Robot#2 calculated 14.76% shorter trajectory.

"Scenario 3" shows that Robot#1 has a loop in the beginning of the trajectory (Figure 36 d) substantiated with the path recalculation after data exchange with Robot#2 and Robot#3. However, the trajectory length is 17.3% less the original one.

The present article offers the original solution able to improve the collaboration inside the robotic group. All tasks discussed in the paper have a common point, guided only by information obtained in local interactions with the environment by our TVS, a group of robots needs to reach the goal using motion planning and communication algorithms.

Method of ranging the striking distances and angle saving equivalent based on Garcia-Cruz et al. (2014) shows the data reduction in 32.9 times for improvement of data transfer. The implemented method gives an additional option of the resolution adjustment according to specifics of the environment or current task.

Figure 33. Reaching the goal

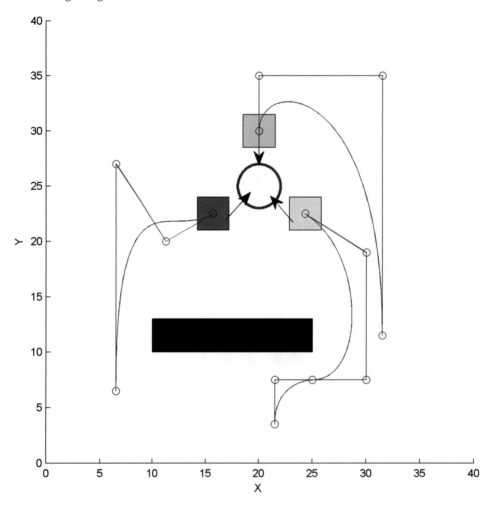

Figure 34. Object search (scenario 1)

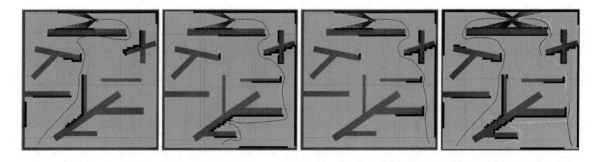

Figure 35. Object search (scenario 2)

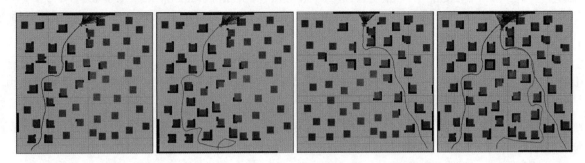

Figure 36. Object search (scenario 3)

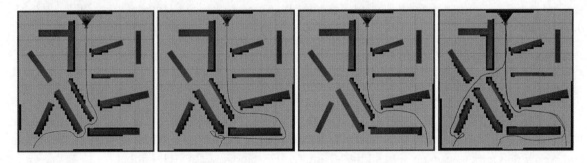

The proposed use of Bezier curves based on four control point's polygon showed good improvement. By increasing the general length by 5% (comparing to 10 point curves in previous research (Básaca-Preciado, et al., 2014)) results in banding energy saving the decreased of 71%. Moreover, such approach gives an ability to build continuous trajectories.

REFERENCES

Ali, A. A., Rashid, A. T., Frasca, M., & Fortuna, L. (2016). An algorithm for multi-robot collision-free navigation based on shortest distance. *Robotics and Autonomous Systems*, *75*, 119–128. doi:10.1016/j.robot.2015.10.010

Balcılar, M., Yavuz, S., & Amasy, M. F. (2017). R-SLAM: Resilient localization and mapping in challenging environments. *Robotics and Autonomous Systems*, *87*, 66–80. doi:10.1016/j.robot.2016.09.013

Básaca-Preciado, L. C., Sergiyenko, O. Y., Rodríguez-Quinonez, J. C., García, X., Tyrsa, V. V., Rivas-Lopez, M., ... Starostenko, O. (2014). Optical 3D laser measurement system for navigation of autonomous mobile robot. *Optics and Lasers in Engineering*, *54*, 159–169. doi:10.1016/j.optlaseng.2013.08.005

Benet, G., Blanes, F., Simo, J., & Perez, P. (2002, September 30). Using infrared sensors for distance measurement in mobile robots. *Robotics and Autonomous Systems*, *40*(4), 255–266. doi:10.1016/S0921-8890(02)00271-3

Bezier, P. E. (1971). Example of an Existing System in the Motor Industry: The Unisurf System. *Proceedings of the Royal Society of London. Series A, Mathematical and Physical Sciences, 321*(1545), 207–218. doi:10.1098/rspa.1971.0027

Castro-Toscano, M. J. (2017). A methodological use of inertial navigation systems for strapdown navigation task. In *26th International Symposium on Industrial Electronics (ISIE)* (pp. 1589-1595). Edinburgh, UK: IEEE. 10.1109/ISIE.2017.8001484

Correal, R., Pajares, G., & Ruz, J. (2014, March). Automatic expert system for 3D terrain reconstruction based on stereo vision and histogram matching. *Expert Systems with Applications, 41*(4), 2043–2051. doi:10.1016/j.eswa.2013.09.003

Dubins, L. E. (1957). On curves of minimal length with a constraint on average curvature, and with prescribed initial and terminal positions and tangents. *American Journal of Mathematics, 79*(3), 497–516. doi:10.2307/2372560

Dumont, T., & Le Corff, S. (2014). Simultaneous localization and mapping in wireless sensor networks. *Signal Processing, 101*, 192–203. doi:10.1016/j.sigpro.2014.02.011

Flores-Fuentes, W., Rivas-Lopez, M., Hernandez-Balbuena, D., Sergiyenko, O., Rodríguez-Quiñonez, J., Rivera-Castillo, J., . . . Basaca-Preciado, L. (2016). Applying Optoelectronic Devices Fusion in Machine Vision: Spatial Coordinate Measurement. In O. Sergiyenko, & J. Rodríguez-Quiñonez (Eds.), Developing and Applying Optoelectronics in Machine Vision (p. 37). Hershey, PA: IGI Global.

Flores-Fuentes, W., Sergiyenko, O., Gonzalez-Navarro, F., Rivas-Lopez, M., Hernandez-Balbuena, D., Rodríguez-Quiñonez, J., ... Lindner, L. (2017). Optoelectronic instrumentation enhancement using data mining feedback for a 3D measurement system. *Optical Review, 23*(6), 891–896. doi:10.100710043-016-0265-z

Flores-Fuentes, W., Sergiyenko, O., Gonzalez-Navarro, F., Rivas-Lopez, M., Rodríguez-Quiñonez, J., Hernandez-Balbuena, D., ... Lindner, L. (2016, August). Multivariate outlier mining and regression feedback for 3D measurement improvement in opto-mechanical system. *Optical and Quantum Electronics, 48*(8), 21. doi:10.100711082-016-0680-1

Forrest, A. R. (1972). Interactive Interpolation and Approximation by Bezier Polynomials. *Comp. J.*, 71-79.

Gamini Dissanayake, P. N.-W. (2001). A solution to the simultaneous localization and map building (SLAM) problem. *IEEE Transactions on Robotics and Automation, 17*(3), 229–241. doi:10.1109/70.938381

Garcia-Cruz, X., Sergiyenko, O., Tyrsa, V., Rivas-Lopez, M., Hernandez-Balbuena, D., Rodriguez-Quiñonez, J., ... Mercorelli, P. (2014). Optimization of 3D laser scanning speed by use of combined variable step. *Optics and Lasers in Engineering, 54*, 141–151. doi:10.1016/j.optlaseng.2013.08.011

Gordon, W. J., & Riesenfeld, R. F. (1974). Bernstein-Bezier Methods for the Computer Aided Design of Free-from Curves and Surfaces. *Journal of the Association for Computing Machinery, 21*(2), 293–310. doi:10.1145/321812.321824

Grymin, D. J., Neas, C. B., & Farhood, M. (2014). A hierarchical approach for primitive-based motion planning and control of autonomous vehicles. *Robotics and Autonomous Systems, 62*(2), 214–228. doi:10.1016/j.robot.2013.10.003

Hiremath, S., van der Heijden, G., van Evert, F., Stein, A., & ter Braak, C. (2014, January). Laser range finder model for autonomous navigation of a robot in a maize field using a particle filter. *Computers and Electronics in Agriculture, 100*, 41–50. doi:10.1016/j.compag.2013.10.005

Kawabata, Ma, Xue, Zhu, & Zheng. (2015). A path generation for automated vehicle based on Bezier curve and via-points. *Robotics and Autonomous Systems, 74*(A), 243-252.

Kovács, B., Szayer, G., Tajti, F., Burdelis, M., & Korondi, P. (2016, August). A novel potential field method for path planning of mobile robots by adapting animal motion attributes. *Robotics and Autonomous Systems, 82*, 24–34. doi:10.1016/j.robot.2016.04.007

Kumar, P., McElhinney, C., Lewis, P., & McCarthy, T. (2013, November). An automated algorithm for extracting road edges from terrestrial mobile LiDAR data. *ISPRS Journal of Photogrammetry and Remote Sensing, 85*, 44–55. doi:10.1016/j.isprsjprs.2013.08.003

Latombe, J.-C. (1991). *Robot Motion Planning* (Vol. 124). Springer, US. doi:10.1007/978-1-4615-4022-9

Lindner, L. (2016). Laser Scanners. In O. Sergiyenko, J. Rodríguez-Quiñonez, O. Sergiyenko, & J. Rodríguez-Quiñonez (Eds.), Developing and Applying Optoelectronics in Machine Vision (p. 38). Hershey, PA: IGI Global.

Lindner, L., Sergiyenko, O., Rivas-Lopez, M., Hernandez-Balbuena, D., Flores-Fuentes, W., Rodríguez-Quiñonez, J., ... Basaca, L. (2017). Exact laser beam positioning for measurement of vegetation vitality. *Industrial Robot: An International Journal, 44*(4), 532–541. doi:10.1108/IR-11-2016-0297

Lindner, L., Sergiyenko, O., Rodríguez-Quiñonez, J., Rivas-Lopez, M., Hernandez-Balbuena, D., Flores-Fuentes, W., ... Tyrsa, V. (2016). Mobile robot vision system using continuous laser scanning for industrial application. *The Industrial Robot, 43*(4), 360–369. doi:10.1108/IR-01-2016-0048

Lindner, L., Sergiyenko, O., Rodríguez-Quiñonez, J., Tyrsa, V., Mercorelli, P., Fuentes-Flores, W., . . . Nieto-Hipolito, J. (2015). Continuous 3D scanning mode using servomotors instead of stepping motors in dynamic laser triangulation. In *Industrial Electronics (ISIE), 2015 IEEE 24th International Symposium on* (pp. 944-949). Buzios: IEEE.

Lindner, L., Sergiyenko, O., Tyrsa, V., & Mercorelli, P. (2014, June 01-04). An approach for dynamic triangulation using servomotors. In *Industrial Electronics (ISIE), 2014 IEEE 23rd International Symposium on* (pp. 1926-1931). Istanbul: IEEE.

Meyer, Y., Isaiah, P., & Shima, T. (2015, March). On Dubins paths to intercept a moving target. *Automatica, 53*, 256–263. doi:10.1016/j.automatica.2014.12.039

Murrieta-Rico, F., Sergiyenko, O., Petranovskii, V., Hernandez-Balbuena, D., Lindner, L., Tyrsa, V., ... Karthashov, V. (2016, May). Pulse width influence in fast frequency measurements using rational approximations. *Measurement, 86*, 67–78. doi:10.1016/j.measurement.2016.02.032

Ohnishi, N., & Imiya, A. (2013, June - July). Appearance-based navigation and homing for autonomous mobile robot. *Image and Vision Computing*, *31*(6-7), 511–532. doi:10.1016/j.imavis.2012.11.004

Rodríguez-Quiñonez, J., Sergiyenko, O., Basaca-Preciado, L., Tyrsa, V., Gurko, A., Podrygalo, M., ... Hernandez-Balbuena, D. (2014, March). Optical monitoring of scoliosis by 3D medical laser scanner. *Optics and Lasers in Engineering*, *54*, 175–186. doi:10.1016/j.optlaseng.2013.07.026

Rodríguez-Quiñonez, J., Sergiyenko, O., Gonzalez-Navarro, F., Basaca-Preciado, L., & Tyrsa, V. (2013, February). Surface recognition improvement in 3D medical laser scanner using Levenberg–Marquardt method. *Signal Processing*, *93*(2), 378–386. doi:10.1016/j.sigpro.2012.07.001

Sergiyenko, O. (2010). Optoelectronic System for Mobile Robot Navigation. *Optoelectronics, Instrumentation and Data Processing*, *46*(5), 414–428. doi:10.3103/S8756699011050037

Sergiyenko, O., Ivanov, M., Tyrsa, V., Kartashov, V., Rivas-López, M., Hernández-Balbuena, D., ... Tchernykh, A. (2016). Data transferring model determination in robotic group. *Robotics and Autonomous Systems*, *83*, 251–260. doi:10.1016/j.robot.2016.04.003

Sergiyenko, O., Tyrsa, V., Basaca-Preciado, L., Rodríguez-Quiñonez, J., Hernandez, W., Nieto-Hipolito, J., . . . Starostenko, O. (2011). Electromechanical 3D Optoelectronic Scanners: Resolution Constraints and Possible Ways of Improvement. In Optoelectronic Devices and Properties. InTech.

Sergiyenko, O. Yu., Ivanov, M. V., Kartashov, V. M., Tyrsa, V. V., Hernández-Balbuena, D., & Nieto-Hipólito, J. I. (2016). Transferring model in robotic group. In *2016 IEEE 25th International Symposium on Industrial Electronics (ISIE)* (pp. 946-952). Santa Clara, CA: IEEE.

Volos, C., Kyprianidis, I., & Stouboulos, I. (2013, December). Experimental investigation on coverage performance of a chaotic autonomous mobile robot. *Robotics and Autonomous Systems*, *61*(12), 1314–1322. doi:10.1016/j.robot.2013.08.004

Xidias, E. K., & Azariadis, P. N. (2016, August). Computing collision-free motions for a team of robots using formation and non-holonomic constraints. *Robotics and Autonomous Systems*, *82*, 15–23. doi:10.1016/j.robot.2016.04.008

Chapter 13
Methods and Algorithms for Technical Vision in Radar Introscopy

Oleg Sytnik
A. Usikov Institute for Radio Physics and Electronics of the National Academy of Sciences of Ukraine, Ukraine

Vladimir Kartashov
Kharkiv's National University of Radio and Electronics, Ukraine

ABSTRACT

Optimization of technical characteristics of radio vision systems is considered in the radars with ultra-wideband sounding signals. Highly noisy conditions, in which such systems operate, determine the requirements that should be met by the signals being studied. The presence of the multiplicative noise makes it difficult to design optimal algorithms of echo-signal processing. Consideration is being given to the problem of discriminating objects hidden under upper layers of the ground at depths comparable to the probing pulse duration. Based upon the cepstrum and textural analysis, a subsurface radar signal processing technique has been suggested. It is shown that, however the shape of the probing signal spectrum might be, the responses from point targets in the cepstrum images of subsurface ground layers make up the texture whose distinctive features enable objects to be detected and identified.

INTRODUCTION

One of the main goals of technical vision systems, based on radar technology is to detect and identify target covered by obstacle. It is important that signal can penetrate throughout and return to receiver. For example, the problems that arise with probing earth subsurface layers using video pulse signals is the ability of discriminating the objects hidden under upper ground layers at depths comparable to the probing pulse duration (Grinev, 2005). The discrimination criteria for two targets are based on personal opinions and feelings. The following criteria are of considerable current use: a) Rayleigh resolution (Born, 1973) and b) statistical criteria (Levin, 1966). In accordance with the first one two point targets

DOI: 10.4018/978-1-5225-5751-7.ch013

the returned signals from which has identical power at the receiver input and the signals delayed relative to a probing pulse for τ_1 and τ_2 respectively can be regarded as defined ones using both a certain system of observation and an algorithm. The latter helps determine signal intensities if the total signals power measured at $\tau = (\tau_1 + \tau_2)/2$ is equal to 74% against the intensities of received signal at τ_1 or τ_2. This resolution criterion is usually employed where to ensure target delay resolution it is the power of received signals is solely utilized but not all their possible characteristics. This criterion is referred to as the Rayleigh resolution criteria. It is evident that, when in use, this criterion makes it fairly difficult to deals with objects' resolution problems, the objects featuring a significant difference in radar scattering cross section (RCS). According to the Rayleigh criterion the resolving power of the observation system involving such an algorithm implies the minimal delay time of signals returned from two point targets having an identical RCS, this time ensuring that the condition of this criterion is satisfied. The shorter is this time, the higher is the resolution of the observation system.

The statistical criterion formulation can be defined in the general form in the following way: two point targets described by identical probabilistic models and the delays of signal arrival from which are τ_1 and τ_2 can be considered to be resolvable with the observation system and as well as with the algorithm that processes the received signal-interference mix if the probability of correct indication of two sources is equal to P_r. In this instance the resolving power of the returned-signal processing algorithm under the statistical criterion is taken to be the minimal angular distance between sources, at which the prescribed probability of correct detection of two sources is ensured. It should be noted that in this particular case no constraints are imposed upon the resolution algorithm structure (in contrast to the Rayleigh criterion). Similar to the Rayleigh criterion it is necessary that the algorithm in question should be able to form the estimates of τ_1 and τ_2 delays. In the statistical criterion the probability P_r is the criterion parameter in the same way as for the Rayleigh criterion the number 74% is the basic parameter.

Thus, one can suggest a great number of resolution algorithms for one and the same observation system. Clearly, of all possible algorithms one might as well refer to the one which in fact provides for the best resolving power.

THE CEPSTRUM METHOD OF DELAY MEASUREMENT

The concept "cepstrum" (Noll, 1967) has been known since the middle of the last century and is defined by the following expression:

$$C_s(q) = \frac{1}{2\pi} \cdot \int_{-\infty}^{\infty} \ln\left[S(\omega)\right]^2 \cdot e^{j\omega q} d\omega \tag{1}$$

where $S(\omega)$ is the amplitude spectrum of signal $s(t)$.

However, expression (1) is not valid for a totally arbitrary signal. For a finite-energy signal the condition $\int_{-\infty}^{\infty} S^2(\omega) d\omega < \infty$ should be met, from which it follows that at $|\omega| \to \infty$, then $S^2(\omega) \to 0$.

Function $\left| \ln \left(S^2 \left(\omega \right) \right) \right|$ at $S^2 \left(\omega \right) \to 0$ becomes infinite and integral (1) diverges. Therefore, for practical purposes, expression (1) can be used over a frequency range of $\pm \Delta \omega$ only, where the main energy of the signal spectrum is concentrated, and, accordingly integration in (1) is performed not within the indefinite limits, but rather over the limits $\pm \Delta \omega$.

It is evident that argument q in (1) has the time dimensionality. An important property of cepstrum representation of signals is that it is possible to obtain a direct estimate of the signal delay regardless of the shape of its spectrum. This can be demonstrated as follows. Let represent signal $s \left(t \right)$ observed on the time interval $\left[0, T \right]$ in the discreet form

$$s \left(m \right) = s_1 \left(m \right) + \alpha s_1 \left(m - m_0 \right), \quad \alpha < 1 \tag{2}$$

where $\left\{ s_1 \left(m \right) \right\}$, $m = 0, 1, ..., N - 1$ is the probing signal sampled by step Δt according to the Nyquist theorem (Wiener, 1949; Openheim, 1979); N is the number of signal samples on the observation interval $\left[0, T \right]$; $\alpha s_1 \left(m - m_0 \right)$ is the target signal delayed for $\tau = m_0 \Delta t$.

It is evident that if the initial signal $s_1 \left(m \right)$ corresponds to z-transformation $S_1 \left(z \right)$ then z-transformation of the second component in (2) can be written as $\alpha S_1 \left(z \right) z^{-m_0}$. Now z-transformation of total signal $s \left(m \right)$ takes the following form:

$$S \left(z \right) = S_1 \left(z \right) + \alpha S_1 \left(z \right) z^{-m_0} = S_1 \left(z \right) \cdot \left(1 + \alpha z^{-m_0} \right) = S_1 \left(z \right) \cdot S_2 \left(z \right) \tag{3}$$

Based upon the known properties of z-transformation (Openheim, 1979), according to (3), the initial signal (2) can be considered as convolution

$$s \left(m \right) = s_1 \left(m \right) \otimes s_2 \left(m \right) \tag{4}$$

where $s_2 \left(m \right)$ is some signal whose -transformation has the form $\left(1 + \alpha z^{-m_0} \right)$.

Hence, in a time domain the signal $s_2 \left(t \right) = \delta \left(t \right) + \alpha \delta \left(t - \tau \right)$ is the sum of two delta-functions. Now substitute variable $e^{j \omega \Delta t}$ into (3) instead of variable z, whereas for the second cofactor one can write

$$\left| S_2 \left(e^{j \omega \Delta t} \right) \right|^2 = \left| 1 + \alpha e^{-j \omega \tau} \right|^2 = \left(1 + \alpha e^{-j \omega \tau} \right) \left(1 + \alpha e^{j \omega \tau} \right) = 1 + \alpha^2 + 2 \alpha \cos \left(\omega \tau \right) \tag{5}$$

The square of the signal spectral density module (3) including (5) can be written as

$$\left| S \left(e^{j \omega \Delta t} \right) \right|^2 = \left| S_1 \left(e^{j \omega \Delta t} \right) \right|^2 \cdot \left[1 + \alpha^2 + 2 \alpha \cos \left(\omega \tau \right) \right] \tag{6}$$

375

From expression (5) it follows that the mix of the probing signal and the target signal constitutes the amplitude-modulated signal spectrum under the law $1 + \alpha^2 + 2\alpha \cos(\omega\tau)$. At the same time the depth of modulation is defined by the coefficient $2\alpha / (1 + \alpha^2)$, and the modulation period is equal to τ. To find the cepstrum, take the logarithm of expression (6)

$$\ln\left|S\left(e^{j\omega\Delta t}\right)\right|^2 = \ln\left|S_1\left(e^{j\omega\Delta t}\right)\right|^2 + \ln\left[1 + \alpha^2 + 2\alpha\cos(\omega\tau)\right] \tag{7}$$

In keeping with definition (1), substitute (7) into (1) and change the integration variable

$$C_s(m) = \frac{1}{2\pi}\int_{-\pi}^{\pi}\ln\left|S_1\left(e^{j\omega\Delta t}\right)\right|^2 d(\omega\Delta t) + \frac{1}{2\pi}\int_{-\pi}^{\pi}\ln\left|1 + \alpha e^{-j\omega\tau}\right|^2 d(\omega\Delta t) = C_{S_1} + C_{S_2} \tag{8}$$

It is apparent that perfect information on signal delay is contained in the second term of (8). By expanding the sub integral function of the second component into a series it is easy to make sure that this function will be something other than 0 at points $m = \pm m_0 \pm 2m_0, \dots$. The true delay of the returned signal is determined from the position of the first peak of this function. Figure 1 shows the probing signal model obtained as a Gaussian video pulse passes through the radiating structure.

Figure 2 presents the cepstrum of initial and time-delayed at $\tau = 0,029$ ns signals.

For comparison Figure 3 and Figure 4 show the cross-correlation function of this particular signal delayed for $\tau = 0,029$ ns and $\tau = 0,29$ ns.

Figure 1. Probing video pulse signal model of the ground-penetrating radar (GPR)

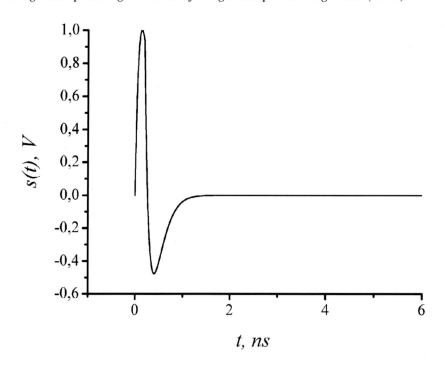

Figure 2. The cepstrum of initial and time-delayed at $\tau = 0,029$ *ns signals*

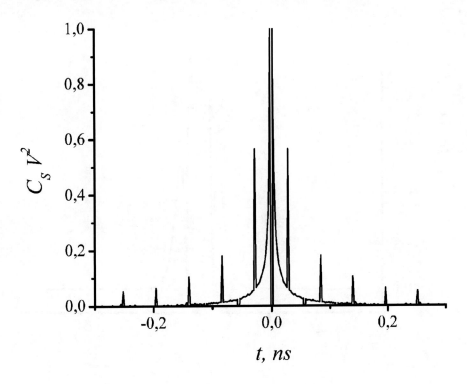

Figure 3. Correlation function at signal delay $\tau = 0,029$ *ns*

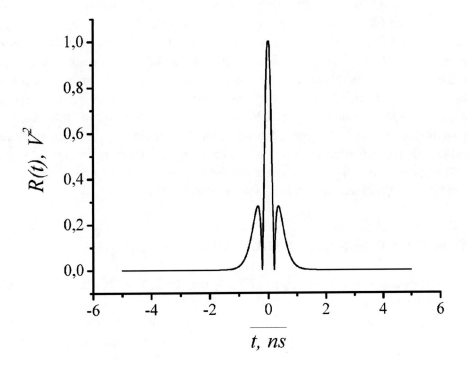

Figure 4. Correlation function at signal delay $\tau = 0,29$ *ns*

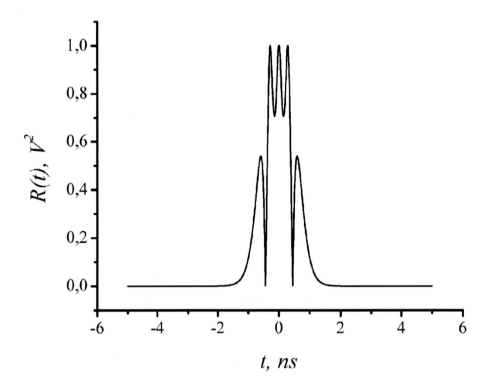

As it is seen from Figure 4, the signals are resolvable according to the Rayleigh criterion. With a signal delayed for $\tau = 0,029$ ns (Figure 3) the signals are not resolvable in Rayleigh criterion. The cepstrum criterion (see Figure 2) makes it possible to determine the target signal delay and, consequently, to resolve point targets at distances that by an order of magnitude smaller than in using correlation processing and the Rayleigh criterion. It can be shown that the limiting resolution for the cepstrum criterion will be the magnitude of the time delay equal to signal time step sampling and the resolving power of the observation system is not dependent upon the shape of the probing signal spectrum.

The limiting resolution as complied with the Rayleigh criterion is determined by the quantity inversely proportional to that of the effective signal spectrum width. The influence of noise upon resolution using the cepstrum criterion reveals itself as a threshold effect using the cepstrum criterion. At a signal/noise ratio of 20 dB and more the delays do not affect the estimation error. If the real S/N ratio is lower than a threshold value, then the cepstrum criterion is not usable.

FORMATION OF SUBSURFACE STRUCTURE IMAGES

The video pulse subsurface radar (Harmouth, 1985; Sytnik, 2007,2006, 2012) is capable of producing radar images (see Figure 5) as it moves relative to a point subsurface reflector. Coordinate x (Figure 5) is directed along the line on which a radar is moving. Arrival time of the returned-signal is registered along coordinate y. The total extent over the earth surface along coordinate x, as it is seeing in Figure 5, is around 50 ns respectively. At the top part of the image one can observe a uniform strip arising from the

Figure 5. Radar image formed by video pulse GPR

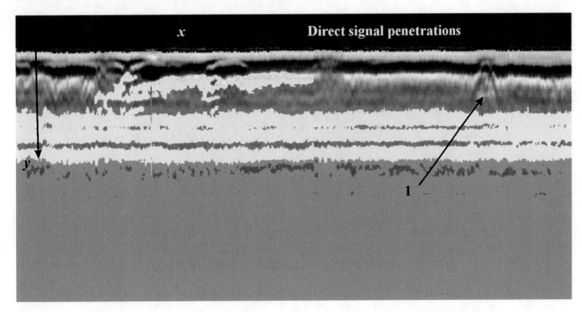

direct signal penetration of the transmitter signal to the receiver. It is a result of limiting the minimally possible probing depth. Furthermore, alternating horizontal light- and dark-colored strips are visible, which correspond to soil inhomogeneities.

The responses from point targets are shown up as more pronounced geometric structures (1) of hyperbolic shape (Figure 5), which are clearly seen against the background of soil heterogeneousness. This allows one to identify objects by means of the cepstrum criterion. A two-dimensional (2-D) cepstrum of an object (a point target – the hyperbole in the right-hand side of Figure 5) is shown in Figure 6. Figure 7 illustrates a 2-D cepstrum of the point-target response model.

Figure 6. The 2-D cepstrum of the real point target

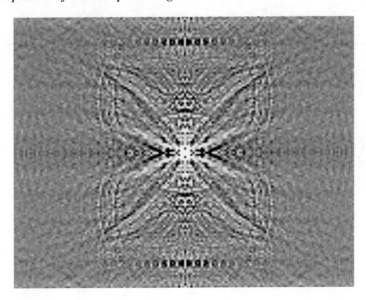

Figure 7. The 2-D cepstrum of the model point target

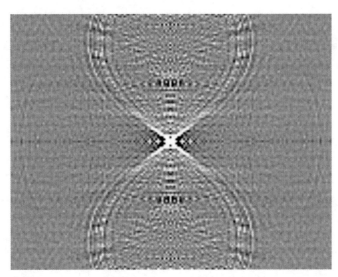

Identification of point targets in a radar image of subsurface structures was performed by comparing the textures of real objects to a model intern of training sample from m peculiar points. The false alarm probabilities plotted versus the volume of training sample m are shown in Figure 8. Curve 1 corresponds to SN ratio = 14 dB; curve 2 – to SN ratio = 16 dB; curve 3 is for SN ratio = 18 dB; curve 4 is for SN ratio = 20 dB; curve 5 corresponds to SN ratio = 22 dB.

As can be seen from the examination of Figure 8, the false alarm probability rapidly decreases with a rising volume of a training sample. The impact of the training sample volume is particularly dramatic in the low-value domain $m \leq 40$. Specifically, an increase in the sample volume from 5 to 40 leads to a decrease of the error probability from 0,08 to 0,003 whereas an increase of m from 40 to 80 tends to

Figure 8. The false alarm probabilities as a function of the training sample in the space of signal parameter spectra

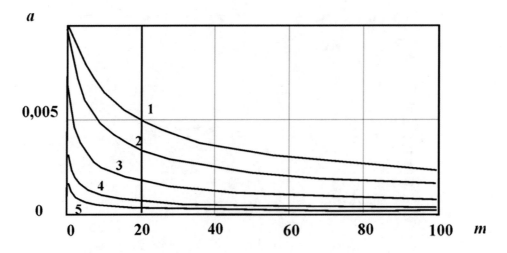

reduce the false alarm probability from 0,003 down to 0,0025. The reason is that at high values of *m* the solving statistics has been formed, and further training makes an insignificant contribution to the description of the probability density distribution. Moreover, with a rising S/N ratio and reducing the value of α the volume of the training sample can be diminished. An increase of S/N ratio by 7-8 dB at $m = 20$ results in decreasing the error probability from 0,005 down to 0,0001. Thus, since S/N ratio can be thought of as a kind of a universal measure of differences in signal and interference statistics, it is evident that the quality of estimates obtained in identifying an object will depend not only upon the latter, but also upon the nature of the set of the training points.

HOMOMORPHIC SIGNAL PROCESSING ALGORITHM OF SUBSURFACE TECHNICAL VISION SYSTEMS

Most part of the ground topsoil is formed by the media having significant values of dispersion and absorption, which vary with variation of the moisture content in the layers along the signal propagation path. Besides, to obtain high resolution upon the sensing depth in ground penetrating radars it is required to use picosecond pulses for impact excitation of the antenna by the voltage drop like the pulse shown at figure 1. But as the result of resonance effects, the antenna forms up a variable-sign attenuating train of electromagnetic waves (Figure 9), which is significantly (by several times) exceeding the duration of exciting video pulse that confines the ground penetrating radar resolution in terms of the depth of subsurface sensing.

Figure 9. Approximate shape of oscillations in the ground penetrating radar antenna at its excitation by a video pulse

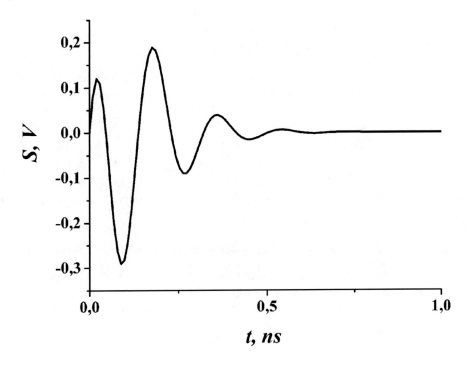

For a number of practically important applications this particularity of a ground penetrating radars creates the problems, which are almost impossible to overcome. The problem of probing pavements of motor-car roads where the thicknesses of the layers might amount from tens of centimeters to several centimeters is a typical example of the above. Availability of two closely positioned boundaries between the media forms the interference image (Figure 10), on the basis of which it is practically impossible to perform measurements of the ranges to each of the boundaries.

Under the above conditions the radar image contains multiple overlapping dark and bright stripes (Figure 11) hampering identification of the objects under investigation.

It is evident that at the extreme correlation processing (Noll, 1967) of the reflected signals the radar image of such a part of the motor-car road makes no practical sense because measurements of layer thickness or detection of defects of the pavement are impossible at such small ranges.

Figure 10. The result of reflection of the signals from closely positioned subsurface layers of the ground

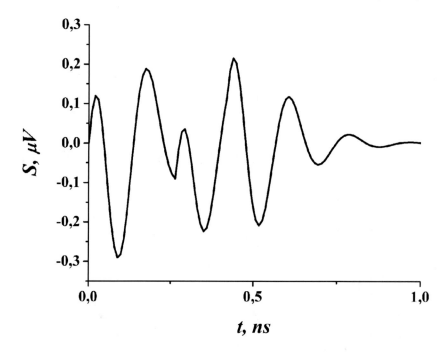

Figure 11. A fragment of the near-surface domain of radar image of the two-layer part at correlational-extremal processing of reflected signals

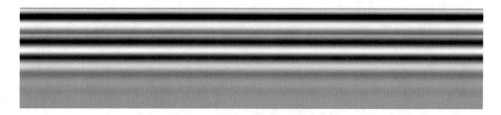

The solution to the problem for a number of particular cases can be found if it is taken into consideration that the range from the asphalt surface (if a motor-car road is selected as the object for probing) to its lower boundary is not large (about several centimeters or tens of centimeters). As the result, the reflected signal level at the input of the ground penetrating radar receiver exceeds the level of eigen noise of the receiver by 20...30 dB. At such values of the signal-to-noise ratio the algorithms of non-linear homomorphic signal processing (Kemerait, 1972; Hassah, 1994; Sytnik, 2011) turn out to be efficient. In order to avoid bulky mathematical correlations and to preserve, at the same time, generality of the analysis, let us consider the signal of one row of the video pulse ground penetrating radar (Grinev, 2005) depth profile

$$y(t) = x(t) + \alpha x(t - \tau) + n(t) \tag{9}$$

where $x(t)$ is the probing signal with the spectral density $\dot{F}_x(\omega)$; $\alpha < 1$ is the attenuation factor of the signal at its propagating in the medium; τ is the signal delay for the time necessary for propagation of the probing signal to the reflecting subsurface boundary and backwards; $n(t)$ is realization of the noise with the spectral density $N(\omega)$ and mathematical expectation equal to 0. So, it is required to estimate the parameter τ. If the layer thickness in the probed medium permits fulfilling of the condition

$$\tau >> \tau_p \tag{10}$$

where τ_p is the total probing signal duration in the propagation medium considering the resonant effects in the antenna, then the parameter τ can be calculated with the help of a number of well-known in the statistical radio physics methods (Levin, 1974-1976; Repin, 1977; Van Trees, 1972-1977). However, for a number of practically important cases when the condition (10) is not fulfilled and the inequality sign is changed to its opposite, the classical time delay estimation methods are non-applicable.

The signal (9) observed in the spectral domain can be expressed in a following way:

$$\dot{F}_y(\omega) = \dot{F}_x(\omega)\left(1 + \alpha e^{j\omega\tau}\right) + \dot{N}(\omega) \tag{11}$$

The squared absolute value (11) will have the following representation

$$\left|\dot{F}_y(\omega)\right| = \left(1 + \alpha^2\right)\left|\dot{F}_x(\omega)\right|^2 + 2\alpha\left|\dot{F}_x(\omega)\right|^2 +$$

$$+\left|\dot{N}(\omega)\right|^2 + \left(1 + \alpha e^{-j\omega\tau}\right)\dot{F}_x(\omega)\dot{N}^*(\omega) + \left(1 + \alpha e^{-j\omega\tau}\right)\dot{F}_x^*(\omega)\dot{N}(\omega) \tag{12}$$

where * is the complex conjugation symbol.

As it is evident from (12), the spectrum of the observed process contains the crossspectrum components $\dot{F}_x(\omega)$ and $\dot{N}(\omega)$, which represent an interference at signal processing. Let us designate the cross-spectrum components as $\dot{F}_{xn}(\omega)$ and express them via the real and imaginary parts:

$$\dot{F}_{xn}(\omega) = 2\left[\operatorname{Re}\left\{\dot{F}_x(\omega)\right\} + \alpha \operatorname{Re}\left\{\dot{F}_x(\omega)\right\}\cos(\omega\tau) + \alpha \operatorname{Im}\left\{\dot{F}_x(\omega)\right\}\sin(\omega\tau)\right]\operatorname{Re}\left\{\dot{N}(\omega)\right\} + \quad (13)$$

$$+\left[\operatorname{Im}\left\{\dot{F}_x(\omega)\right\} + \alpha \operatorname{Im}\left\{\dot{F}_x(\omega)\right\}\cos(\omega\tau) - \alpha \operatorname{Re}\left\{\dot{F}_x(\omega)\right\}\sin(\omega\tau)\right]\operatorname{Im}\left\{\dot{N}(\omega)\right\}.$$

By means of apparent transforms we simplify (13), and at the same time designate

$$A = \operatorname{Re}\left\{\dot{F}_x(\omega)\right\} + \alpha \operatorname{Re}\left\{\dot{F}_x(\omega)\right\}\cos(\omega\tau) + \alpha \operatorname{Im}\left\{\dot{F}_x(\omega)\right\}\sin(\omega\tau)$$

$$B = \operatorname{Im}\left\{\dot{F}_x(\omega)\right\} + \alpha \operatorname{Im}\left\{\dot{F}_x(\omega)\right\}\cos(\omega\tau) - \alpha \operatorname{Re}\left\{\dot{F}_x(\omega)\right\}\sin(\omega\tau).$$

It is evident that

$$A^2 + B^2 = \left|\dot{F}_x(\omega)\right|^2\left(1 + \alpha^2 + 2\alpha\cos(\omega\tau)\right)$$

$$\beta = 2\alpha\left(1 + \alpha^2\right)^{-1}$$

$$\left|\dot{R}(\omega)\right|^2 = \left\{1 + \left|\dot{N}(\omega)\right|^2\left[\left(1 + \alpha^2\right)\left|\dot{F}_x(\omega)\right|^2\right]^{-1}\right\}^{-1}$$

Then

$$\left|\dot{F}_y(\omega)\right|^2 = \left(1 + \alpha^2\right)\left|\dot{F}_x(\omega)\right|^2\left[1 + \left|\dot{N}(\omega)\right|^2\left[\left(1 + \alpha^2\right)\left|\dot{F}_x\right|^2\right]^{-1}\right] \times \left[1 + \beta\left|\dot{R}(\omega)\right|^2\cos(\omega\tau)\right]\left[1 + \left|\dot{Z}(\omega)\right|^2\right] \quad (14)$$

Taking the logarithms from the right-hand and the left-hand sides of (6), it is obtained

$$\ln\left(\left|\dot{F}_y(\omega)\right|^2\right) = \ln\left\{\left(1 + \alpha^2\right)\left|\dot{F}_x(\omega)\right|^2\right\} + \ln\left\{1 + \left|\dot{N}(\omega)\right|^2\left[\left(1 + \alpha^2\right)\left|\dot{F}_x(\omega)\right|^2\right]^{-1}\right\} +$$

$$+\ln\left\{1 + \beta\left|\dot{R}(\omega)\right|^2\cos(\omega\tau)\right\} + \ln\left\{1 + \left|\dot{Z}(\omega)\right|^2\right\} \quad (15)$$

The analysis of (15) shows that the first summand is represented by a determinate member, which is non-dependent upon τ. The second and the last summands of (15) contain cross-spectra of the signal and noise that is a direct interference in separating the information about the time delay. In the summand $\ln\left\{1 + \beta\left|\dot{R}(\omega)\right|^2\cos(\omega\tau)\right\}$ the function $\dot{R}(\omega)$ is nothing else but modulation of the time delay cepstrum by the noise. Let us transform (15) by means of decomposing the last two summands upon the formula (Bronshtein, 1980): $\ln(1+x) = \sum\limits_{k=1}^{\infty}(-1)^{k+1}\dfrac{x^k}{k}$. As the result we obtain

$$\ln\left(\left|\dot{F}_y(\omega)\right|^2\right) = \ln\left\{\left(1+\alpha^2\right)\left|\dot{F}_x(\omega)\right|^2\right\} + \ln\left\{1 + \left|\dot{N}(\omega)\right|^2\left[\left(1+\alpha^2\right)\left|\dot{F}_x\right|^2\right]^{-1}\right\} +$$

$$+\sum_{k=1}^{\infty}\left[(-1)^{k+1}D_k(\omega)\cos(k\omega\tau)\right] + \sum_{l=1}^{\infty}\frac{(-1)^{l+1}}{l}\left|\dot{Z}(\omega)\right|^{2l} \qquad (16)$$

where

$$D_0(\omega) = \frac{\left\{\beta\left|\dot{R}(\omega)\right|^2\right\}^2}{4} + \frac{\left\{\beta\left|\dot{R}(\omega)\right|^2\right\}^4}{32} + \cdots$$

$$D_1(\omega) = \beta\left|\dot{R}(\omega)\right|^2 + \frac{\left\{\beta\left|\dot{R}(\omega)\right|^2\right\}}{4} + \frac{\left\{\beta\left|\dot{R}(\omega)\right|^2\right\}^5}{8} + \cdots$$

The inverse Fourier transform from (16) provides for

$$\Im^{-1}\left\{\ln\left|\dot{F}_y(\omega)\right|^2\right\} = \Im^{-1}\left\{\ln\left(1+\alpha^2\right)\left|\dot{F}_x(\omega)\right|^2\right\} + \Im^{-1}\left\{\ln\left\{1 + \left|\dot{N}(\omega)\right|^2\left[\left(1+\alpha^2\right)\left|\dot{F}_x(\omega)\right|^2\right]^{-1}\right\}\right\} + \sum_{l=1}^{\infty}\frac{(-1)^{l+1}}{l}d_l(t) -$$

$$-\left\{\frac{\beta^2}{4}C_2(t) + \frac{3}{32}\beta^4 C_4(t) + \frac{5}{96}\beta^6 C_6(t) + \cdots\right\} + 0,5\left\{\beta b(t-\tau) + \frac{\beta^3}{4}C_3(t-\tau) + \frac{\beta^5}{8}C_5(t-\tau) + \cdots\right\} -$$

$$-0,5\left\{\frac{\beta^2}{4}C_2(t-2\tau) + \frac{\beta^2}{8}C_4(t-2\tau) + \frac{5}{64}\beta^6 C_6(t-2\tau) + \cdots\right\} + 0,5\left\{\frac{\beta^3}{12}C_3(t-3\tau) +$$

$$+\frac{\beta^5}{16}C_5(t-3\tau) + \cdots\right\} - 0,5\left\{\frac{\beta^6}{192}C_6(t-6\tau) + \cdots\right\}\cdots \qquad (17)$$

where $\Im^{-1}\left\{\dot{R}\left(\omega\right)\right\} = b\left(t\right)C_i\left(t\right)$, while $d_i\left(t\right)$ and $C_i\left(t\right)$ represent relevant convolutions of the process $x\left(t\right)$ with $b\left(t\right)$ and $C\left(t\right)$ correspondingly.

Analysis of the correlation (17) shows that unambiguous identification of the position on the time axis of the cepstrum peak, which corresponds to the true time delay of the signal τ, is rather complicated under availability of the noise. Actually, after transformation of (17) the next processing stage will be solving of the statistical problem of separation of the determinate signal at the fluctuation interference background, whereas it is necessary to perform a preliminary statistical estimation of the interference concerned.

ESTIMATION OF STATISTICAL PROPERTIES OF THE NOISE

Initially, in (9) the noise $n(t)$ was represented by a stochastic Gaussian process with the known spectral density and zero mathematical expectation. Apparently that the process probability density $n(t)$ can be represented in the form of (Levin, 1974-1976)

$$W\left(t\right) = \frac{1}{\sqrt{2\pi\sigma^2}}\, e^{-\frac{n^2\left(t\right)}{2\sigma^2}}$$

After applying the Fourier transform to the correlation (9) the interference probability density can be put down as follows

$$W\left(N_r\left(\omega\right), N_i\left(\omega\right)\right) = \frac{\exp\left\{-\frac{\left[N_r^2\left(\omega\right) + N_i^2\left(\omega\right)\right]}{\left|\dot{N}\left(\omega\right)\right|^2}\right\}}{\pi\left|\dot{N}\left(\omega\right)\right|^2} \tag{18}$$

where $N_r\left(\omega\right), N_i\left(\omega\right)$ are the real and imaginary components of the noise spectrum correspondingly.

As it follows from (18) that the resulting interference process in the spectral domain is represented by a process with the distribution χ^2 (Van Trees, 1972-1977) and two degrees of freedom. After the non-linear logarithmic transform the initial noise process acquired the following form of the noise

$$\dot{N}\left(\omega\right) = \left(1 + \alpha^2\right)\dot{F}_x\left(\omega\right)\left\{e^{\dot{C}\left(\omega\right)} - 1\right\}$$

where $\dot{C}\left(\omega\right) = \ln\left\{1 + \dot{N}\left(\omega\right)\left[\left(1 + \alpha^2\right)\dot{F}_x\left(\omega\right)\right]^{-1}\right\}$

The probability density of such a noise is

$$W\left(\dot{N}\left(\omega\right)\right) = \begin{cases} \dot{\chi}\left(\omega\right)\exp\left[\dot{C}\left(\omega\right) - \dot{\chi}\left(\omega\right)\exp\left(\dot{C}\left(\omega\right) - 1\right)\right] & C \geq 0 \\ 0 & C < 0 \end{cases} \tag{19}$$

where $\dot{\chi}\left(\omega\right) = \left(1 + \alpha^2\right)\dot{F}_x\left(\omega\right)\left[\dot{N}\left(\omega\right)\right]^{-1}$ is the weighed with the coefficient $\left(1 + \alpha^2\right)$ ratio between the signal spectrum density and the spectrum density of the noise.

The probability density of combinational components of the cepstrum has the following form

$$W\left(R\left(\omega\right)\right) = \begin{cases} \dfrac{\chi\left(\omega\right)}{R^2\left(\omega\right)}e^{\left\{-\chi\left(\omega\right)\left(R^{-1}\left(\omega\right) - 1\right)\right\}} \end{cases} \tag{20}$$

The expressions (19) and (20) are represented by the non-Gaussian probability density distributions of the processes, each of which can be used for point estimation of signal-to-noise ratios within the limits of determining the function $\chi\left(\omega\right)$ in (19). The family of functions $W\left(R\left(\omega\right)\right)$ at different values of $\chi\left(\omega\right)$ is shown in Figure 12.

It should be noted that at low values of the signal-to-noise ratio main values of $W\left(R\left(\omega\right)\right)$ are concentrated near the zero values of the argument. Due to dependence upon the frequency of all the functions included into the correlation (20) the highfrequency components of the cross-cepstrum are subject to masking by the interference to the greatest extent. This phenomenon imposes serious restrictions upon

Figure 12. Family of distributions of the function W(R) at different values of $\chi\left(\omega\right)$

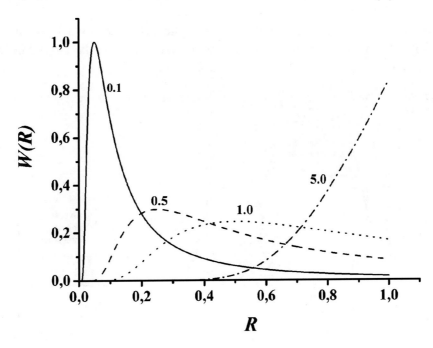

the possibility of applying a cepstrum-based approach to 'super-resolution' of the surface layers in the case with the clearly expressed multilayer structures with closely positioned boundaries, because the periodic cepstrum of the reflected signal delays is rapidly decreasing along the signal propagation path. However, for simple structures with two separation boundaries of separation the method of cepstrum-processing produces better results.

STRUCTURE OF SIGNAL PROCESSING ALGORITHM

Figure 13 shows the generalized block diagram of the homomorphic processing algorithm for ground penetrating radar signals. The radar output signal is subject to the Fourier transform, then it is performed calculation of the square of the spectrum absolute value and taking the logarithms from it. The result of the above operation is subject to the inverse Fourier transform, thus providing for shaping of the signal cepstrum at the input of the threshold device. The threshold device is used for solving of the problem of detection of the first peak of a discrete cepstrum of the signal delayed for the time of its propagation to the lower boundary of the probed layer and backwards to the receiver. Threshold selection is performed in accordance with the method of maximum trustworthiness (Levin, 1974-1976) at the fixed value of the probability distribution function (20).

The result of image processing (Figure 10) performed with the help of the above procedure is shown in Figure 14.

Comparison of Figsures 10 and 14 demonstrates that the boundaries of the near-surface layer are resolved, that is practically unattainable at applying extreme correlation signal processing algorithms.

CONCLUSION

The processing results provide an apparent illustration of the fact that at relatively high signal-to-noise ratios the non-linear homomorphic signal processing allows increasing the ground penetrating radar

Figure 13. Structure of the homomorphic signal processing algorithm of the ground penetrating

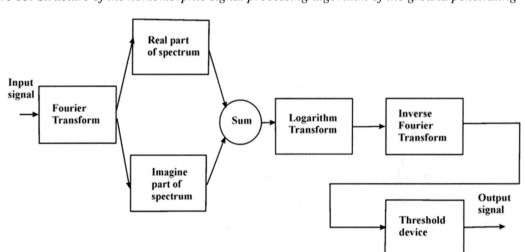

Figure 14. Fragment of the near-surface domain of radar image of the two-layer part at using cepstrum processing of the reflected signals

resolution upon depth by up to 5% of the pulse duration that cannot be attained while using traditional spectrum correlation algorithms. However, it must be born in mind that it is attainable at a small depth in the near surface layers. The restrictions in terms of the resolution are stipulated by the fact that the rate of decreasing of the signal cepstrum envelope curve is a finite value and at the times of the reflected signal delay of less than 5% of the probing pulse duration the first peak of the delay cepstrum is hidden under the envelope curve of the probing signal cepstrum.

Another aspect, which is limiting the scope of application for the above algorithm, is stipulated by the dependence upon frequency of all the functions included into the correlation (20). Evidently, the high-frequency components of the cross-cepstrum are subject to masking by the interference to the greatest extent. This factor imposes serious restrictions on the possibility of application of the cepstrum-based approach to 'super-resolution' of the near-surface layers in the case of explicitly expressed multilayer structures with closely positioned boundaries, because the periodic cepstrum of the reflected signal delays is rapidly decreasing along the signal propagation path.

One of the possible ways to decrease the influence of the fluctuation noise upon the procedure of identification of the boundaries of reflecting subsurface structures is in applying a preliminary, the so-called 'window', weighing of the squared spectrum absolute value logarithm function of the basic signal. This procedure is used before applying the inverse Fourier transform. However, in that case, it occurs decreasing of the potential resolution capacity depending upon the properties of the selected window.

Despite the fact that the result is attained due to a substantial complication of the processing procedure this approach is of a great importance for a number of practical problems. Using of relatively powerful in modern meaning of the term computing systems allows compensation of the above drawback and performing processing of signals in a real time scale. Using of the ground penetrating radars with software embedded homomorphic signal processing would allow saving costs while performing road repair works, road building, at diagnostics and non-destructive control over civil construction structures and solving a number of other practically important problems.

REFERENCES

Born, M., & Wolf, E. (1973). *Fundamentals of optics*. Moscow: Nauka. (in Russian)

Bronshtein, I. N., & Semendyayev, K. A. (1980). *Handbook of mathematics*. Moscow: Nauka. (in Russian)

Grinev, A. Yu. (Ed.). (2005). *Problems of subsurface radiolocation*. Moscow: Radiotehnika. (in Russian)

Harmouth, Kh. F. (1985). *Nonsinusoidal waves in radio location and radio communication*. Moscow: Radio and Svyaz. (in Russian)

Hassah, J. C. (1994). Time delay processing near the ocean surface. *Journal of Sound and Vibration, 35*(4), 489–501.

Kemerait, R. C., & Childers, D. G. (1972). Signal detection and extraction by Cepstrum techniques. *IEEE Transactions on Information Theory, IT-18*(1), 745–759. doi:10.1109/TIT.1972.1054926

Levin, B. R. (1966). *Theoretical foundations of statistical radioengineering*. Moscow: Sov. Radio. (in Russian)

Levin, B.R. (1974-1976). *Theoretical foundations of the statistical radio engineering: in 3 vol.* Sov. radio. (in Russian)

Masalov, S. A., Sytnik, O. V., & Ruban, V. P. (2012). Wavelet UWB signal processing for underground sounding systems. *Proc. Int. Conf. "Ultrawideband and Ultrashort Impulse Signals*, 123–125. 10.1109/UWBUSIS.2012.6379754

Noll, A.M. (1967). Cepstrum pitch determination. *Trans of the Acoustical Soc. of America, 41*(2).

Openheim, A. V., & Shafer, R. V. (1979). *Digital signal processing*. Moscow: Svyaz. (in Russian)

Repin, V. G., & Tartakovsky, G. P. (1977). *Statistical synthesis with a priori indeterminacy and adaptation of informational systems*. Moscow: Sov. Radio. (in Russian)

Sytnik, O. V. (2006). Ground-penetrating radar data preprocessing. *Telecommunications and Radio Engineering, 65*(7), 621–631. doi:10.1615/TelecomRadEng.v65.i7.40

Sytnik, O. V. (2011). Textural Analysis of Cepstrum Images of Subsurface Structure. *Telecommunications and Radio Engineering, 70*(1), 87–94. doi:10.1615/TelecomRadEng.v70.i1.90

Sytnik, O. V., & Gorokhovatsky, A. V. (2007). Algorithm for signal processing in identifying subsurface objects, *Izv. VUZov.* [in Russian]. *Radioelektronika, 50*(10), 43–52.

Van Trees, H. (1972-1977). *Detection estimation and modulation theory: in 3 vol.* Sov. radio. (in Russian)

Wiener, N. (1949). *Extrapolation, interpolation and smoothing of stationary time series*. New York: John Willey.

KEY TERMS AND DEFINITIONS

Cepstrum: Is the inverse Fourier transform of the logarithm of the square of the signal spectrum.

False Alarm: Is the deceptive or erroneous report of an emergency, causing unnecessary panic and/or bringing resources (such as emergency services) to a place where they are not needed.

Ground-Penetrating Radar: Is special radar which radiates UWB signal to soil and receive reflated signal.

Inhomogeneities: Is something that is not homogeneous or uniform.

Rayleigh Resolution Criteria: Describes the ability of any image-forming device such as an optical or radio telescope, rarar, a microscope, a camera, or an eye, to distinguish small details of an object, thereby making it a major determinant of image resolution.

S/N Ratio: Is a ratio of signal power to dispersion of noise.

Statistical Criterion: Two point targets described by identical probabilistic models and the delays of signal arrival from which are τ_1 and τ_2 can be considered to be resolvable with the bservation system and as well as with the algorithm that processes the received signal-interference mix if he probability of correct indication of two sources is equal to P_r.

Compilation of References

Abdul Aziz, Z. A., & Marzuki, A. (2010). Residual Folding Technique adopting switched capacitor residue amplifiers and folded cascode amplifier with novel pmos isolation for high speed pipelined ADC applications. *3rd AUN/SEED-Net Regional Conference in Electrical and Electronics Engineering: International Conference on System on Chip Design Challenges (ICoSoC 2010)*, 14–17.

Acosta, D., Garcia, O., & Aponte, J. (2006). Laser Triangulation for shape acquisition in a 3D Scanner Plus Scan. In *Electronics, Robotics and Automotive Mechanics Conference* (pp. 14-19). IEEE.

Aggarwal, J. K., & Xia, L. (2014). Human activity recognition from 3D data: A review. *Pattern Recognition Letters*, *48*, 70–80. doi:10.1016/j.patrec.2014.04.011

Ahola, R., & Millylä, R. (1986). A Time-of-Flight Laser Receiver for Moving Objects. *IEEE Transactions on Instrumentation and Measurement*, *IM-35*(2), 216–221. doi:10.1109/TIM.1986.6499093

Aizawa, K., Nakamura, Y., & Satoh, S. (2004). *Advances in Multimedia Information Processing - PCM 2004: 5th Pacific Rim Conference on Multimedia, Tokyo, Japan, November 30 - December 3, 2004, Proceedings*. Springer Berlin Heidelberg.

Aksenov, V. p., & Pogutsa, Ch. (2012). The effect of an optical vortex on the random shifts of a Laguerre-Gaussian laser beam propagating in the turbulent atmosphere. *Atmospheric and Oceanic Optics*, *25*(7), 561–565.

Alazrai, R., Mowafi, Y., & Lee, C. (2015). Anatomical-plane-based representation for human–human interactions analysis. *Pattern Recognition*, *48*(8), 2346–2363. doi:10.1016/j.patcog.2015.03.002

Aldrich, C., Marais, C., Shean, B. J., & Cilliers, J. J. (2010). Online monitoring and control of froth flotation systems with machine vision: A review. *International Journal of Mineral Processing*, *96*(1), 1–13. doi:10.1016/j.minpro.2010.04.005

Ali, A. A., Rashid, A. T., Frasca, M., & Fortuna, L. (2016). An algorithm for multi-robot collision-free navigation based on shortest distance. *Robotics and Autonomous Systems*, *75*, 119–128. doi:10.1016/j.robot.2015.10.010

Alivisatos, P. (2004). The use of nanocrystals in biological detection. *Nature Biotechnology*, *22*(1), 47–52. doi:10.1038/nbt927 PMID:14704706

Aptina. (2010). *An Objective Look at FSI and BSI*. Aptina White Paper.

Ardeshirpour, Y., Deen, M. J., Shirani, S., West, M. S., & Ls, O. N. (2004). 2-D CMOS based image sensor system for fluorescent detection. In *Canadian Conference on Electrical and Computer Engineering* (pp. 1441–1444). IEEE.

Artikis, A., Sergot, M., & Paliouras, G. (2010). A logic programming approach to activity recognition. In *International workshop on events in multimedia (EiMM 2010)* (pp. 3–8). ACM. 10.1145/1877937.1877941

Baader, F., Calvanese, D., McGuinness, D., Nardi, D., & Patel-Schneider, P. (2002). *The Description Logic Handbook*. Cambridge University Press.

Badica, C., Braubach, L., & Paschke, A. (2011). Rule-based distributed and agent systems. In *International Workshop on Rules and Rule Markup Languages for the Semantic Web* (pp. 3–28). Heidelberg, Germany: Springer.

Baerlocher, Ch., & McCusker, L. B. (2017). *Database of Zeolite Structures*. Retrieved from http://www.iza-structure.org/databases/

Baker, R. J. (2010). *CMOS: Circuit Design, Layout, and Simulation* (3rd ed.). Wiley-IEEE Press. doi:10.1002/9780470891179

Balcılar, M., Yavuz, S., & Amasy, M. F. (2017). R-SLAM: Resilient localization and mapping in challenging environments. *Robotics and Autonomous Systems*, *87*, 66–80. doi:10.1016/j.robot.2016.09.013

Baldoni, M., Baroglio, C., Mascardi, V., Omicini, A., & Torroni, P. (2010). Agents, multiagent systems and declarative programming: What, when, where, why, who, how? In A. Dovier & E. Pontelli (Eds.), *A 25-year perspective on logic programming* (pp. 204–230). Heidelberg, Germany: Springer. doi:10.1007/978-3-642-14309-0_10

Banbara, M., Tamura, N., & Inoue, K. (2006). Prolog Cafe: A Prolog to Java translator system. In M. Umeda, A. Wolf, O. Bartenstein, U. Geske, D. Seipel, & O. Takata (Eds.), *Declarative programming for knowledge management* (pp. 1–11). Heidelberg, Germany: Springer. doi:10.1007/11963578_1

Baral, C., Gelfond, G., Son, T. C., & Pontelli, E. (2010). Using answer set programming to model multi-agent scenarios involving agents' knowledge about other's knowledge. In *18th International Joint Conference on Artificial Intelligence* (pp. 259–266). Toronto, Canada: Academic Press.

Barmpoutis, A. (2013). Tensor body: Real-time reconstruction of the human body and avatar synthesis from RGB-D. *IEEE Transactions on Cybernetics*, *43*(5), 1347–1356. doi:10.1109/TCYB.2013.2276430 PMID:23974673

Barnum, P., Sheikh, Y., Datta, A., & Kanade, T. (2009). Dynamic seethroughs: Synthesizing hidden views of moving objects. In *Mixed and augmented reality* (pp. 111–114). ISMAR. doi:10.1109/ISMAR.2009.5336483

Barone, P. W., Parker, R. S., & Strano, M. S. (2005). In vivo fluorescence detection of glucose using a single-walled carbon nanotube optical sensor: Design, fluorophore properties, advantages, and disadvantages. *Analytical Chemistry*, *77*(23), 7556–7562. doi:10.1021/ac0511997 PMID:16316162

Basaca, L., Rodruiguez, J., Sergiyenko, O. Y., Tyrsa, V. V., Hernadez, W., Nieto-Hipolito, J. I., & Starostenko, O. (2010). Resolution improvement of Dynamic Triangulation method for 3D Vision System in Robot Navigation task. In *IECON 2010-36th Annual Conference on IEEE Industrial Electronics Society* (págs. 2886-2891). IEEE.

Básaca-Preciado, L. C., Sergiyenko, O. Y., Rodríguez-Quinonez, J. C., García, X., Tyrsa, V. V., Rivas-Lopez, M., ... Starostenko, O. (2014). Optical 3D laser measurement system for navigation of autonomous mobile robot. *Optics and Lasers in Engineering*, *54*, 159–169. doi:10.1016/j.optlaseng.2013.08.005

Basu, B. J., Anandan, C., & Rajam, K. S. (2003). Study of the mechanism of degradation of pyrene-based pressure sensitive paints. *Sensors and Actuators. B, Chemical*, *94*(3), 257–266. doi:10.1016/S0925-4005(03)00450-7

Baybakov, A. N., Ladigin, B. I., Pastushenko, A. I., Plotnikov, S. V., Tukubaev, N. T., & Yunoshev, S. P. (2004). Laser triangulation position sensors in industrial monitoring and diagnostics. *Optoelectronics, Instrumentation and Data Processing*, *2*(40), 105–113.

Bay, H., Ess, A., Tuytelaars, T., & Van Gool, L. (2008, June). Speeded-Up Robust Features (SURF). *Computer Vision and Image Understanding*, *110*(3), 346–359. doi:10.1016/j.cviu.2007.09.014

Beattie, C., Leibo, J. Z., Teplyashin, D., Ward, T., Wainwright, M., Küttler, H., . . . Schrittwieser, J. (2016). *Deepmind lab*. arXiv preprint arXiv:1612.03801

Beltrán Ortega, J., Gila, M., Diego, M., Aguilera Puerto, D., Gámez García, J., & Gómez Ortega, J. (2016). Novel technologies for monitoring the in-line quality of virgin olive oil during manufacturing and storage. *Journal of the Science of Food and Agriculture, 96*(14), 4644–4662. doi:10.1002/jsfa.7733 PMID:27012363

Benalia, S., Cubero, S., Prats-Montalbán, J. M., Bernardi, B., Zimbalatti, G., & Blasco, J. (2016). Computer vision for automatic quality inspection of dried figs (Ficus carica L.) in real-time. *Computers and Electronics in Agriculture, 120,* 17–25. doi:10.1016/j.compag.2015.11.002

Benedek, C. (2014). 3D people surveillance on range data sequences of a rotating Lidar. *Pattern Recognition Letters, 50,* 149–158. doi:10.1016/j.patrec.2014.04.010

Benet, G., Blanes, F., Simo, J., & Perez, P. (2002, September 30). Using infrared sensors for distance measurement in mobile robots. *Robotics and Autonomous Systems, 40*(4), 255–266. doi:10.1016/S0921-8890(02)00271-3

Benounis, M., Jaffrezic-Renault, N., Dutasta, J. P., Cherif, K., & Abdelghani, A. (2005). Study of a new evanescent wave optical fibre sensor for methane detection based on cryptophane molecules. *Sensors and Actuators. B, Chemical, 107*(1), 32–39. doi:10.1016/j.snb.2004.10.063

Beraldin, J. A., Latouche, C., El-Hakim, S. F., & Filiatrault, A. (2004). Applications of photogrammetric and computer vision techniques in shake table testing. *Proceedings of the 13th World Conference on Earthquake Engineering.*

Berardi, R., Longhi, G., & Rinaldis, D. (1991). Qualification of the European strong motion data bank: Influence of the accelerometric station response and pre-processing techniques. *European Earthquake Engineering, 2,* 38–53.

Bezier, P. E. (1971). Example of an Existing System in the Motor Industry: The Unisurf System. *Proceedings of the Royal Society of London. Series A, Mathematical and Physical Sciences, 321*(1545), 207–218. doi:10.1098/rspa.1971.0027

Bhanse & Vasekar. (2017). Survey on Classification for Moving Object Detection and Tracking Using Multiple Sensor Fusion Architecture. *International Journal of Innovative Research in Computer and Communication Engineering, 5*(2), 2054–2060.

Bigas, M., Cabruja, E., Forest, J., & Salvi, J. (2006). Review of CMOS image sensors. *Microelectronics Journal, 37*(5), 433–451. doi:10.1016/j.mejo.2005.07.002

Biggs, G. L., Blackmer, T. M., Demetriades-Shah, T. H., Holland, K. H., Schepers, J. S., & Wurm, J. H. (2002). *Method and apparatus for real-time determination and application of nitrogen fertilizer using rapid, non-destructive crop canopy measurements.* Academic Press.

Bityukov, S. I., Maksimushkina, A. V., & Smirnova, V. V. (2016). Comparison of histograms in physical research. *Nuclear Energy and Technology, 2*(2), 108–113. doi:10.1016/j.nucet.2016.05.007

Blum, A., & Rivest, R. L. (1989). Training a 3-node neural network is NP-complete. In Advances in neural information processing systems (pp. 494-501). Academic Press.

Blunsden, S.J., & Fisher, R.B. (2010). The BEHAVE video dataset: Ground truthed video for multi-person behavior classification. *Annals of the BMVA,* (4), 1–11.

Bol'basova, L. A., Lukin, V. P., & Nosov, V. V. (2009). Image jitter of a laser guide star in a monostatic formation scheme. *Optics and Spectroscopy, 107*(6), 993–999. doi:10.1134/S0030400X09120236

Bordini, R. H., Hübner, J. F., & Wooldridge, M. (2007). Programming multi-agent systems in AgentSpeak using Jason. Chichester, UK: John Wiley & Sons.

Borges, P. V. K. (2013). Pedestrian detection based on blob motion statistics. *IEEE Transactions on Circuits and Systems for Video Technology, 23*(2), 224–235. doi:10.1109/TCSVT.2012.2203217

Borges, P. V. K., Conci, N., & Cavallaro, A. (2013). Video-based human behavior understanding: A survey. *IEEE Transactions on Circuits and Systems for Video Technology, 23*(11), 1993–2008. doi:10.1109/TCSVT.2013.2270402

Born, M., & Wolf, E. (1973). *Fundamentals of optics.* Moscow: Nauka. (in Russian)

Boukhriss, Fendri, & Hammami. (2016). Moving object classification in infrared and visible spectra. *Proc. SPIE 10341, Ninth International Conference on Machine Vision (ICMV 2016).*

Boyd, S. P., & Vandenberghe, L. (2004). *Convex Optimization.* Cambridge University Press. doi:10.1017/CBO9780511804441

Bradshaw, G. (1999). *Non-Contact Surface Geometry Measurements Techniques.* Dublin: Trinity College Dublin, Department of Computer Science.

Bragagnolo, J., Amado, T. J. C., Nicoloso, R. D. S., Jasper, J., Kunz, J., & Teixeira, T. D. G. (2013). Optical crop sensor for variable-rate nitrogen fertilization in corn: I - plant nutrition and dry matter production. *Revista Brasileira de Ciência do Solo, 37*(5), 1288–1298. doi:10.1590/S0100-06832013000500018

Bretscher, O. (1995). *Linear Algebra With Applications* (3rd ed.). Upper Saddle River, NJ: Prentice Hall.

Bronshtein, I. N., & Semendyayev, K. A. (1980). *Handbook of mathematics.* Moscow: Nauka. (in Russian)

Brookes, M. (1997). *VOICEBOX: Speech Processing Toolbox for MATLAB.* Retrieved from http://www.ee.ic.ac.uk/hp/staff/dmb/voicebox/voicebox.html

Burke, M. W. (2012). *Image Acquisition: Handbook of machine vision engineering.* Springer Netherlands.

Canny, J. (1987). A computational approach to edge detection. In Readings in Computer Vision (pp. 184-203). Academic Press. doi:10.1016/B978-0-08-051581-6.50024-6

Castro-Toscano, M. J. (2017). A methodological use of inertial navigation systems for strapdown navigation task. In *26th International Symposium on Industrial Electronics (ISIE)* (pp. 1589-1595). Edinburgh, UK: IEEE. 10.1109/ISIE.2017.8001484

Chaaraoui, A. A., Padilla-López, J. R., & Flórez-Revuelta, F. (2015). Abnormal gait detection with RGB-D devices using joint motion history features. In Automatic face and gesture recognition (FG) (Vol. 7, pp. 1–6). Academic Press. doi:10.1109/FG.2015.7284881

Chae, Y., Cheon, J., Lim, S., Kwon, M., Yoo, K., Jung, W., & Han, G. (2011). A 2.1 M Pixels, 120 Frame/s CMOS Image Sensor ADC Architecture with Column-Parallel Delta Sigma ADC Architecture. *IEEE Journal of Solid-State Circuits, 46*(1), 236–247. doi:10.1109/JSSC.2010.2085910

Chakrabarti, P. (1998). *Optical Fiber Communication.* Mc Graw Hill.

Chang, C.-L., & Lee, R. (1973). *Symbolic logic and mechanical theorem proving.* New York: Academic Press.

Chan, P. K., Ng, L. S., Siek, L., & Lau, K. T. (2001). Designing CMOS folded-cascode operational amplifier with flicker noise minimisation. *Microelectronics Journal, 32*(1), 69–73. doi:10.1016/S0026-2692(00)00105-1

Chauhan, Parmar, Parmar, & Chauhan. (2013). Hybrid Approach for Video Compression Based on Scene Change Detection. *IEEE International conference on Signal Processing, Computing and Control (ISPCC)*, 1-5.

Chen, F., Brown, G. M., & Song, M. (2000). Overview of 3-D shape measurement using optical methods. *Optical Engineering (Redondo Beach, Calif.)*, *39*(1), 10–23. doi:10.1117/1.602438

Chen, J. G., Wadhwa, N., Cha, Y. J., Durand, F., Freeman, W. T., & Büyüköztürk, O. (2014). Structural Modal Identification through High Speed Camera Video: Motion Magnification. *Proceedings of the 32nd International Modal Analysis Conference*, 7, 191-197. 10.1007/978-3-319-04753-9_19

Chen, L., Wei, H., & Ferryman, J. (2013). A survey of human motion analysis using depth imagery. *Pattern Recognition Letters*, *34*(15), 1995–2006. doi:10.1016/j.patrec.2013.02.006

Chhasatia, N. J., Trivedi, C. U., & Shah, K. A. (2013). Performance Evaluation of Localized Person Based Scene Detection and Retrieval in Video. *IEEE International Conference on Image Information Processing (ICIIP)*, 78 – 83. 10.1109/ICIIP.2013.6707559

Chong, K. P., Carino, N. J., & Washer, G. (2003). Health monitoring of civil infrastructures. *Structural Health Monitoring*, *12*(3), 483–493.

Cinti, F. R. (2008). The 1997-1998 Umbria-Marche post-earthquakes investigation: Perspective from a decade of analyses and debates on the surface fractures. *Annals of Geophysics*, *51*(2-3), 361–381.

Claes, K., & Bruyninckx, H. (2007, August). Robot positioning using structured light patterns suitable for self calibration and 3D tracking. *Proceedings of the 2007 International Conference on Advanced Robotics*.

Clocksin, W. F., & Mellish, C. S. (2003). *Programming in Prolog: Using the ISO standard*. Heidelberg, Germany: Springer. doi:10.1007/978-3-642-55481-0

Codognet, P., & Diaz, D. (1995). wamcc: Compiling Prolog to C. In L. Sterling (Ed.), ICLP 1995 (pp. 317–331). MIT Press.

Comparison, H. (2016). Retrieved from http://docs.opencv.org/2.4/doc/tutorials/imgproc/ histograms/histogram_comparison/histogram_comparison.html

Cook, J. J. (2004). Optimizing P#: Translating Prolog to more idiomatic C#. In CICLOPS 2004 (pp. 59–70). Academic Press.

Correal, R., Pajares, G., & Ruz, J. (2014, March). Automatic expert system for 3D terrain reconstruction based on stereo vision and histogram matching. *Expert Systems with Applications*, *41*(4), 2043–2051. doi:10.1016/j.eswa.2013.09.003

Cubero, S., Lee, W. S., Aleixos, N., Albert, F., & Blasco, J. (2016). Automated systems based on machine vision for inspecting citrus fruits from the field to postharvest—a review. *Food and Bioprocess Technology*, *9*(10), 1623–1639. doi:10.100711947-016-1767-1

Cuevas-Jimenez, E. V. (2006). *Intelligent Robotic Vision* (Doctoral dissertation). Freie Universität Berlin.

Dang, C. T., Kumar, M., & Radha, H. (2012). Key frame extraction from consumer videos using epitome. *19th IEEE International Conference on Image processing (ICIP)*, 93 – 96. 10.1109/ICIP.2012.6466803

Dang, C. T., & Radha, H. (2014, June). Heterogeneity Image Patch Index and its Application to Consumer Video Summarization. *IEEE Transactions on Image Processing*, *23*(6), 2704–2718. doi:10.1109/TIP.2014.2320814 PMID:24801112

Dastani, M. (2008). 2APL: A practical agent programming language. *Autonomous Agents and Multi-Agent Systems*, *16*(3), 214–248. doi:10.100710458-008-9036-y

David, Y. (1999). *Digital pixel cmos image sensors*. Stanford University.

Davison, A. (1992, January). *A survey of logic programming-based object oriented languages* (Tech. Rep. No. 92/3). Melbourne, Australia: Department of Computer Science, University of Melbourne.

De Canio, G., Andersen, P., Roselli, I., Mongelli, M., & Esposito, E. (2011). Displacement Based approach for a robust operational modal analysis. In Sensors, Instrumentation and Special Topics, 6, 187-195. doi:10.1007/978-1-4419-9507-0_19

De Canio, G., de Felice, G., De Santis, S., Giocoli, A., Mongelli, M., Paolacci, F., & Roselli, I. (2016). Passive 3D motion optical data in shaking table tests of a SRG-reinforced masonry wall. *Earthquakes and Structures*, *40*(1), 53–71. doi:10.12989/eas.2016.10.1.053

De Canio, G., Mongelli, M., & Roselli, I. (2013). 3D Motion capture application to seismic tests at ENEA Casaccia Research Center: 3D optical motion capture system and DySCo virtual lab. *WIT Transactions on the Built Environment*, *134*, 803–814. doi:10.2495/SAFE130711

De Fauw, J., Keane, P., Tomasev, N., Visentin, D., van den Driessche, G., Johnson, M., ... Peto, T. (2017). Automated analysis of retinal imaging using machine learning techniques for computer vision. *F1000 Research*, 5. PMID:27830057

DeCamp, P., Shaw, G., Kubat, R., & Roy, D. (2010). An immersive system for browsing and visualizing surveillance video. In *International conference on multimedia* (pp. 371–380). Academic Press. 10.1145/1873951.1874002

DeCost, B. L., & Holm, E. A. (2015). A computer vision approach for automated analysis and classification of microstructural image data. *Computational Materials Science*, *110*, 126–133. doi:10.1016/j.commatsci.2015.08.011

Dedeoglu, Y., Toreyin, B. U., Gudukbay, U., & Cetin, A. E. (2006). Silhouette Based Method for Object Classification and Human Action Recognition in Video. *European Conference on Computer vision*, 64 - 77. 10.1007/11754336_7

Degarmo, E. P., Black, J. T., & Kohser, R. A. (2003). Materials and Processes in Manufacturing (9th ed.). Wiley.

Deguchi, J., Tachibana, F., Morimoto, M., Chiba, M., Miyaba, T., & Tanaka, H. K. T. (2013). A 187.5μVrms -Read-Noise 51mW 1.4Mpixel CMOS Image Sensor with PMOSCAP Column CDS and 10b Self-Differential Offset-Cancelled Pipeline SAR-ADC. ISSCC 2013, 494–496.

Demoen, B., & Tarau, P. (1997). *jProlog home page*. Retrieved October 27, 2017, from https://people.cs.kuleuven. be/~bart.demoen/PrologInJava/

Deng, Y., Yang, Q., Lin, X., & Tang, X. (2007). Stereo Correspondence with Occlusion Handling in a Symmetric Patch-Based Graph-Cuts Model. *IEEE Transactions on Pattern Analysis and Machine Intelligence*, *29*(6), 1068–1079. doi:10.1109/TPAMI.2007.1043 PMID:17431303

Deponte, Ed. (2004). *European Patent N° EP1619466A1*. Munich, Germany: European Patent Office.

Di Leo, G., Liguori, C., Pietrosanto, A., & Sommella, P. (2016). A vision system for the online quality monitoring of industrial manufacturing. *Optics and Lasers in Engineering*, *89*, 162–168. doi:10.1016/j.optlaseng.2016.05.007

Diederich, J. (2008). *Rule Extraction from Support Vector Machines*. Springer Berlin Heidelberg. doi:10.1007/978-3-540-75390-2

Diraco, G., Leone, A., & Siciliano, P. (2013). Human posture recognition with a time-of-flight 3D sensor for in-home applications. *Expert Systems with Applications*, *40*(2), 744–751. doi:10.1016/j.eswa.2012.08.007

Doerr, K. U., Hutchinson, T. C., & Kuester, F. (2005). A methodology for image-based tracking of seismic-induced motions. *Proceedings of the Society for Photo-Instrumentation Engineers*, *5758*, 321–332. doi:10.1117/12.597679

Draper, N., & Smith, H. (1998). *Applied regression analysis*. Wiley-Interscience. doi:10.1002/9781118625590

Dubins, L. E. (1957). On curves of minimal length with a constraint on average curvature, and with prescribed initial and terminal positions and tangents. *American Journal of Mathematics*, *79*(3), 497–516. doi:10.2307/2372560

Dumont, T., & Le Corff, S. (2014). Simultaneous localization and mapping in wireless sensor networks. *Signal Processing*, *101*, 192–203. doi:10.1016/j.sigpro.2014.02.011

Durou, J. D., Falcone, M., & Sagona, M. (2008). Numerical methods for shape-from-shading: A new survey with benchmarks. *Computer Vision and Image Understanding*, *109*(1), 22–43. doi:10.1016/j.cviu.2007.09.003

Dvojnishnikov, S. V., Bakakin, V. G., Glavnij, V. G., Kabardin, I. K., & Meledin, V. G. (2013). *RU Patent N° 2537522*. Moscow: FIPS.

Dvoynishnikov, S. V., Bakakin, G. V., Glavnyj, V. G., Kabardin, I. K., & Meledin, V. G. (2015). *RU Patent N° 1826697*. Moscow: FIPS.

Dvoynishnikov, S., & Rakhmanov, V. (2015). Power installations geometrical parameters optical control method steady against thermal indignations. *EPJ Web of Conferences*, *82*, 01035-3.

Dvoynishnikov, S. V., Rakhmanov, V. V., Meledin, V. G., Kulikov, D. V., Anikin, Yu. A., & Kabardin, I. K. (2015). Experimental Assessment of the Applicability of Laser Triangulators for Measurements of the Thickness of Hot Rolled Product. *Measurement Techniques*, *57*(12), 1378–1385. doi:10.100711018-015-0638-x

Eichberg, M. (2011). Compiling Prolog to idiomatic Java. In J. P. Gallagher & M. Gelfond (Eds.), *ICLP 2011* (pp. 84–94). Saarbrücken, Wadern: Dagstuhl Publishing.

Eisenbeiss, H., Lambers, K., Sauerbier, M., & Li, Z. (2005). Photogrammetric documentation of an archaeological site (Palpa, Peru) using an autonomous model helicopter. In CIPA 2005 (pp. 238-243). Academic Press.

Ejaz, N., Manzoor, U., Nefti, S., & Baik, S. (2012). A collaborative multi-agent framework for abnormal activity detection in crowded areas. *International Journal of Innovative Computing, Information, & Control*, *8*, 4219–4234.

El Gamal, A. (2002). Trends in CMOS Image Sensor Technology and Design. *Electron Devices Meeting*, 805–808. 10.1109/IEDM.2002.1175960

El-Gamal, F. E. Z. A., Elmogy, M., & Atwan, A. (2016). Current trends in medical image registration and fusion. *Egyptian Informatics Journal*, *1*(17), 99–124. doi:10.1016/j.eij.2015.09.002

Ellenberg, A., Kontsos, A., Moon, F., & Bartoli, I. (2016). Bridge related damage quantification using unmanned aerial vehicle imagery. *Structural Control and Health Monitoring*, *23*(9), 1168–1179. doi:10.1002tc.1831

Endo, T., Yanagida, Y., & Hatsuzawa, T. (2007). Colorimetric detection of volatile organic compounds using a colloidal crystal-based chemical sensor for environmental applications. *Sensors and Actuators. B, Chemical*, *125*(2), 589–595. doi:10.1016/j.snb.2007.03.003

Escalera, S., Athitsos, V., & Guyon, I. (2016). Challenges in multimodal gesture recognition. *Journal of Machine Learning Research*, *17*(72), 1–54.

Eyas, E. (2003). Scene Change Detection Schemes for Video Indexing in Uncompressed Domain. *INFORMATICA*, *14*(1), 19–36.

Faugeras, O. (1993). *Three-dimensional computer vision: a geometric viewpoint*. MIT Press.

Feng, D., & Feng, M. Q. (2018). Computer vision for SHM of civil infrastructure: From dynamic response measurement to damage detection – A review. *Engineering Structures*, *156*, 105–117. doi:10.1016/j.engstruct.2017.11.018

Feng, Z. (2014). *Méthode de simulation rapide de capteur d'image CMOS prenant en compte les paramètres d'extensibilité et de variabilité*. Ecole Centrale de Lyon.

Fernández-Corazza, M., Turovets, S., Luu, P., Anderson, E., & Tucker, D. (2016). Transcranial electrical neuromodulation based on the reciprocity principle. *Frontiers in Psychiatry*, 7. PMID:27303311

Fernando, W. A. C., Canagarajah, C. N., & Bull, D. R. (2000). A Unified Approach To Scene Change Detection In Uncompressed And Compressed Video, Digest of Technical Papers. *International Conference on Consumer Electronics*, 350-351.

Ferryman, J., Hogg, D., Sochman, J., Behera, A., Rodriguez-Serrano, J., Worgan, S., ... Dose, M. (2013). Robust abandoned object detection integrating wide area visual surveillance and social context. *Pattern Recognition Letters*, *34*(7), 789–798. doi:10.1016/j.patrec.2013.01.018

Fillard, J. P. (1992). Sub-pixel accuracy location estimation from digital signals. *Optical Engineering (Redondo Beach, Calif.)*, *31*(11), 2465. doi:10.1117/12.59956

Fioriti, V., Roselli, I., Tatì, A., & De Canio, G. (2017). Historic masonry monitoring by motion magnification analysis. *WIT Transactions on Ecology and the Environment*, *223*, 367–375. doi:10.2495/SC170321

Fischer, R. (2008). *Optical System Design* (2nd ed.). McGraw-Hill Education.

Fisher, R. (2007). *CAVIAR test case scenarios. The EC funded project IST 2001 37540.* Retrieved October 27, 2017, from http://homepages.inf.ed.ac.uk/rbf/CAVIAR/

Fisher, R. (2013). *BEHAVE: Computer-assisted prescreening of video streams for unusual activities. The EPSRC project GR/S98146.* Retrieved October 27, 2017, from http://groups.inf.ed.ac.uk/vision/BEHAVEDATA/INTERACTIONS/

Flores-Fuentes, W., Rivas-Lopez, M., Hernandez-Balbuena, D., Sergiyenko, O., Rodríguez-Quiñonez, J., Rivera-Castillo, J., . . . Basaca-Preciado, L. (2016). Applying Optoelectronic Devices Fusion in Machine Vision: Spatial Coordinate Measurement. In O. Sergiyenko, & J. Rodríguez-Quiñonez (Eds.), Developing and Applying Optoelectronics in Machine Vision (p. 37). Hershey, PA: IGI Global.

Flores-Fuentes, W., Sergiyenko, O., Gonzalez-Navarro, F., Rivas-Lopez, M., Hernandez-Balbuena, D., Rodríguez-Quiñonez, J., ... Lindner, L. (2017). Optoelectronic instrumentation enhancement using data mining feedback for a 3D measurement system. *Optical Review*, *23*(6), 891–896. doi:10.100710043-016-0265-z

Flores-Fuentes, W., Sergiyenko, O., Gonzalez-Navarro, F., Rivas-Lopez, M., Rodríguez-Quiñonez, J., Hernandez-Balbuena, D., ... Lindner, L. (2016, August). Multivariate outlier mining and regression feedback for 3D measurement improvement in opto-mechanical system. *Optical and Quantum Electronics*, *48*(8), 21. doi:10.100711082-016-0680-1

Flores-Fuentes, W., Sergiyenko, O., Rodriguez-Quiñonez, J. C., Rivas-López, M., Hernández-Balbuena, D., Básaca-Preciado, L. C., ... González-Navarro, F. F. (2016). Optoelectronic scanning system upgrade by energy center localization methods. *Optoelectronics, Instrumentation and Data Processing*, *52*(6), 592–600. doi:10.3103/S8756699016060108

Ford, R. M., Robson, C., Temple, D., & Gerlach, M. (1997). Metrics for Scene Change Detection in Digital Video Sequences. *IEEE International Conference on Multimedia Computing and Systems*, 610-611. 10.1109/MMCS.1997.609780

Forrest, A. R. (1972). Interactive Interpolation and Approximation by Bezier Polynomials. *Comp. J.*, 71-79.

Fowler, B., Balicki, J., How, D., & Godfrey, M. (2001). Low FPN High Gain Capacitive Transimpedance Amplifier for Low Noise CMOS Image Sensors. *Sensor and Camera Systems for Scientific, Industrial and Digital Photography Applications II, Proceedings of the SPIE*, 4306. 10.1117/12.426991

França, J. G., Gazziro, M. A., Ide, A. N., & Saito, J. H. (2005). A 3D scanning system based on laser triangulation and variable field of view. In *Image Processing, 2005. ICPI 2005. IEEE International Conference on* (pp. I-425). IEEE.

Fu, G., Corradi, P., Menciassi, A., & Dario, P. (2011). An Integrated Triangulation Laser Scanner for Obstacle Detection of Miniature Mobile Robots in Indoor Environment. *IEEE/ASME Transactions on Mechatronics, 16*(4), 778–783. doi:10.1109/TMECH.2010.2084582

Fujise, T., Chikayama, T., Rokusava, K., & Nakase, A. (1994, December). KLIC: A portable implementation of KL1. In *FGCS 1994* (pp. 66–79). Tokyo: ICOT.

Fukuda, Y., Feng, M. Q., & Shinozuka, M. (2010). Cost-effective vision-based system for monitoring dynamic response of civil engineering structures. *Structural Control and Health Monitoring, 17*(8), 918–936. doi:10.1002tc.360

Fukunaga, K. (2013). *Introduction to Statistical Pattern Recognition*. Elsevier Science.

Gade, R., & Moeslund, T. B. (2014). Thermal cameras and applications: A survey. *Machine Vision and Applications, 25*(1), 245–262. doi:10.100700138-013-0570-5

Gamini Dissanayake, P. N.-W. (2001). A solution to the simultaneous localization and map building (SLAM) problem. *IEEE Transactions on Robotics and Automation, 17*(3), 229–241. doi:10.1109/70.938381

Garcia-Cruz, X., Sergiyenko, O., Tyrsa, V., Rivas-Lopez, M., Hernandez-Balbuena, D., Rodriguez-Quiñonez, J., ... Mercorelli, P. (2014). Optimization of 3D laser scanning speed by use of combined variable step. *Optics and Lasers in Engineering, 54*, 141–151. doi:10.1016/j.optlaseng.2013.08.011

Gascueña, J., & Fernández-Caballero, A. (2011). On the use of agent technology in intelligent, multisensory and distributed surveillance. *The Knowledge Engineering Review, 26*(2), 191–208. doi:10.1017/S0269888911000026

Gavrila, D. M. (1999, January). The visual analysis of human movement: A survey. *Computer Vision and Image Understanding, 73*(1), 82–98. doi:10.1006/cviu.1998.0716

Germain, M. E., & Knapp, M. J. (2009). Optical explosives detection: From color changes to fluorescence turn-on. *Chemical Society Reviews, 38*(9), 2543–2555. doi:10.1039/b809631g PMID:19690735

Goldman, E. R., Medintz, I. L., Whitley, J. L., Hayhurst, A., Clapp, A. R., Uyeda, H. T., ... Mattoussi, H. (2005). A hybrid quantum dot-antibody fragment fluorescence resonance energy transfer-based TNT sensor. *Journal of the American Chemical Society, 127*(18), 6744–6751. doi:10.1021/ja043677l PMID:15869297

Gomes, P. (2011). Surgical robotics: Reviewing the past, analysing the present, imagining the future. *Robotics and Computer-integrated Manufacturing, 27*(2), 261–266. doi:10.1016/j.rcim.2010.06.009

Gonzalez, W. (1992). *Derivation of HSI to RGB and RGB to HSI conversion equations, Digital Image Processing* (1st ed.). Addison Wesley.

Gooch, B. (2017). *Silhouette Extraction*. Retrieved from https://www.cs.rutgers.edu/~decarlo/readings/gooch-sg03c.pdf

Goodman, J. W. (1985). *Statistical Optics*. Wiley.

Goodman, J. W. (1996). *Introduction to Fourier Optics*. McGraw-Hill.

Goos, F., & Lindberg-Hänchen, H. (1949). Neumessung des strahlversetzungseffektes bei totalreflexion. *Annalen der Physik, 440*(3-5), 251–252. doi:10.1002/andp.19494400312

Göran, S. (1989). Roentgen stereophotogrammetry: A method for the study of the kinematics of the skeletal system. *Acta Orthopaedica Scandinavica*, 1–51. PMID:2686344

Gordon, K., & Jacob, S. (1969). *U.S. Patent N° 3565531*. Washington, DC: U.S. Patent and Trademark Office.

Gordon, W. J., & Riesenfeld, R. F. (1974). Bernstein-Bezier Methods for the Computer Aided Design of Free-from Curves and Surfaces. *Journal of the Association for Computing Machinery, 21*(2), 293–310. doi:10.1145/321812.321824

Goy, J., Courtois, B., Karam, J. M., & Pressecq, F. (2001). Design of an APS CMOS Image Sensor for Low Light Level Applications Using Standard CMOS Technology. *Analog Integrated Circuits and Signal Processing, 29*(1/2), 95–104. doi:10.1023/A:1011286415014

Gray, P. R., Hurst, P. J., Lewis, S. H., & Meyer, R. G. (2009). *Analysis and Design of Analog Integrated Circuits.* JohnWiley & Sons, Inc.

Gregory, J. W., Asai, K., Kameda, M., Liu, T., & Sullivan, J. P. (2008). A review of pressure-sensitive paint for high-speed and unsteady aerodynamics. *Proceedings of the Institution of Mechanical Engineers. Part G, Journal of Aerospace Engineering, 222*(2), 249–290. doi:10.1243/09544100JAERO243

Grinev, A. Yu. (Ed.). (2005). *Problems of subsurface radiolocation.* Moscow: Radiotehnika. (in Russian)

Groppa, S., Bergmann, T. O., Siems, C., Mölle, M., Marshall, L., & Siebner, H. R. (2010). Slow-oscillatory transcranial direct current stimulation can induce bidirectional shifts in motor cortical excitability in awake humans. *Neuroscience, 166*(4), 1219–1225. doi:10.1016/j.neuroscience.2010.01.019 PMID:20083166

Gruber, M., & Hausler, G. (1992). Simple, robust and accurate phase-measuring triangulation. *Optik (Stuttgart), 3,* 118–122.

Grymin, D. J., Neas, C. B., & Farhood, M. (2014). A hierarchical approach for primitive-based motion planning and control of autonomous vehicles. *Robotics and Autonomous Systems, 62*(2), 214–228. doi:10.1016/j.robot.2013.10.003

Gul, M., & Catbas, N. (2009). Statistical pattern recognition for Structural Health Monitoring using time series modeling: Theory and experimental verifications. *Mechanical Systems and Signal Processing, 23*(7), 2192–2204. doi:10.1016/j.ymssp.2009.02.013

Gurvich, A. S., Gorbunov, M. E., Fedorova, O. V., Fortus, M. I., Kirchengast, G., Proschek, V., & Tereszchuk, K. A. (2014). Spatiotemporal structure of a laser beam at a path length of 144 km: Comparative analysis of spatial and temporal spectra. *Applied Optics, 53*(12), 2625–2631. doi:10.1364/AO.53.002625 PMID:24787588

Gutschoven, B., & Verlinde, P. (2000, July). Multi-modal identity verification using support vector machines (SVM). In *Information Fusion, 2000. FUSION 2000. Proceedings of the Third International Conference on* (Vol. 2, pp. THB3-3). IEEE.

Halevy, A., Norvig, P., & Pereira, F. (2009). The unreasonable effectiveness of data. *IEEE Intelligent Systems, 24*(2), 8–12. doi:10.1109/MIS.2009.36

Hamami, S., Fleshel, L., Yadid-pecht, O., & Driver, R. (2004). CMOS Aps Imager Employing 3.3V 12 bit 6.3 ms/s pipelined ADC. *Proceedings of the 2004 International Symposium on Circuits and Systems.*

Han & Kamber. (2006). Data Mining: Concepts and Techniques (2nd ed.). Morgan Kaufmann Publishers.

Han, F., Reily, B., Hoff, W., & Zhang, H. (2016). *Space-time representation of people based on 3D skeletal data: A review.* arXiv preprint arXiv:1601.01006v2 [cs.CV]

Han, J., Shao, L., Xu, D., & Shotton, J. (2013, October). Enhanced computer vision with Microsoft Kinect sensor: A review. *IEEE Transactions on Cybernetics, 43*(5), 1318–1334. doi:10.1109/TCYB.2013.2265378 PMID:23807480

Hansen, D. W., Hansen, M. S., Kirschmeyer, M., Larsen, R., & Silvestre, D. (2008). Cluster tracking with time-of-flight cameras. In Computer vision and pattern recognition workshops (CVPRW'08) (pp. 1–6). Academic Press.

Haritaoglu, I., Harwood, D., & Davis, L. S. (1998, April). W4: Who? When? Where? What? A real time system for detecting and tracking people. In FG 1998 (pp. 222–227). Nara, Japan: Academic Press.

Harmouth, Kh. F. (1985). *Nonsinusoidal waves in radio location and radio communication*. Moscow: Radio and Svyaz. (in Russian)

Harrington, J., & Cassidy, S. (2012). *Techniques in Speech Acoustics*. Springer Netherlands.

Hassah, J. C. (1994). Time delay processing near the ocean surface. *Journal of Sound and Vibration, 35*(4), 489–501.

Havaei, M., Davy, A., Warde-Farley, D., Biard, A., Courville, A., Bengio, Y., ... Larochelle, H. (2016). Brain tumor segmentation with deep neural networks. *Medical Image Analysis, 35*, 18–31. doi:10.1016/j.media.2016.05.004 PMID:27310171

Hayes, M. H. (1999). *Schaum's Outline of Digital Signal Processing*. McGraw-Hill Companies,Incorporated.

He, K., Zhang, X., Ren, S., & Sun, J. (2015). *Deep Residual Learning for Image Recognition*. Retrieved from https://arxiv.org/abs/1512.03385

Hecht-Nielsen, R. (1988). Theory of the backpropagation neural network. *Neural Networks, 1*(Supplement-1), 445–448. doi:10.1016/0893-6080(88)90469-8

Hefele, J., & Brenner, C. (2001, February). Robot pose correction using photogrammetric tracking. In *Machine Vision and Three-Dimensional Imaging Systems for Inspection and Metrology* (Vol. 4189, pp. 170–179). International Society for Optics and Photonics. doi:10.1117/12.417194

He, G., Yan, N., Yang, J., Wang, H., Ding, L., Yin, S., & Fang, Y. (2011). Pyrene-containing conjugated polymer-based fluorescent films for highly sensitive and selective sensing of TNT in aqueous medium. *Macromolecules, 44*(12), 4759–4766. doi:10.1021/ma200953s

Henderson, F., & Somogyi, Z. (2002). Compiling Mercury to high-level C code. In CC 2002. Grenoble, France: Academic Press. doi:10.1007/3-540-45937-5_15

Higashia, K., Fukuib, S., Iwahoria, Y., & Adachia, Y. (2016). New Feature for Shadow Detection by Combination of Two Features Robust to Illumination Changes. *Procedia Computer Science, 96*, 896–903. doi:10.1016/j.procs.2016.08.268

Hiremath, S., van der Heijden, G., van Evert, F., Stein, A., & ter Braak, C. (2014, January). Laser range finder model for autonomous navigation of a robot in a maize field using a particle filter. *Computers and Electronics in Agriculture, 100*, 41–50. doi:10.1016/j.compag.2013.10.005

Homola, J. (2008). Surface plasmon resonance sensors for detection of chemical and biological species. *Chemical Reviews, 108*(2), 462–493. doi:10.1021/cr068107d PMID:18229953

Hong, K., Lee, C., Jung, K., & Oh, K. (2008). Real-time 3D Feature Extraction without Explicit 3D Object Reconstruction. *International Journal of Computer, Electrical, Automation. Control and Information Engineering, 2*(8), 2613–2618.

HSI Color Space - Color Space Conversion. (2016). Retrieved from https://www.blackice.com/colorspaceHSI.htm

Hsieh, C.-T., Wang, H.-C., Wu, Y.-K., Chang, L.-C., & Kuo, T.-K. (2012). A Kinect-based people-flow counting system. In Intelligent signal processing and communication systems (ISPACS 2012) (pp. 146–150). IEEE. doi:10.1109/ISPACS.2012.6473470

Huang, H. W., Sun, Y., Xue, Y. D., & Wang, F. (2017). Inspection equipment study for subway tunnel defects by grey-scale image processing. *Advanced Engineering Informatics, 32*, 188–201. doi:10.1016/j.aei.2017.03.003

Huang, M., Xia, L., Zhang, J., & Dong, H. (2013). An Integrated Scheme for Video Key Frame Extraction. *2nd International Symposium on Computer, Communication, Control and Automation*, 258-261. 10.2991/3ca-13.2013.64

Huber, C., Klimant, I., Krause, C., & Wolfbeis, O. S. (2001). Dual lifetime referencing as applied to a chloride optical sensor. *Analytical Chemistry*, *73*(9), 2097–2103. doi:10.1021/ac9914364 PMID:11354496

Hu, F. (2011). An automatic measuring method and system using a light curtain for the thread profile of a ballscrew. *Measurement Science & Technology*, *22*(8), 085106. doi:10.1088/0957-0233/22/8/085106

Hulme, K. F., Collins, B. S., Constant, G. D., & Pinson, J. T. (1981). A CO2 laser rangefinder using heterodyne detection and chirp pulse compression. *Optical and Quantum Electronics*, *13*(1), 35–45. doi:10.1007/BF00620028

Hutchinson, T. C., & Kuester, F. (2004). Monitoring global earthquake-induced demands using vision-based sensors. *IEEE Transactions on Instrumentation and Measurement*, *53*(1), 31–36. doi:10.1109/TIM.2003.821481

Hu, X., Wang, B., & Ji, H. (2013). A wireless sensor network-based structural health monitoring system for highway bridges. *Computer-Aided Civil and Infrastructure Engineering*, *28*(3), 193–209. doi:10.1111/j.1467-8667.2012.00781.x

Hu, Z., Deibert, B. J., & Li, J. (2014). Luminescent metal–organic frameworks for chemical sensing and explosive detection. *Chemical Society Reviews*, *43*(16), 5815–5840. doi:10.1039/C4CS00010B PMID:24577142

Ibañez, R., Soria, Á., Teyseyre, A., & Campo, M. (2014). Easy gesture recognition for Kinect. *Advances in Engineering Software*, *76*, 171–180. doi:10.1016/j.advengsoft.2014.07.005

Ibrahim, D. (2013). *Practical Digital Signal Processing Using Microcontrollers*. Elektor International Media.

Ingle, V. K., & Proakis, J. G. (2011). *Digital Signal Processing Using MATLAB*. Cengage Learning.

Innocent, M. (n.d.). *General introduction to CMOS image sensors*. Academic Press.

Intaravanne, Y., & Sumriddetchkajorn, S. (2015). Android-based rice leaf color analyzer for estimating the needed amount of nitrogen fertilizer. *Computers and Electronics in Agriculture*, *116*, 228–233. doi:10.1016/j.compag.2015.07.005

Ishimaru, A. (1999). *Wave Propagation and Scattering in Random Media*. Wiley. doi:10.1109/9780470547045

Jahromi, S., Jansson, J. P., & Kostamovaara, J. (2015). Pulsed TOF Laser Rangefinding with a 2D SPAD-TDC Receiver. In SENSORS (pp. 1-4). IEEE.

Jain, R., Kasturi, R., & Schunk, B. G. (1995). *Machine Vision*. New York: McGraw-Hill.

Jazebi, F., & Rashidi, A. (2013). An automated procedure for selecting project managers in construction firms. *Journal of Civil Engineering and Management*, *19*(1), 97–106. doi:10.3846/13923730.2012.738707

Jin, Z., Su, Y., & Duan, Y. (2000). An improved optical pH sensor based on polyaniline. *Sensors and Actuators. B, Chemical*, *71*(1-2), 118–122. doi:10.1016/S0925-4005(00)00597-9

Johanna, M. (2013). *Recognizing activities with the Kinect. A logic-based approach for the support room* (Unpublished master's thesis). Radboud University Nijmegen.

Joshi, U., & Patel, K. (2016). Object tracking and classification under illumination variations. *International Journal of Engineering Development and Research*, *4*(1), 667–670.

Jui-Lin, L., Ting-You, L., Cheng-Fang, T., Yi-Te, L., & Rong-Jian, C. (2010). Design a low-noise operational amplifier with constant-gm. *Proceedings of SICE Annual Conference*, 322-326.

Kandidov, V. P., Tamarov, M. P., & Shlenov, S. A. (1998). The influence of the outer scale of atmospheric turbulence on the dispersion of the center of gravity of the laser beam. *Atmospheric and Oceanic Optics, 12*(1), 27–33.

Kannel, P. R., Lee, S., Lee, Y. S., Kanel, S. R., & Khan, S. P. (2007). Application of water quality indices and dissolved oxygen as indicators for river water classification and urban impact assessment. *Environmental Monitoring and Assessment, 132*(1), 93–110. doi:10.100710661-006-9505-1 PMID:17279460

Katzouris, N., Artikis, A., & Paliouras, G. (2014). Event recognition for unobtrusive assisted living. In *Hellenic conference on artificial intelligence* (pp. 475–488). Academic Press.

Kavitha & Kavitha. (2017). Abnormal Moving Object Detection Using Sparse Based Graph K Nearest Neighbour (SGk-NN). *International Journal of Innovative Research in Computer and Communication Engineering, 5*(4), 7901–7906.

Kawabata, Ma, Xue, Zhu, & Zheng. (2015). A path generation for automated vehicle based on Bezier curve and via-points. *Robotics and Autonomous Systems, 74*(A), 243-252.

Kelm, P., Schmiedeke, S., & Sikora, T. (2009). Feature-based video key frame extraction for low quality video sequences, Image Analysis for Multimedia Interactive Services. *10th Workshop on Image Analysis for Multimedia Interactive Services*, 25 – 28.

Kemerait, R. C., & Childers, D. G. (1972). Signal detection and extraction by Cepstrum techniques. *IEEE Transactions on Information Theory, IT-18*(1), 745–759. doi:10.1109/TIT.1972.1054926

Kilpela, A. (2002). *Pulsed time-of-flight laser range finder techniques for fast high-precision measurement applications.* Academic Press.

Kimmig, A., Demoen, B., Raedt, L. D., Costa, V. S., & Rocha, R. (2011). On the implementation of the probabilistic logic programming language ProbLog. *Theory and Practice of Logic Programming, 11*(11), 235–262. doi:10.1017/S1471068410000566

Kind, T., & Fiehn, O. (2007). Seven golden rules for heuristic filtering of molecular formulas obtained by accurate mass spectrometry. *BMC Bioinformatics, 8*(1), 105. doi:10.1186/1471-2105-8-105 PMID:17389044

Kleinfelder, S., Lim, S., Liu, X., & El Gamal, A. (2001). A 10 000 Frames/s CMOS Digital Pixel Sensor. *IEEE Journal of Solid-State Circuits, 36*(12), 2049–2059. doi:10.1109/4.972156

Koenderink, J. J., van Doorn, A. J., Chalupa, L. M., & Werner, J. S. (2003). Shape and Shading. *Visual Neuroscience*, 1090–1105.

Kolb, A., Barth, E., Koch, R., & Larsen, R. (2009, March). Time-of-Flight Sensors in Computer Graphics. In Eurographics (STARs) (pp. 119-134). Academic Press.

Komissarov, A. G. (1990). *RU Patent Nº 1826697.* Moscow: FIPS.

Kostamovaara, J., Huikari, J., Hallman, L., Nissinen, L., Nissinen, J., Rapakko, H., ... Ryvkin, B. (2015). On Laser Ranging Based on High-Speed/Energy Laser Diode Pulses and Single-Photon Detection Techniques. *IEEE Photonics Journal, 7*(2), 1–15. doi:10.1109/JPHOT.2015.2402129

Kovács, B., Szayer, G., Tajti, F., Burdelis, M., & Korondi, P. (2016, August). A novel potential field method for path planning of mobile robots by adapting animal motion attributes. *Robotics and Autonomous Systems, 82*, 24–34. doi:10.1016/j.robot.2016.04.007

Kowalski, R., & Sergot, M. (1986). A logic-based calculus of events. *New Generation Computing, 1*(4), 67–96. doi:10.1007/BF03037383

Kravari, K., & Bassiliades, N. (2015). A survey of agent platforms. *Journal of Artificial Societies and Social Simulation*, *18*(1), 191–208. doi:10.18564/jasss.2661

Kravcov, Y. A., Feizulin, Z. I., & Vinogradov, A. G. (1983). *Passage of radio waves through Earth atmosphere*. Moscow: Radio i svyaz.

Krishnan, P., & Naveen, S. (2015). RGB-D face recognition system verification using Kinect and FRAV3D databases. *Procedia Computer Science*, *46*, 1653–1660. doi:10.1016/j.procs.2015.02.102

Krizhevsky, A., Sutskever, I., & Hinton, G. E. (2012). Imagenet classification with deep convolutional neural networks. In Advances in neural information processing systems (pp. 1097-1105). Academic Press.

Krotkov, E. (1988). Focusing. *International Journal of Computer Vision*, 223-237.

Kumar, P., McElhinney, C., Lewis, P., & McCarthy, T. (2013, November). An automated algorithm for extracting road edges from terrestrial mobile LiDAR data. *ISPRS Journal of Photogrammetry and Remote Sensing*, *85*, 44–55. doi:10.1016/j.isprsjprs.2013.08.003

Kuramochi, S. (1999). *KLIJava home page*. Retrieved October 27, 2017, from http://www.ueda.info.waseda.ac.jp/~satoshi/klijava/klijava-e.html

Kuzmany, H. (2009). *Solid-State Spectroscopy: An Introduction*. Springer Berlin Heidelberg. doi:10.1007/978-3-642-01479-6

Lal, K., & Arif, K. M. (2016). Feature extraction for moving object detection in a non-stationary background. *12th IEEE/ASME International Conference on Mechatronic and Embedded Systems and Applications (MESA)*, 1-6. 10.1109/MESA.2016.7587172

Lan, X., Huang, J., Han, Q., Wei, T., Gao, Z., Jiang, H., ... Xiao, H. (2012). Fiber ring laser interrogated zeolite-coated singlemode-multimode-singlemode structure for trace chemical detection. *Optics Letters*, *37*(11), 1998–2000. doi:10.1364/OL.37.001998 PMID:22660100

Lao, W., Han, J., & With, P. H. N. (2010). Flexible human behavior analysis framework for video surveillance applications. *International Journal of Digital Multimedia Broadcasting*.

Lapray, P. J., Heyrman, B., & Ginhac, D. (2016). HDR-ARtiSt: An adaptive real-time smart camera for high dynamic range imaging. *Journal of Real-Time Image Processing*, *12*(4), 747–762. doi:10.100711554-013-0393-7

Latombe, J.-C. (1991). *Robot Motion Planning* (Vol. 124). Springer, US. doi:10.1007/978-1-4615-4022-9

Lau, T. B., Ong, A. C., & Putra, F. A. (2014, June). Non-invasive monitoring of people with disabilities via motion detection. *International Journal of Signal Processing Systems*, *2*(1), 37–41.

Lauwerys, R. R., & Hoet, P. (2001). *Industrial chemical exposure: guidelines for biological monitoring*. CRC Press.

Lee, D. (2001). *U.S. Patent Nº 20030007161*. Washington, DC: U.S. Patent and Trademark Office.

Lee, J., Farha, O. K., Roberts, J., Scheidt, K. A., Nguyen, S. T., & Hupp, J. T. (2009). Metal-organic framework materials as catalysts. *Chemical Society Reviews*, *38*(5), 1450–1459. doi:10.1039/b807080f PMID:19384447

Lee, J., Lee, K. C., Cho, S., & Sim, S. H. (2017). Computer Vision-Based Structural Displacement Measurement Robust to Light-Induced Image Degradation for In-Service Bridges. *Sensors (Basel)*, *17*(10), 2317. doi:10.339017102317 PMID:29019950

Lee, Y. S., Joo, B. S., Choi, N. J., Lim, J. O., Huh, J. S., & Lee, D. D. (2003). Visible optical sensing of ammonia based on polyaniline film. *Sensors and Actuators. B, Chemical*, *93*(1-3), 148–152. doi:10.1016/S0925-4005(03)00207-7

Lee, Y.-S., & Chung, W.-Y. (2012). Visual sensor based abnormal event detection with moving shadow removal in home healthcare applications. *Sensors (Basel)*, *12*(1), 573–584. doi:10.3390120100573 PMID:22368486

Leightley, D., Yap, M. H., Hewitt, B. M., & McPhee, J. S. (2016). Sensing behaviour using the Kinect: Identifying characteristic features of instability and poor performance during challenging balancing tasks. In Measuring behavior. Academic Press.

Levin, B.R. (1974-1976). *Theoretical foundations of the statistical radio engineering: in 3 vol.* Sov. radio. (in Russian)

Levin, B. R. (1966). *Theoretical foundations of statistical radioengineering*. Moscow: Sov. Radio. (in Russian)

Lewis, J. M., Lakshmivarahan, S., & Dhall, S. (2006). *Dynamic Data Assimilation: A Least Squares Approach*. Cambridge University Press. doi:10.1017/CBO9780511526480

Liang, C.-W., & Juang, C.-F. (2015). Moving object classification using local shape and HOG features in wavelet-transformed space with hierarchical SVM classifiers. *Applied Soft Computing*, *28*, 483–497. doi:10.1016/j.asoc.2014.09.051

Li, B., Jiang, L., Wang, S., Zhou, L., Xiao, H., & Tsai, H. L. (2011). Ultra-abrupt tapered fiber Mach-Zehnder interferometer sensors. *Sensors (Basel)*, *11*(6), 5729–5739. doi:10.3390110605729 PMID:22163923

Lim, R., & Yang, X. Gao, & Lin. (2017). *Real-Time Optical flow-based Video Stabilization for Unmanned Aerial Vehicles*. Cornel University Library. arXiv:1701.03572 [cs.CV]

Lindner, L. (2016). Laser Scanners. In O. Sergiyenko, J. Rodríguez-Quiñonez, O. Sergiyenko, & J. Rodríguez-Quiñonez (Eds.), Developing and Applying Optoelectronics in Machine Vision (p. 38). Hershey, PA: IGI Global.

Lindner, L., Sergiyenko, O., Rodríguez-Quiñonez, J., Tyrsa, V., Mercorelli, P., Fuentes-Flores, W., . . . Nieto-Hipolito, J. (2015). Continuous 3D scanning mode using servomotors instead of stepping motors in dynamic laser triangulation. In *Industrial Electronics (ISIE), 2015 IEEE 24th International Symposium on* (pp. 944-949). Buzios: IEEE.

Lindner, L., Sergiyenko, O., Tyrsa, V., & Mercorelli, P. (2014, June 01-04). An approach for dynamic triangulation using servomotors. In *Industrial Electronics (ISIE), 2014 IEEE 23rd International Symposium on* (pp. 1926-1931). Istanbul: IEEE.

Lindner, L., Sergiyenko, O., Rivas-Lopez, M., Hernandez-Balbuena, D., Flores-Fuentes, W., Rodríguez-Quiñonez, J., ... Basaca, L. (2017). Exact laser beam positioning for measurement of vegetation vitality. *Industrial Robot: An International Journal*, *44*(4), 532–541. doi:10.1108/IR-11-2016-0297

Lindner, L., Sergiyenko, O., Rodríguez-Quiñonez, J., Rivas-Lopez, M., Hernandez-Balbuena, D., Flores-Fuentes, W., ... Tyrsa, V. (2016). Mobile robot vision system using continuous laser scanning for industrial application. *The Industrial Robot*, *43*(4), 360–369. doi:10.1108/IR-01-2016-0048

Ling, Ho, Fu, & Ming Kai. (2011). *Home-Made 3-D image measuring instrument data process and analysis*. Paper presented at the 2011 International Conference on Multimedia Technology.

Lin, N., Jiang, L., Wang, S., Yuan, L., Xiao, H., Lu, Y., & Tsai, H. (2010). Ultrasensitive chemical sensors based on whispering gallery modes in a microsphere coated with zeolite. *Applied Optics*, *49*(33), 6463–6471. doi:10.1364/AO.49.006463 PMID:21102672

Li, R., Fan, K., Miao, J., Huang, Q., Tao, S., & Gong, E. (2014). An analogue contact probe using a compact 3D optical sensor for micro-nano coordinate measuring machines. *Measurement Science & Technology*, *25*(9), 094008. doi:10.1088/0957-0233/25/9/094008

Liu, J. Y., & Li, Y. F. (2014). A Whole-Field 3D Shape Measurement Algorithm Based on Phase-Shifting Projected Fringe Profilometry. In *Information and Automation (ICIA), 2014 IEEE International Conference* (pp. 495-500). IEEE.

Liu, M., Ma, J., Lin, L., Ge, M., Wang, Q., & Liu, C. (2017). Intelligent assembly system for mechanical products and key technology based on internet of things. *Journal of Intelligent Manufacturing*, 28(2), 271–299. doi:10.100710845-014-0976-6

Liu, N., Hui, J., Sun, C., Dong, J., Zhang, L., & Xiao, H. (2006). Nanoporous zeolite thin film-based fiber intrinsic Fabry-Perot interferometric sensor for detection of dissolved organics in water. *Sensors (Basel)*, 6(8), 835–847. doi:10.33906080835

Liu, Y. F., Cho, S. J., Spencer, B. F., & Fan, J. S. (2014). Automated assessment of cracks on concrete surfaces using adaptive digital image processing. *Smart Structures and Systems*, 14(10), 719–741. doi:10.12989ss.2014.14.4.719

Livni, R., Shalev-Shwartz, S., & Shamir, O. (2014). *On the Computational Efficiency of Training Neural Networks*. Retrieved from https://arxiv.org/abs/1410.1141

Li, Y., Ruichek, Y., & Capelle, C. (2013). Optimal Extrinsic Calibration Between a Stereoscopic System and a LIDAR. *IEEE Transactions on Instrumentation and Measurement*, 62(8), 2258–2269. doi:10.1109/TIM.2013.2258241

Loinaz, M. J., Singh, K. J., Blanksby, A. J., Member, S., Inglis, D. A., Azadet, K., & Ackland, B. D. (1998). A 200-mW, 3.3-V, CMOS color Camera IC producing 352x288 24-b Video at 30 Frames/s. *IEEE Journal of Solid-State Circuits*, 33(12), 2092–2103. doi:10.1109/4.735552

Lopez, M. (2010). Optoelectronic Method for Structural Health Monitoring. *Structural Health Monitoring*, 9(2), 105–120. doi:10.1177/1475921709340975

Lou, J., Tong, L., & Ye, Z. (2005). Modeling of silica nanowires for optical sensing. *Optics Express*, 13(6), 2135–2140. doi:10.1364/OPEX.13.002135 PMID:19495101

Lowe, D. G. (1999). Object recognition from local scale-invariant features. *Proceedings of the Seventh International Conference on Computer Vision*, 1150–1157. 10.1109/ICCV.1999.790410

Lu, W. S., & Antoniou, A. (2000). *Design of digital filters and filter banks by optimization: A state of the art review*. Paper presented at the 2000 10th European Signal Processing Conference.

Luan, L., Evans, R. D., Jokerst, N. M., & Fair, R. B. (2008). Integrated optical sensor in a digital microfluidic platform. *IEEE Sensors Journal*, 8(5), 628–635. doi:10.1109/JSEN.2008.918717

Luhmann, T., Robson, S., Kyle, S. A., & Harley, I. A. (2006). *Close range photogrammetry: principles, techniques and applications*. Whittles.

Lulé, T., Benthien, S., Keller, H., Mütze, F., Rieve, P., Seibel, K., & Böhm, M. (2000). Sensitivity of CMOS Based Imagers and Scaling Perspectives. *IEEE Transactions on Electron Devices*, 47(11), 2110–2122. doi:10.1109/16.877173

Lunghi, F., Pavese, A., Peloso, S., Lanese, I., & Silvestri, D. (2012). Computer Vision System for Monitoring in Dynamic Structural Testing. *Geotechnical, Geological, and Earthquake Engineering*, 22, 159–176. doi:10.1007/978-94-007-1977-4_9

Lun, R., & Zhao, W. (2015, March). A survey of applications and human motion recognition with Microsoft Kinect. *International Journal of Pattern Recognition and Artificial Intelligence*, 29(5), 1555008. doi:10.1142/S0218001415550083

Luo, P., Zhang, M., Liu, Y., Han, D., & Li, Q. (2012). *A Moving Average Filter Based Method of Performance Improvement for Ultraviolet Communication System*. Academic Press.

Lyness, J. N., & Moler, C. B. (1967). Numerical Differentiation of Analytic Functions. *SIAM Journal on Numerical Analysis*, 4(2), 202–210. doi:10.1137/0704019

Lyu, T., Yao, S., Nie, K., & Xu, J. (2014). A 12-bit high-speed column-parallel two-step single-slope analog-to-digital converter (ADC) for CMOS image sensors. *Sensors (Basel)*, *14*(11), 21603–21625. doi:10.3390141121603 PMID:25407903

Machot, F., Kyamakya, K., Dieber, B., & Rinner, B. (2011). *Real time complex event detection for resource-limited multimedia sensor networks*. AMMCSS. doi:10.1109/AVSS.2011.6027378

Mahmoodi, A., & Joseph, D. (2008). Pixel-Level Delta-Sigma ADC with Optimized Area and Power for Vertically-Integrated Image Sensors. *51st Midwest Symposium on Circuits and Systems*, 41–44. 10.1109/MWSCAS.2008.4616731

Mahmud, M., Joannic, D., Roy, M., Isheil, A., & Fontaine, J. F. (2011). 3D part inspection path planning of a laser scanner with control on the uncertainty. *Computer Aided Design*, *43*(4), 345–355. doi:10.1016/j.cad.2010.12.014

Mangal, S., & Kumar, A. (2017). Real time moving object detection for video surveillance based on improved GMM. *International Journal of Advanced Technology and Engineering Exploration*, *4*(26), 17–22. doi:10.19101/IJATEE.2017.426004

Mangini, F., Dinia, L., Frezza, F., Beccarini, A., Del Muto, M., Federici, E., . . . Segneri, A. (2017). New crack measurement methodology: tag recognition. In *Proceedings of IMEKO International Conference on Metrology for Archaeology and Cultural Heritage* (pp. 433-436). Lecce, Italy: Academic Press.

Man, Z., Lee, K., Wang, D., Cao, Z., & Khoo, S. (2013). An optimal weight learning machine for handwritten digit image recognition. *Signal Processing*, *93*(6), 1624–1638. doi:10.1016/j.sigpro.2012.07.016

Markus-Christian, A., Thierry Bosch, M. L., Risto, M., & Marc, R. (2001). Laser ranging: A critical review of usual techniques for distance measurement. *Optical Engineering (Redondo Beach, Calif.)*, 10–19.

Marzuki, A. (2016). CMOS Image Sensor: Analog and Mixed-Signal Circuit. In O. Sergiyenko & J. C. Rodriguez-Quiñonez (Eds.), Developing and Applying Optoelectronics in Machine Vision. Hershey, PA: IGI Global.

Marzuki, A., Aziz, Z. A. A., & Manaf, A. A. (2011, May). A review of CMOS analog circuits for image sensing application. In *2011 IEEE International Conference on Imaging Systems and Techniques (IST)* (pp. 180-184). IEEE. 10.1109/IST.2011.5962187

Masalov, S. A., Sytnik, O. V., & Ruban, V. P. (2012). Wavelet UWB signal processing for underground sounding systems. *Proc. Int. Conf. "Ultrawideband and Ultrashort Impulse Signals*, 123–125. 10.1109/UWBUSIS.2012.6379754

Mastorakis, G., & Makris, D. (2014). Fall detection system using Kinect's infrared sensor. *Journal of Real-Time Image Processing*, *9*(4), 635–646. doi:10.100711554-012-0246-9

Mayorga-Ortiz, P., Druzgalski, C., Miranda Vega, J. E., & Zeljkovic, V. (2016). Determinación del Tamaño Óptimo de Modelos HMM-GMM para Clasificación de las Señales Bioacústicas. *Revista Mexicana de Ingeniería Biomédica*, *37*, 63–79.

McConnell, R. K. (1986). *Method of and apparatus for pattern recognition*. U.S. Patent No. 4,567,610.

McCoy, A. D., Thomsen, B. C., Ibsen, M., & Richardson, D. J. (2004). Filtering effects in a spectrum-sliced WDM system using SOA-based noise reduction. *IEEE Photonics Technology Letters*, *16*(2), 680–682. doi:10.1109/LPT.2003.821077

McDonagh, C., Burke, C. S., & MacCraith, B. D. (2008). Optical chemical sensors. *Chemical Reviews*, *108*(2), 400–422. doi:10.1021/cr068102g PMID:18229950

Meier, B., Werner, T., Klimant, I., & Wolfbeis, O. S. (1995). Novel oxygen sensor material based on a ruthenium bipyridyl complex encapsulated in zeolite Y: Dramatic differences in the efficiency of luminescence quenching by oxygen on going from surface-adsorbed to zeolite-encapsulated fluorophores. *Sensors and Actuators. B, Chemical*, *29*(1), 240–245. doi:10.1016/0925-4005(95)01689-9

Meledin, V. (2011). Optoelectronic Measurements in Science and Innovative Industrial Technologies. *Optoelectronic Devices and Properties*, *18*, 373–399.

Meng, C., Xiao, Y., Wang, P., Zhang, L., Liu, Y., & Tong, L. (2011). Quantum-dot-doped polymer nanofibers for optical sensing. *Advanced Materials*, *23*(33), 3770–3774. PMID:21766349

Meyer, Y., Isaiah, P., & Shima, T. (2015, March). On Dubins paths to intercept a moving target. *Automatica*, *53*, 256–263. doi:10.1016/j.automatica.2014.12.039

Micheletti, N., Chandler, J. H., & Lane, S. N. (2015). *Structure from motion (SFM) photogrammetry. In Geomorphological Techniques*. London, UK: British Society for Geomorphology.

Miller, J. M. (1920). Dependence of the input impedance of a three-electrode vacuum tube upon the load in the plate circuit. *Scientific Papers of the Bureau of Standards*, 367-385.

Mishra, P. K., & Saroha, G. P. (2016). A Study on Classification for Static and Moving Object in Video Surveillance System. *International Journal of Image, Graphics & Signal Processing*, *8*(5), 76–82.

Mithlesh, C. S., & Shukla, D. (2016). Detection of Scene Change in Video. *International Journal of Science and Research*, *5*(3), 1187–1201.

Moghadam, P., Sardha, W., & Feng, D. J. (2008). Improving Path Planning and Mapping Based on Stereo Vision and Lidar. In *Control, Automation, Robotics and Vision, 2008. ICARV 2008, 10th International Conference* (pp. 384-389). IEEE.

Mohamed, A. N., Ahmed, H. N., Elkhatib, M., & Shehata, K. A. (2013). A low power low noise capacitively coupled chopper instrumentation amplifier in 130 nm CMOS for portable biopotential acquisiton systems. In *International Conference on Computer Medical Applications*, 1-5. 10.1109/ICCMA.2013.6506168

Mongelli, M., De Canio, G., Roselli, I., Malena, M., Nacuzi, A., & de Felice, G. (2017). 3D photogrammetric reconstruction by drone scanning for FE analysis and crack pattern mapping of the "Bridge of the Towers", Spoleto. *Key Engineering Materials*, *747*, 423–430. doi:10.4028/www.scientific.net/KEM.747.423

Mongelli, M., Roselli, I., De Canio, G., & Ambrosino, F. (in press). Quasi real-time FEM calibration by 3D displacement measurements of large shaking table tests using HPC resources. *Advances in Engineering Software*.

Morgenthal, G., & Hallermann, N. (2014). Quality assessment of unmanned aerial vehicle (UAV) based visual inspection of structures. *Advances in Structural Engineering*, *17*(3), 289–302. doi:10.1260/1369-4332.17.3.289

Mori, S., & Barth, H. G. (2013). *Size exclusion chromatography*. Springer Science & Business Media.

Morozov, A. A. (1994). The Prolog with actors. *Programmirovanie*, (5), 66–78. (in Russian)

Morozov, A. A. (1999, September). Actor Prolog: an object-oriented language with the classical declarative semantics. In K. Sagonas & P. Tarau (Eds.), IDL 1999 (pp. 39–53). Paris, France: Academic Press.

Morozov, A. A. (2002, September). On semantic link between logic, object-oriented, functional, and constraint programming. In Proc. of MultiCPL'02 workshop (pp. 43–57). Ithaca, NY: Academic Press.

Morozov, A. A. (2007, September). Operational approach to the modified reasoning, based on the concept of repeated proving and logical actors. In V. S. C. Salvador Abreu (Ed.), CICLOPS 2007 (pp. 1–15). Porto, Portugal: Academic Press.

Morozov, A. A. (2018). *A GitHub repository containing source codes of Actor Prolog built-in classes*. Retrieved February 12, 2018, from https://github.com/Morozov2012/actor-prolog-java-library

Morozov, A. A., & Sushkova, O. S. (2018). *The intelligent visual surveillance logic programming Web Site.* Retrieved February 12, 2018, from http://www.fullvision.ru

Morozov, A. A., Sushkova, O. S., & Polupanov, A. F. (2017a). Towards the distributed logic programming of intelligent visual surveillance applications. In O. Pichardo-Lagunas & S. Miranda-Jimenez (Eds.), *Advances in Soft Computing: 15th Mexican International Conference on Artificial Intelligence, MICAI 2016, Cancun, Mexico, Proceedings, Part II* (pp. 42–53). Cham: Springer International Publishing. 10.1007/978-3-319-62428-0_4

Morozov, A. A., Sushkova, O. S., & Polupanov, A. F. (2017b, June). Object-oriented logic programming of 3D intelligent video surveillance: The problem statement. In *IEEE 26th International Symposium on Industrial Electronics (ISIE), 2017* (pp. 1631–1636). IEEE Xplore Digital Library.

Morozov, A. A. (2003). Logic object-oriented model of asynchronous concurrent computations. *Pattern Recognition and Image Analysis, 13*(4), 640–649.

Morozov, A. A. (2015). Development of a method for intelligent video monitoring of abnormal behavior of people based on parallel object-oriented logic programming. *Pattern Recognition and Image Analysis, 25*(3), 481–492. doi:10.1134/S1054661815030153

Morozov, A. A., & Polupanov, A. F. (2014, June). Intelligent visual surveillance logic programming: Implementation issues. In T. Ströder & T. Swift (Eds.), *CICLOPS-WLPE 2014* (pp. 31–45). RWTH Aachen University.

Morozov, A. A., & Polupanov, A. F. (2015). 5). Development of the logic programming approach to the intelligent monitoring of anomalous human behaviour. In D. Paulus, C. Fuchs, & D. Droege (Eds.), *OGRW 2014* (pp. 82–85). Koblenz: University of Koblenz-Landau.

Morozov, A. A., & Sushkova, O. S. (2016). Real-time analysis of video by means of the Actor Prolog language. *Computer Optics, 40*(6), 947–957. doi:10.18287/2412-6179-2016-40-6-947-957

Morozov, A. A., Sushkova, O. S., & Polupanov, A. F. (2015a). 8). An approach to the intelligent monitoring of anomalous human behaviour based on the Actor Prolog object-oriented logic language. In N. Bassiliades & ... (Eds.), *RuleML 2015 DC and Challenge*. Berlin: CEUR.

Morozov, A. A., Sushkova, O. S., & Polupanov, A. F. (2015b). 8). A translator of Actor Prolog to Java. In N. Bassiliades & ... (Eds.), *RuleML 2015 DC and Challenge*. Berlin: CEUR.

Morozov, A. A., Sushkova, O. S., & Polupanov, A. F. (2017c). Object-oriented logic programming of 3D intelligent video surveillance systems: The problem statement. *RENSIT, 9*(2), 205–214. doi:10.17725/rensit.2017.09.205

Morozov, A. A., Vaish, A., Polupanov, A. F., Antciperov, V. E., Lychkov, I. I., Alfimtsev, A. N., & Deviatkov, V. V. (2014). Development of concurrent object-oriented logic programming system to intelligent monitoring of anomalous human activities. In A. Cliquet Jr, G. Plantier, T. Schultz, A. Fred, & H. Gamboa (Eds.), *BIODEVICES 2014* (pp. 53–62). SCITEPRESS.

Morozov, A. A., Vaish, A., Polupanov, A. F., Antciperov, V. E., Lychkov, I. I., Alfimtsev, A. N., & Deviatkov, V. V. (2015). Development of concurrent object-oriented logic programming platform for the intelligent monitoring of anomalous human activities. In G. Plantier, T. Schultz, A. Fred, & H. Gamboa (Eds.), *BIOSTEC 2014* (Vol. 511, pp. 82–97). Heidelberg, Germany: Springer. doi:10.1007/978-3-319-26129-4_6

Murphy, C. J. (2002). Optical sensing with quantum dots. *Analytical Chemistry, 74*(19), 520A–526 A. doi:10.1021/ac022124v PMID:12380801

Murrieta-Rico, F. N., Mercorelli, P., Sergiyenko, O. Y., Petranovskii, V., Hernandez-Balbuena, D., & Tyrsa, V. (2015). Mathematical modelling of molecular adsorption in zeolite coated frequency domain sensors. *IFAC-PapersOnLine, 48*(1), 41–46. doi:10.1016/j.ifacol.2015.05.060

Murrieta-Rico, F., Sergiyenko, O., Petranovskii, V., Hernandez-Balbuena, D., Lindner, L., Tyrsa, V., ... Karthashov, V. (2016, May). Pulse width influence in fast frequency measurements using rational approximations. *Measurement, 86,* 67–78. doi:10.1016/j.measurement.2016.02.032

Mustafah, Y. M., Zainuddin, N. A., Rashidan, M. A., Aziz, N. N. A., & Saripan, M. I. (2017). Intelligent Surveillance System for Street Surveillance. *Pertanika Journal of Science & Technology, 25*(1), 181–190.

Nair, K. (2007). Time series based structural damage detection algorithm using Gaussian mixtures modeling. *Journal of Dynamic Systems, Measurement, and Control, 129*(3), 285–293. doi:10.1115/1.2718241

Nayar, S. K., & Nakagawa, Y. (1990). Shape from Focus: An Effective Approach for Rough Surfaces. In *IEEE International Conference* (pp. 218-225). IEEE. 10.1109/ROBOT.1990.125976

Nelli, F. (2017). *OpenCV & Python – Harris Corner Detection – a method to detect corners in an image.* Retrieved from http://www.meccanismocomplesso.org/en/opencv-python-harris-corner-detection-un-metodo-per-rilevare-i-vertici-in-unimmagine/

Nemirovsky, Y., Brouk, I., & Jakobson, C. G. (2001). 1/f noise in CMOS transistors for analog applications. *IEEE Transactions on Electron Devices, 48*(5), 921–927. doi:10.1109/16.918240

Nierstrasz, O., & Dami, L. (1995). Component-oriented software technology. In O. Nierstrasz & D. Tsichritzis (Eds.), *Object-Oriented Software Composition* (pp. 3–28). Prentice Hall.

Noll, A.M. (1967). Cepstrum pitch determination. *Trans of the Acoustical Soc. of America, 41*(2).

Norouzi, A., Rahim, M. S. M., Altameem, A., Saba, T., Rad, A. E., Rehman, A., & Uddin, M. (2014). Medical image segmentation methods, algorithms, and applications. *IETE Technical Review, 31*(3), 199–213. doi:10.1080/02564602.2014.906861

Norsworthy, S. R., Schreier, R., & Temes, G. C. (Eds.). (1997). *Delta-sigma data converters: theory, design, and simulation* (Vol. 97). New York: IEEE press.

Nosov, V. V., Lukin, V. P., Nosov, E. V., Torgaev, A. V., Grigoriev, V. M., & Kovadlo, P. G. (2009). *Coherent structures in the turbulent atmosphere. In Mathematical models of nonlinear phenomena* (pp. 120–154). Nova Science Publishers.

O'Hara, S. (2008). *VERSA – video event recognition for surveillance applications* (Unpublished master's thesis). University of Nebraska at Omaha.

Odell, J. (2002). Objects and agents compared. *Journal of Object Technology, 1*(1), 41–53. doi:10.5381/jot.2002.1.1.c4

Ohnishi, N., & Imiya, A. (2013, June - July). Appearance-based navigation and homing for autonomous mobile robot. *Image and Vision Computing, 31*(6-7), 511–532. doi:10.1016/j.imavis.2012.11.004

Oh, S.-I., & Kang, H.-B. (2017). Object Detection and Classification by Decision-Level Fusion for Intelligent Vehicle Systems. *Sensors (Basel), 17*(1), 207. doi:10.339017010207 PMID:28117742

Openheim, A. V., & Shafer, R. V. (1979). *Digital signal processing.* Moscow: Svyaz. (in Russian)

Ott, R., Gutiérrez, M., Thalmann, D., & Vexo, F. (2006). Advanced virtual reality technologies for surveillance and security applications. *International conference on virtual reality continuum and its applications,* 163–170. 10.1145/1128923.1128949

Pages, J., Collewet, C., Chaumette, F., & Salvi, J. (2006, June). A camera-projector system for robot positioning by visual servoing. In *Computer Vision and Pattern Recognition Workshop, 2006. CVPRW'06. Conference on* (pp. 2-2). IEEE. 10.1109/CVPRW.2006.9

Pajares, G., García-Santillán, I., Campos, Y., Montalvo, M., Guerrero, J., Emmi, L., & Gonzalez-de-Santos, P. (2016). Machine-Vision Systems Selection for Agricultural Vehicles: A Guide. *Journal of Imaging*, 2(4), 34. doi:10.3390/jimaging2040034

Palojärvi, P., Määttä, K., & Kostamovaara, J. (1997). Integrated Time-of-Flight Laser Radar. *IEEE Transactions on Instrumentation and Measurement*, 969–999.

Pan, B., Qian, K., Xie, H., & Asundi, A. (2009). Two-dimensional digital image correlation for in-plane displacement and strain measurement: A review. *Measurement Science & Technology*, 20(6), 1–17. doi:10.1088/0957-0233/20/6/062001 PMID:20463843

Patra, S., Bhowmick, B., Banerjee, S., & Kalra, P. (2012). High Resolution Point Cloud Generation from Kinect and HD Cameras using Graph Cut. *VISAPP*, 12(2), 311–316.

Patwardhan, A., & Knapp, G. (2016). *Aggressive actions and anger detection from multiple modalities using Kinect.* arXiv preprint arXiv:1607.01076.

Pavia, D. L., Lampman, G. M., Kriz, G. S., & Vyvyan, J. A. (2008). *Introduction to spectroscopy*. Cengage Learning.

Payne, J. (1973). An Optical Distance Measuring Instrument. *The Review of Scientific Instruments*, 44(3), 304–306. doi:10.1063/1.1686113

Payra, P., & Dutta, P. K. (2003). Development of a dissolved oxygen sensor using tris (bipyridyl) ruthenium (II) complexes entrapped in highly siliceous zeolites. *Microporous and Mesoporous Materials*, 64(1), 109–118. doi:10.1016/j.micromeso.2003.06.002

Pearson, K. (1901). On Lines and Planes of Closest Fit to Systems of Points in Space. *Philosophical Magazine*, 2(11), 559–572.

Penney, C. M., & Thomas, B. (1989). High performance laser triangulation ranging. *Journal of Laser Applications*, 1(2), 51–58. doi:10.2351/1.4745229

Penza, M., Cassano, G., Aversa, P., Cusano, A., Cutolo, A., Giordano, M., & Nicolais, L. (2005). Carbon nanotube acoustic and optical sensors for volatile organic compound detection. *Nanotechnology*, 16(11), 2536–2547. doi:10.1088/0957-4484/16/11/013

Plotnikov, S. V. (1995). Comparison of methods for signal processing in the triangulation measurement systems. *Optoelectronics, Instrumentation and Data Processing*, 6, 58–63.

Pollinger, F., Hieta, T., Vainio, M., Doloca, N. R., Abou-Zeid, A., Meiners-Hagen, K., & Merimaa, M. (2012). Effective humidity in length measurements: Comparison of three approaches. *Measurement Science & Technology*, 23(2), 025503. doi:10.1088/0957-0233/23/2/025503

Ponti, G., Palombi, F., Abate, D., Ambrosino, F., Aprea, G., Bastianelli, T., . . . Vita, A. (2014). The role of medium size facilities in the HPC ecosystem: the case of the new CRESCO4 cluster integrated in the ENEAGRID infrastructure. In *Proceedings of the 2014 International Conference on High Performance Computing and Simulation, HPCS 2014*, (pp. 1030-1033). Bologna, Italy: Academic Press. 10.1109/HPCSim.2014.6903807

Poonam, D. (2017). Image Stitching Based on Corner Detection. *International Journal on Recent and Innovation Trends in Computing and Communication, Volume*, 5(4), 351–354.

Poorani, M., Prathiba, T., & Ravindran, G. (2013). Integrated Feature Extraction for Image Retrieval. *International Journal of Computer Science and Mobile Computing, 2*(2), 28–35.

Popa, M., Koc, A. K., Rothkrantz, L. J., Shan, C., & Wiggers, P. (2011). Kinect sensing of shopping related actions. In *International joint conference on ambient intelligence* (pp. 91–100). Academic Press.

Pramanik, S., Zheng, C., Zhang, X., Emge, T. J., & Li, J. (2011). New microporous metal–organic framework demonstrating unique selectivity for detection of high explosives and aromatic compounds. *Journal of the American Chemical Society, 133*(12), 4153–4155. doi:10.1021/ja106851d PMID:21384862

Preis, J., Kessel, M., Werner, M., & Linnhoff-Popien, C. (2012). Gait recognition with Kinect. In *International workshop on Kinect in pervasive computing*. Newcastle, UK: Academic Press.

Presti, L. L., & Cascia, M. L. (2016). 3D skeleton-based human action classification: A survey. *Pattern Recognition, 53*, 130–147. doi:10.1016/j.patcog.2015.11.019

Raheja, J., Minhas, M., Prashanth, D., Shah, T., & Chaudhary, A. (2015). Robust gesture recognition using Kinect: A comparison between DTW and HMM. *Optik – International Journal for Light and Electron Optics, 126*(11), 1098–1104.

Rashidi, A., Sigari, M. H., Maghiar, M., & Citrin, D. (2016). An analogy between various machine-learning techniques for detecting construction materials in digital images. *KSCE Journal of Civil Engineering, 20*(4), 1178–1188. doi:10.100712205-015-0726-0

Raymond, M. (1984). *Laser Remote Sensing: Fundamentals and Applications*. Wiley.

Real-Moreno, O., Rodriguez-Quiñonez, J. C., Sergiyenko, O., Basaca-Preciado, L. C., Hernandez-Balbuena, D., Rivas-Lopez, M., & Flores-Fuentes, W. (2017). Accuracy Improvement in 3D Laser Scanner Based on Dynamic Triangulation for Autonomous Navigation System. In *Industrial Electronics (ISIE), 2017 IEEE 26th International Symposium on* (pp. 1602-1608). IEEE.

Regtien, P., van der Heijden, F., Korsten, M. J., & Otthius, W. (2004). *Measurement Science for Engineers*. Elsevier Science.

Remillard, J. T., Jones, J. R., Poindexter, B. D., Narula, C. K., & Weber, W. H. (1999). Demonstration of a high-temperature fiber-optic gas sensor made with a sol-gel process to incorporate a fluorescent indicator. *Applied Optics, 38*(25), 5306–5309. doi:10.1364/AO.38.005306 PMID:18324032

Remillard, J. T., Poindexter, B. D., & Weber, W. H. (1997). Fluorescence characteristics of Cu-ZSM-5 zeolites in reactive gas mixtures: Mechanisms for a fiber-optic-based gas sensor. *Applied Optics, 36*(16), 3699–3707. doi:10.1364/AO.36.003699 PMID:18253395

Remondino, F., Barazzetti, L., Nex, F., Scaioni, M., & Sarazzi, D. (2011). UAV photogrammetry for mapping and 3D modeling - Current status and future perspectives. *Int. Archives of Photogrammetry, Remote Sensing and Spatial Information Sciences, 38*(1/C22).

Repin, V. G., & Tartakovsky, G. P. (1977). *Statistical synthesis with a priori indeterminacy and adaptation of informational systems*. Moscow: Sov. Radio. (in Russian)

Resendiz, M. J., Noveron, J. C., Disteldorf, H., Fischer, S., & Stang, P. J. (2004). A self-assembled supramolecular optical sensor for Ni (II), Cd (II), and Cr (III). *Organic Letters, 6*(5), 651–653. doi:10.1021/ol035587b PMID:14986941

Richter, S. R., & Roth, S. (2015). Discriminative Shape from Shading in Uncalibrated Illumination. In *Proceedings of the IEEE Conference on Computer Vision and Pattern Recognition* (pp. 1128-1136). IEEE. 10.1109/CVPR.2015.7298716

Rivas-Lopez, M., Sergiyenko, O., & Tyrsa, V. (2008). Machine Vision: Approaches and Limitations. In Computer Vision. InTech.

Rodríguez-Iznaga, I., Rodríguez-Fuentes, G., & Petranovskii, V. (2018). Ammonium modified natural clinoptilolite to remove manganese, cobalt and nickel ions from wastewater: Favorable conditions to the modification and selectivity to the cations. *Microporous and Mesoporous Materials*, *255*, 200–210. doi:10.1016/j.micromeso.2017.07.034

Rodriguez-Quiñonez, J. C., Sergiyenko, O. Y., Basaca-Preciado, L. C., Tyrsa, V. V., Gurko, A. G., Podrygalo, M. A., . . . Hernandez-Balbuena, D. (2014). Optical monitoring of scoliosis by 3D medical laser scanner. *Optics and Laser in Engineering*, 175-186.

Rodriguez-Quiñonez, J. C., Sergiyenko, O., Flores-Fuentes, W., Rivas-Lopez, M., Hernandez-Balbuena, D., Rascon, R., & Mercorelli, P. (2017). Improve a 3D distance measurement accuracy in stereo vision systems using optimization methods' approach. *Opto-Electronics Review*, *25*(1), 24–32. doi:10.1016/j.opelre.2017.03.001

Rodríguez-Quiñonez, J., Sergiyenko, O., Basaca-Preciado, L., Tyrsa, V., Gurko, A., Podrygalo, M., ... Hernandez-Balbuena, D. (2014, March). Optical monitoring of scoliosis by 3D medical laser scanner. *Optics and Lasers in Engineering*, *54*, 175–186. doi:10.1016/j.optlaseng.2013.07.026

Rodríguez-Quiñonez, J., Sergiyenko, O., Gonzalez-Navarro, F., Basaca-Preciado, L., & Tyrsa, V. (2013, February). Surface recognition improvement in 3D medical laser scanner using Levenberg–Marquardt method. *Signal Processing*, *93*(2), 378–386. doi:10.1016/j.sigpro.2012.07.001

Roselli, I., Paolini, D., Mongelli, M., De Canio, G., & de Felice, G. (2017). Processing of 3D optical motion data of shaking table tests: filtering optimization and modal analysis. In *Proceedings of the 6th International Conference on Computational Methods in Structural Dynamics and Earthquake Engineering (COMPDYN)* (*vol. 2*, pp. 4174-4183). Rhodes, Greece: Academic Press. 10.7712/120117.5714.18115

Roselli, I., Mencuccini, G., Mongelli, M., Beone, F., De Canio, G., Di Biagio, F., & Rocchi, A. (2010). The DySCo virtual lab for seismic and vibration tests at the ENEA Casaccia Research Center. *Proceedings of the 14th European Conference On Earthquake Engineering (14ECEE)*.

Rougier, C., Auvinet, E., Rousseau, J., Mignotte, M., & Meunier, J. (2011). Fall detection from depth map video sequences. In *International conference on smart homes and health telematics* (pp. 121–128). Academic Press. 10.1007/978-3-642-21535-3_16

Rouse, M. (2016). *Image metadata*. Retrieved from http://whatis.techtarget.com/definition/image-metadata

Rublee, E., Rabaud, V., Konolige, K., & Bradski, G. (2011). ORB: An efficient alternative to SIFT or SURF. *Proceedings of the 2011 International Conference on Computer Vision*, 2564-2571. 10.1109/ICCV.2011.6126544

Russakovsky, O., Deng, J., Su, H., Krause, J., Satheesh, S., Ma, S., . . . Fei-Fei, L. (2015). *ImageNet Large Scale Visual Recognition Challenge*. Retrieved from https://arxiv.org/abs/1409.0575

Russell, S., & Norvig, P. (1995). *Artificial intelligence. A modern approach*. London: Prentice-Hall.

Rusu, R. B., & Cousins, S. (2011, May). 3d is here: Point cloud library (pcl). In *Robotics and automation (ICRA), 2011 IEEE International Conference on* (pp. 1-4). IEEE.

Ruta, M., Scioscia, F., Summa, M. D., Ieva, S., Sciascio, E. D., & Sacco, M. (2014). Semantic matchmaking for Kinect-based posture and gesture recognition. *International Journal of Semantic Computing*, *8*(4), 491–514. doi:10.1142/S1793351X14400169

Saberioon, M., Gholizadeh, A., Cisar, P., Pautsina, A., & Urban, J. (2016). Application of machine vision systems in aquaculture with emphasis on fish: State-of-the-art and key issues. *Reviews in Aquaculture.*

Sakarya, U., & Telatar, Z. (2010). Video scene detection using graph-based representations. *Signal Processing Image Communication*, 25(10), 774–783. doi:10.1016/j.image.2010.10.001

Salehi, H., Burgueño, R., Das, S., Biswas, S., & Chakrabartty, S. (2016). Structural Health Monitoring from Discrete Binary Data through. *Pattern Recognition.*

Salvi, J. (1998). *An approach to coded structured light to obtain three dimensional information.* Universitat de Girona.

Salvi, J., Armangué, X., & Batlle, J. (2002). A comparative review of camera calibrating methods with accuracy evaluation. *Pattern Recognition*, 35(7), 1617–1635. doi:10.1016/S0031-3203(01)00126-1

Sargano, A. B., Angelov, P., & Habib, Z. (2016). Human Action Recognition from Multiple Views Based on View-Invariant Feature Descriptor Using Support Vector Machines. *Applied Sciences*, 6(10), 309. doi:10.3390/app6100309

Satrughan, K. U. M. A. R., & Jigyendra Sen, Y. A. D. A. V. (2016, June). Segmentation of Moving Objects using Background Subtraction Method in Complex Environments. *Wuxiandian Gongcheng*, 25(2), 399–408.

Satta, R., Pala, F., Fumera, G., & Roli, F. (2013). *Real-time appearance-based person re-identification over multiple Kinect™ cameras.* VISAPP.

Savage, R., Clarke, N., & Li, F. (2013). Multimodal biometric surveillance using a Kinect sensor. In *Proceedings of the 12th annual security conference.* Las Vegas, NV: Academic Press.

Savitzky, A., & Golay, M. J. E. (1964). Smoothing and Differentiation of Data by Simplified Least Squares Procedures. *Analytical Chemistry*, 36(8), 1627–1639. doi:10.1021/ac60214a047

Savory, S. J. (2008). Digital filters for coherent optical receivers. *Optics Express*, 16(2), 804–817. doi:10.1364/OE.16.000804 PMID:18542155

Scheffer, D., Dierickx, B., & Meynants, G. (1997). Random addressable 2048x2048 active pixel image sensor. *IEEE Transactions on Electron Devices*, 44(10), 1716–1720. doi:10.1109/16.628827

Schwitzke, M., Bachmann, V., Noack, M., Freyer, T., & Launert, B. (2010). Concrete-crack monitoring on structural elements by automatic digital image analysis. *Bauingenieur*, 85, 455–459.

Sebe, I. O., Hu, J., You, S., & Neumann, U. (2003). 3D video surveillance with augmented virtual environments. In *First ACM SIGMM international workshop on video surveillance* (pp. 107–112). ACM.

Semenov, D. V., Sidorov, I. S., Nippolainen, E., & Kamshilin, A. A. (2010). Speckle-based sensor system for real-time distance and thickness monitoring of fast moving objects. *Measurement Science & Technology*, 21(4), 045304. doi:10.1088/0957-0233/21/4/045304

Sengupta, P., & Li, B. (2014). Hysteresis behavior of reinforced concrete walls. *Journal of Structural Engineering*, 140(7), 1–18. doi:10.1061/(ASCE)ST.1943-541X.0000927

Sergiyenko, O. Y., Tyrsa, V. V., Hernandez, W., Starostenko, O., & Rivas, M. (2009). Dynamic laser scanning method for mobile robot navigation. 제어로봇시스템학회 국제학술대회 논문집, 4884-4889.

Sergiyenko, O. Yu., Ivanov, M. V., Kartashov, V. M., Tyrsa, V. V., Hernández-Balbuena, D., & Nieto-Hipólito, J. I. (2016). Transferring model in robotic group. In *2016 IEEE 25th International Symposium on Industrial Electronics (ISIE)* (pp. 946-952). Santa Clara, CA: IEEE.

Sergiyenko, O., Tyrsa, V., Basaca-Preciado, L., Rodríguez-Quiñonez, J., Hernandez, W., Nieto-Hipolito, J., ... Starostenko, O. (2011). Electromechanical 3D Optoelectronic Scanners: Resolution Constraints and Possible Ways of Improvement. In Optoelectronic Devices and Properties. InTech.

Sergiyenko, O., Tyrsa, V., Hernandez-Balbuena, D., Lopez, M., Lopez, I., & Cruz, L. (2008). Precise Optical Scanning for practical multi-applications. In *Industrial Electronics, 2008. IECON 2008. 34th Annual Conference of IEEE* (pp. 1656-1661). IEEE.

Sergiyenko, O. (2010). Optoelectronic System for Mobile Robot Navigation. *Optoelectronics, Instrumentation and Data Processing*, *46*(5), 414–428. doi:10.3103/S8756699011050037

Sergiyenko, O., Hernandez, W., Tyrsa, V., Cruz, L. F. D., Starostenko, O., & Peña-Cabrera, M. (2009). Remote Sensor for Spatial Measurements by Using Optical Scanning. *Sensors (Basel)*, *9*(7), 5477–5492. doi:10.339090705477 PMID:22346709

Sergiyenko, O., Ivanov, M., Tyrsa, V., Kartashov, V., Rivas-López, M., Hernández-Balbuena, D., ... Tchernykh, A. (2016). Data transferring model determination in robotic group. *Robotics and Autonomous Systems*, *83*, 251–260. doi:10.1016/j.robot.2016.04.003

Sergiyenko, O., & Rodriguez-Quiñonez, J. C. (2016). *Developing and Applying Optoelectronics in Machine Vision*. IGI Global.

Shao, L., & Ji, L. (2009). Motion histogram analysis based key frame extraction for human action/activity representation. *2009 Canadian Conference on Computer and Robot Vision, IEEE CRV'09*, 88 – 92.

Sharkov, E. A. (2003). *Passive Microwave Remote Sensing of the Earth: Physical Foundations*. Springer.

Sharma, S., Katiyal, S., & Arya, L. D. (2015). Performance Comparison of Teaching-Learning-Based Optimization and Differential Evolution Algorithms Applied to the Design of Linear Phase Digital FIR Filter. *IUP Journal of Telecommunications, 7*(1), 23-38.

Shelhamer, E., Long, J., & Darrell, T. (2016). *Fully Convolutional Networks for Semantic Segmentation*. Retrieved from https://arxiv.org/abs/1605.06211

Shen, W., Hao, Q., Yoon, H., & Norrie, D. (2006). Applications of agent-based systems in intelligent manufacturing: An updated review. *Advanced Engineering Informatics*, *20*(4), 415–431. doi:10.1016/j.aei.2006.05.004

Shet, V., Harwood, D., & Davis, L. (2005). VidMAP: Video monitoring of activity with Prolog. In AVSS 2005 (pp. 224–229). IEEE.

Shet, V., Singh, M., Bahlmann, C., Ramesh, V., Neumann, J., & Davis, L. (2011, June). Predicate logic based image grammars for complex pattern recognition. *International Journal of Computer Vision*, *93*(2), 141–161. doi:10.100711263-010-0343-9

Shiang, C. W., Onn, B. T., Tee, F. S., & Khairuddin, M. A. (2016). Developing agent-oriented video surveillance system through agent-oriented methodology (AOM). *CIT. Journal of Computing and Information Technology*, *4*(24), 349–367. doi:10.20532/cit.2016.1002869

Shin, T., Kim, J.-G., Lee, H., & Kim, J. (1998). Hierarchical Scene Change Detection In An Mpeg-2 Compressed Video Sequence. *IEEE International Symposium on Circuits and Systems*, 4, 253 - 256.

Shirasu, M., & Touhara, K. (2011). The scent of disease: Volatile organic compounds of the human body related to disease and disorder. *Journal of Biochemistry*, *150*(3), 257–266. doi:10.1093/jb/mvr090 PMID:21771869

Shukla, Mithlesh, & Sharma. (2015). A Survey on Different Video Scene Change Detection Techniques. *International Journal of Science and Research*, 214 – 219.

Shyamal, M., Maity, S., Mazumdar, P., Sahoo, G. P., Maity, R., & Misra, A. (2017). Synthesis of an efficient Pyrene based AIE active functional material for selective sensing of 2, 4, 6-trinitrophenol. *Journal of Photochemistry and Photobiology A Chemistry*, *342*, 1–14. doi:10.1016/j.jphotochem.2017.03.030

Sicard, J., & Sirohi, J. (2013). Measurement of the deformation of an extremely flexible rotor blade using digital image correlation. *Measurement Science & Technology*, *24*(6), 065203. doi:10.1088/0957-0233/24/6/065203

Silvestre, D. (2007). *Video surveillance using a time-of-light camera* (Unpublished master's thesis). Informatics and Mathematical Modelling, Technical University of Denmark.

Sinha, A., Chakravarty, K., & Bhowmick, B. (2013). Person identification using skeleton information from Kinect. In *The sixth international conference on advances in computer-human interactions (ACHI 2013)* (pp. 101–108). Academic Press.

Skarlatidis, A., Artikis, A., Filippou, J., & Paliouras, G. (2014). A probabilistic logic programming event calculus. *Theory and Practice of Logic Programming*, 1–33.

Smith, J. V. (1984). Definition of a zeolite. *Zeolites*, *4*(4), 309–310. doi:10.1016/0144-2449(84)90003-4

Smith, S. W. (1997). *The scientist and engineer's guide to digital signal processing*. California Technical Publishing.

Sohn, H., Czarnecki, J. A., & Farrar, C. R. (2000). Structural health monitoring using statistical process control. *Journal of Structural Engineering*, *126*(11), 1356–1363. doi:10.1061/(ASCE)0733-9445(2000)126:11(1356)

Sohn, H., Farrar, C. R., Hemez, F. M., Shunk, D. D., Stinemates, D. W., Nadler, B. R., & Czarnecki, J. J. (2003). *A review of structural health monitoring literature: 1996–2001*. Los Alamos National Laboratory.

Sohrabnezhad, S., Pourahmad, A., & Sadjadi, M. A. (2007). New methylene blue incorporated in mordenite zeolite as humidity sensor material. *Materials Letters*, *61*(11), 2311–2314. doi:10.1016/j.matlet.2006.09.006

Song, B. (2000). Nyquist-Rate ADC and DAC. In E. W. Chen (Ed.), VLSI Handbook. Academic Press.

Song, W., Fu, M., Yang, Y., Wang, M., Wang, X., & Kornhauser, A. (2017). Real-Time Lane Detection and Forward Collision Warning System Based on Stereo Vision. In *Intelligent Vehicles Symposium (IV), 2017 IEEE* (pp. 493-498). IEEE. 10.1109/IVS.2017.7995766

Stettner, R., Bailey, H., & Silverman, S. (2008). Three dimensional Flash LADAR focal planes and time dependent imaging. *International Journal of High Speed Electronics and Systems*, *18*(02), 401–406. doi:10.1142/S0129156408005436

Striebel, C., Hoffmann, K., & Marlow, F. (1997). The microcrystal prism method for refractive index measurements on zeolite-based nanocomposites. *Microporous Materials*, *9*(1), 43–50. doi:10.1016/S0927-6513(96)00090-9

Sukegawa, S., Umebayashi, T., Nakajima, T., Kawanobe, H., Koseki, K., Hirota, I., & Fukushima, N. (2013). A 1/4-inch 8Mpixel Back-Illuminated Stacked CMOS Image Sensor. In ISSCC 2013 (pp. 484–486). IEEE.

Sulaiman, S., Hussain, A., Tahir, N., Samad, S. A., & Mustafa, M. M. (2008). Human Silhouette Extraction Using Background Modeling and Subtraction Techniques. *Information Technology Journal*, *7*(1), 155–159. doi:10.3923/itj.2008.155.159

Sun, S., Hu, W., Gao, H., Qi, H., & Ding, L. (2017). Luminescence of ferrocene-modified pyrene derivatives for turn-on sensing of Cu 2+ and anions. *Spectrochimica Acta. Part A: Molecular and Biomolecular Spectroscopy*, *184*, 30–37. doi:10.1016/j.saa.2017.04.073 PMID:28477514

Susperregi, L., Sierra, B., Castrillón, M., Lorenzo, J., Martínez-Otzeta, J. M., & Lazkano, E. (2013). On the use of a low-cost thermal sensor to improve kinect people detection in a mobile robot. *Sensors (Basel)*, *13*(11), 14687–14713. doi:10.3390131114687 PMID:24172285

Suwajanakorn, S., Hernandez, C., & Seitz, S. M. (2015). Depth from Focus with Your Mobile. In *IEEE Conference on Computer Vision and Pattern Recognition* (pp. 3497-3506). IEEE.

Suzuki, A., Shimamura, N., Kainuma, T., Kawazu, N., Okada, C., Oka, T., & Wakabayashi, H. (2015). A 1/1.7 inch 20Mpixel Back illuminated stacked CMOS Image sensor for new Imaging application. In ISSCC 2015 (pp. 110–112). IEEE.

Swartz, B. E. (1998). The advantages of digital over analog recording techniques. *Electroencephalography and Clinical Neurophysiology*, *106*(2), 113–117. doi:10.1016/S0013-4694(97)00113-2 PMID:9741771

Sytnik, O. V. (2006). Ground-penetrating radar data preprocessing. *Telecommunications and Radio Engineering*, *65*(7), 621–631. doi:10.1615/TelecomRadEng.v65.i7.40

Sytnik, O. V. (2011). Textural Analysis of Cepstrum Images of Subsurface Structure. *Telecommunications and Radio Engineering*, *70*(1), 87–94. doi:10.1615/TelecomRadEng.v70.i1.90

Sytnik, O. V., & Gorokhovatsky, A. V. (2007). Algorithm for signal processing in identifying subsurface objects, *Izv. VUZov*. [in Russian]. *Radioelektronika*, *50*(10), 43–52.

Tajbakhsh, N., Shin, J. Y., Gurudu, S. R., Hurst, R. T., Kendall, C. B., Gotway, M. B., & Liang, J. (2016). Convolutional neural networks for medical image analysis: Full training or fine tuning? *IEEE Transactions on Medical Imaging*, *35*(5), 1299–1312. doi:10.1109/TMI.2016.2535302 PMID:26978662

Takayanagi, I., Yoshimura, N., Sato, T., Matsuo, S., Kawaguchi, T., Mori, K., & Nakamura, J. (2013). A 1-inch Optical Format, 80fps, 10. 8Mpixel CMOS Image Sensor Operating in a Pixel-to-ADC Pipelined Sequence Mode. *Proc. Int'l Image Sensor Workshop*, 325–328.

Tang, X., Provenzano, J., Xu, Z., Dong, J., Duan, H., & Xiao, H. (2011). Acidic ZSM-5 zeolite-coated long period fiber grating for optical sensing of ammonia. *Journal of Materials Chemistry*, *21*(1), 181–186. doi:10.1039/C0JM02523B

Tan, L., & Jiang, J. (2007). *Fundamentals of Analog and Digital Signal Processing*. AuthorHouse.

Tao, M. W., Srinivasan, P. P., Hadap, S., Rusinkiewicz, S., Malik, J., & Ramamoorthi, R. (2017). Shape Estimation from Shading, Defocus, and Correspondance Using Light-Field Angular Coherence. *IEEE Transactions on Pattern Analysis and Machine Intelligence*, *39*(3), 546–560. doi:10.1109/TPAMI.2016.2554121 PMID:27101598

Tarau, P. (2012). The BinProlog experience: Architecture and implementation choices for continuation passing Prolog and first-class logic engines. *Theory and Practice of Logic Programming*, *12*(1–2), 97–126. doi:10.1017/S1471068411000433

Tekieli, M., De Santis, S., de Felice, G., Kwiecień, A., & Roscini, F. (2017). Application of Digital Image Correlation to composite reinforcements testing. *Composite Structures*, *160*, 670–688. doi:10.1016/j.compstruct.2016.10.096

The MathWorks, Inc. (2017). *gmdistribution*. Author.

The MathWorks, Inc. (2018). *IIR Filter Method Summary*. Author.

Tittl, A., Mai, P., Taubert, R., Dregely, D., Liu, N., & Giessen, H. (2011). Palladium-based plasmonic perfect absorber in the visible wavelength range and its application to hydrogen sensing. *Nano Letters*, *11*(10), 4366–4369. doi:10.1021/nl202489g PMID:21877697

Trépanier, J., Sawan, M., Audet, Y., & Coulombe, J. (2002). A Wide Dynamic Range CMOS Digital Pixel Sensor. In *The 2002 45th Midwest Symposium on Circuits and Systems*. IEEE. 10.1109/MWSCAS.2002.1186892

Tscharke, M., & Banhazi, T. M. (2016). A brief review of the application of machine vision in livestock behaviour analysis. *Journal of Agricultural Informatics, 7*(1), 23-42.

Tschmelak, J., Proll, G., & Gauglitz, G. (2005). Optical biosensor for pharmaceuticals, antibiotics, hormones, endocrine disrupting chemicals and pesticides in water: Assay optimization process for estrone as example. *Talanta, 65*(2), 313–323. doi:10.1016/j.talanta.2004.07.011 PMID:18969801

Tseng, T. L. B., Aleti, K. R., Hu, Z., & Kwon, Y. J. (2015). E-quality control: A support vector machines approach. *Journal of Computational Design and Engineering, 3*(2), 91–101. doi:10.1016/j.jcde.2015.06.010

Turbo Prolog Owner's Handbook . (1986). Borland International.

Turhan, H., Tukenmez, E., Karagoz, B., & Bicak, N. (2018). Highly fluorescent sensing of nitroaromatic explosives in aqueous media using pyrene-linked PBEMA microspheres. *Talanta, 179*, 107–114. doi:10.1016/j.talanta.2017.10.061 PMID:29310209

Usamentiaga, R., Molleda, J., & Garcia, D. F. (2014). Structured-light sensor using two laser stripes for 3D reconstruction without vibrations. *Sensors (Basel), 14*(11), 20041–20063. doi:10.3390141120041 PMID:25347586

Vaidyanathan, P. P. (2008). *The Theory of Linear Prediction*. Morgan & Claypool.

Vallejo, D., Albusac, J., Castro-Schez, J., Glez-Morcillo, C., & Jiménez, L. (2011). A multiagent architecture for supporting distributed normality-based intelligent surveillance. *Engineering Applications of Artificial Intelligence, 24*(2), 325–340. doi:10.1016/j.engappai.2010.11.005

Van Trees, H. (1972-1977). *Detection estimation and modulation theory: in 3 vol.* Sov. radio. (in Russian)

Vanithamani, S. (2017). Vehicle classification and analyzing motion features. *Indian Journal of Engineering, 14*(36), 89–94.

Villatoro, J., & Monzón-Hernández, D. (2005). Fast detection of hydrogen with nano fiber tapers coated with ultra thin palladium layers. *Optics Express, 13*(13), 5087–5092. doi:10.1364/OPEX.13.005087 PMID:19498497

Volos, C., Kyprianidis, I., & Stouboulos, I. (2013, December). Experimental investigation on coverage performance of a chaotic autonomous mobile robot. *Robotics and Autonomous Systems, 61*(12), 1314–1322. doi:10.1016/j.robot.2013.08.004

Wadhwa, N., Chen, J. G., Sellon, J. B., Wei, D., Rubinstein, M., Ghaffari, R., ... Freeman, W. T. (2017a). Motion microscopy for visualizing and quantifying small motions. *Proceedings of the National Academy of Sciences of the United States of America, 114*(44), 11639–11644. doi:10.1073/pnas.1703715114 PMID:29078275

Wadhwa, N., Wu, H., Davis, A., Rubinstein, M., Shih, E., Mysore, G. J., ... Durand, F. (2017b). Eulerian video magnification and analysis. *Communications of the ACM, 60*(1), 87–95. doi:10.1145/3015573

Wahbeh, A. M., Caffrey, J. P., & Masri, S. F. (2003). A vision-based approach for the direct measurement of displacements in vibrating systems. *Smart Materials and Structures, 12*(5), 785–794. doi:10.1088/0964-1726/12/5/016

Walters, P. (1982). *An Introduction to Ergodic Theory*. Berlin: Springer-Verlag. doi:10.1007/978-1-4612-5775-2

Wang, C., & Liu, H. (2013). Unusual events detection based on multi-dictionary sparse representation using Kinect. In *International conference on image processing* (pp. 2968–2972). Academic Press. 10.1109/ICIP.2013.6738611

Wang, L., Schmidt, B., & Nee, A. Y. (2013). Vision-guided active collision avoidance for human-robot collaborations. *Manufacturing Letters, 1*(1), 5–8. doi:10.1016/j.mfglet.2013.08.001

Warren, D. H. D. (1983, October). *An abstract Prolog instruction set.* Technical Note 309. Menlo Park, CA: SRI International.

Wielemaker, J., Schrijvers, T., Triska, M., & Lager, T. (2012). SWI-Prolog. *Theory and Practice of Logic Programming, 12*(1–2), 67–96. doi:10.1017/S1471068411000494

Wiener, N. (1949). *Extrapolation, interpolation and smoothing of stationary time series.* New York: John Willey.

Wild, W. J., & Giles, C. L. (1982). Goos-Hänchen shifts from absorbing media. *Physical Review A., 25*(4), 2099–2101. doi:10.1103/PhysRevA.25.2099

Wong, H. P. (1997). *CMOS Image sensors - Recent Advances and Device Scaling Considerations.* IEDM.

Worch, J.-H., Bálint-Benczédi, F., & Beetz, M. (2016). Perception for everyday human robot interaction. *KI – Künstliche Intelligenz, 30*(1), 21–27.

Xiao, H., Zhang, J., Dong, J., Luo, M., Lee, R., & Romero, V. (2005). Synthesis of MFI zeolite films on optical fibers for detection of chemical vapors. *Optics Letters, 30*(11), 1270–1272. doi:10.1364/OL.30.001270 PMID:15981503

Xidias, E. K., & Azariadis, P. N. (2016, August). Computing collision-free motions for a team of robots using formation and non-holonomic constraints. *Robotics and Autonomous Systems, 82*, 15–23. doi:10.1016/j.robot.2016.04.008

Xiong, Y., & Shafer, S. A. (1993). Depth from focusing and defocusing. In *IEEE Computer Society Conference* (pp. 68-73). IEEE.

Xu, R., Ng, W. C., Yuan, J., Member, S., Yin, S., & Wei, S. (2014). A 1 / 2. 5 inch VGA 400 fps CMOS Image Sensor with High Sensitivity for Machine Vision. *IEEE Journal of Solid-State Circuits, 49*(10), 2342–2351. doi:10.1109/JSSC.2014.2345018

Xu, Y., Dong, J., Zhang, B., & Xu, D. (2016, January). Background modeling methods in video analysis: A review and comparative evaluation. *CAAI Transactions on Intelligence Technology, 1*(1), 43–60. doi:10.1016/j.trit.2016.03.005

Yalla, V. G., & Hassebrook, L. G. (2005). Very High Resolution 3-D Surface Scanning Using Multi-frequency Phase Measuring Profilometry. *SPIE*, 44-53.

Yamaguchi, T., & Hashimoto, S. (2010). Fast crack detection method for large-size concrete surface images using percolation-based image processing. *Machine Vision and Applications, 21*(5), 797–809. doi:10.100700138-009-0189-8

Yang, M., Sun, Y., Zhang, D., & Jiang, D. (2010). Using Pd/WO$_3$ composite thin films as sensing materials for optical fiber hydrogen sensors. *Sensors and Actuators. B, Chemical, 143*(2), 750–753. doi:10.1016/j.snb.2009.10.017

Ye, J. W., Zhou, H. L., Liu, S. Y., Cheng, X. N., Lin, R. B., Qi, X. L., ... Chen, X. M. (2015). Encapsulating Pyrene in a Metal Organic Zeolite for Optical Sensing of Molecular Oxygen. *Chemistry of Materials, 27*(24), 8255–8260. doi:10.1021/acs.chemmater.5b03955

Ye, X. W., Ni, Y. Q., Wai, T. T., Wong, K. Y., Zhang, X. M., & Xu, F. (2013). A vision-based system for dynamic displacement measurement of long-span bridges: Algorithm and verification. *Smart Structures and Systems, 12*(3-4), 363–379. doi:10.12989ss.2013.12.3_4.363

Ye, X. W., Ni, Y. Q., Wong, K. Y., & Ko, J. M. (2012). Statistical analysis of stress spectra for fatigue life assessment of steel bridges with structural health monitoring data. *Engineering Structures, 45*(15), 166–176. doi:10.1016/j.engstruct.2012.06.016

Yoshino, T., Kurosawa, K., Itoh, K., & Ose, T. (1982). Fiber-optic Fabry-Perot interferometer and its sensor applications. *IEEE Transactions on Microwave Theory and Techniques*, *30*(10), 1612–1621. doi:10.1109/TMTT.1982.1131298

Yu, Y. J., & Lim, C. (2009). Low-Complexity Design of Variable Bandedge Linear Phase FIR Filters With Sharp Transition Band. Academic Press.

Zetie, K. P., Adams, S. F., & Tocknell, R. M. (2000). How does a Mach-Zehnder interferometer work? *Physics Education*, *35*(1), 46–48. doi:10.1088/0031-9120/35/1/308

Zhang, C. S., & Elaksher, A. (2012). An unmanned aerial vehicle-based imaging system for 3D measurement of unpaved road surface distresses. *Computer-Aided Civil and Infrastructure Engineering*, *27*(2), 118–129. doi:10.1111/j.1467-8667.2011.00727.x

Zhang, J., Liu, F., Shao, H., & Wang, G. (2007). An Effective Error Concealment Framework For H.264 Decoder Based on Video Scene Change Detection. *Fourth International Conference on Image and Graphics*, 285 - 290. 10.1109/ICIG.2007.174

Zhang, J., Luo, M., Xiao, H., & Dong, J. (2006). Interferometric study on the adsorption-dependent refractive index of silicalite thin films grown on optical fibers. *Chemistry of Materials*, *18*(1), 4–6. doi:10.1021/cm0525353

Zhang, J., Tang, X., Dong, J., Wei, T., & Xiao, H. (2008). Zeolite thin film-coated long period fiber grating sensor for measuring trace chemical. *Optics Express*, *16*(11), 8317–8323. doi:10.1364/OE.16.008317 PMID:18545545

Zhang, J., Tang, X., Dong, J., Wei, T., & Xiao, H. (2009). Zeolite thin film-coated long period fiber grating sensor for measuring trace organic vapors. *Sensors and Actuators. B, Chemical*, *135*(2), 420–425. doi:10.1016/j.snb.2008.09.033

Zhang, R., Tsai, P.-S., Cryer, J. E., & Shah, M. (1999). Shape from Shading: A Survey. *IEEE Transactions on Pattern Analysis and Machine Intelligence*, *21*(8), 690–706. doi:10.1109/34.784284

Zhao, W., Wang, T., Pham, H., Hu-Guo, C., Dorokhov, A., & Hu, Y. (2014). Development of CMOS Pixel Sensors with digital pixel dedicated to future particle physics experiments. *Journal of Instrumentation: An IOP and SISSA Journal*, *9*(02), C02004–C02004. doi:10.1088/1748-0221/9/02/C02004

Zhao, Z., Sevryugina, Y., Carpenter, M. A., Welch, D., & Xia, H. (2004). All-optical hydrogen-sensing materials based on tailored palladium alloy thin films. *Analytical Chemistry*, *76*(21), 6321–6326. doi:10.1021/ac0494883 PMID:15516124

Zhou, W., Lyu, C., Jiang, X., Zhou, W., Li, P., Chen, H., ... Liu, Y.-H. (2017). Efficient and Fast Implementation of Embedded Time-of-Flight RAnging System Based on FPGAs. *IEEE Sensors Journal*, *17*(18), 5862–5870. doi:10.1109/JSEN.2017.2728724

Zuev, V. V., Zuev, V. E., Makushkin, Y. S., Marichev, V. N., & Mitsel, A. A. (1983). Laser sounding of atmospheric humidity: Experiment. *Applied Optics*, *22*(23), 3742–3746. doi:10.1364/AO.22.003742 PMID:18200259

Zuo-chun, S., Hua-wei, L., & Yan-zhou, Z. (2011). Calibration of Measurement System Based on Phase Measurement Profilometry. In *Optoelectronics and Microelectronics Technology (AISOMT), 2011 Academic International Symposium* (pp. 200-203). IEEE.

About the Contributors

Moises Rivas-Lopez was born in June, 1, 1960. He received the B.S. and M.S. degrees in Autonomous University of Baja California, México, in 1985 and 1991, respectively and the PhD degree in Applied Physics, in the same university, in 2010. He is editor of a book, has written 38 papers and 6 book chapters in optical scanning, 3D coordinates measurement, and structural health monitoring applications. He holds a patent and has presented different works in several international congresses, of IEEE, ICROS, SICE, AMMAC in America and Europe. He was Dean of Engineering Institute of Autonomous University Baja California (1997-2005) and Rector of Polytechnic University of Baja California (2006 -2010). He is member of National Researcher System Dr. Rivas was Head of Engineering Institute of Baja California Autonomous University Since 1997 to 2005; was Rector of Baja California Polytechnic University Since 2006 to 2010 and now is full researcher and the head of physic engineering department, of Engineering Institute of UABC, Mexico.

Oleg Sergiyenko received the B.S., and M.S., degrees in Kharkiv National University of Automobiles and Highways, Kharkiv, Ukraine, in 1991, 1993, respectively. He received the Ph.D. degree in Kharkiv National Polytechnic University on specialty "Tools and methods of non-destructive control" in 1997. He has written 103 papers and holds 1 patent of Ukraine and 1 in Mexico. He is author of 2 books, Editor of 7 books in international Editorials, session chair on many IEEE conferences (IECON, ISIE). He wins the "Best session presentation Award" on IECON-2014 in Dallas, USA, and IECON-2016 in Florence, Italy. In December 2004 was invited by Engineering Institute of Baja California Autonomous University for researcher position. He is currently Head of Applied Physics Department of Engineering Institute of Baja California Autonomous University, Mexico. His scientific interests are in automated metrology & smart sensors, frequency and time measurement, control systems, robot navigation, 3D coordinates measurement.

Wendy Flores-Fuentes received the master's degree in engineering from Technological Institute of Mexicali in 2006, and the Ph.D. degree in science, applied physics, with emphasis on Optoelectronic Scanning Systems for SHM, from Autonomous University of Baja California in June 2014. Until now she is the author of 13 journal articles in Elsevier, IEEE Emerald and Springer, 5 book chapters and 3 books in Intech, IGI global and Springer, 27 proceedings articles in IEEE ISIE 2014-2017, IECON 2014, the World Congress on Engineering and Computer Science (IAENG 2013), IEEE Section Mexico IEEE ROCC2011, and the VII International Conference on Industrial Engineering ARGOS 2014. Recently, she has organized and participated as Chair of Special Session on ''Machine Vision, Control and Navigation'' at IEEE ISIE 2015, 2016 and 2017. She has been a reviewer of 11 articles in Taylor and Francis, IEEE, Elsevier, and EEMJ (Gh. Asachi Technical University of Iasi). Currently, she is a full-time professor-researcher at Universidad Autónoma de Baja California, at the Faculty of Engineering.

Julio Cesar Rodríguez-Quiñonez received the Ph.D. degree from Baja California Autonomous University, México, in 2013. He is currently Professor of Electronic Topics with the Engineering Faculty, Autonomous University of Baja California. His current research interests include automated metrology, stereo vision systems, control systems, robot navigation and 3D laser scanners. He has written over 40 papers, 5 Book Chapters, has been guest editor of Journal of sensors, book editor, and has been reviewer for IEEE Sensors Journal, Optics and Lasers in Engineering, IEEE Transaction on Mechatronics and Neural Computing and Applications of Springer, he participated as a reviewer and session chair of IEEE conferences in 2014, 2015, 2016 and 2017. He is involved in the development of optical scanning prototype in the Applied Physics Department and is currently research head in the development of a new stereo vision system prototype.

* * *

Moises J. Castro-Toscano received the M.S. from Baja California Autonomous University, Mexico, in 2014. He is currently a PhD student of Electronic-Instrumentation Topics with the Engineering Faculty, Autonomous University of Baja California, where he is involved in the development of Inertial Navigation Systems in the Applied Electronic-Instrumentation Department.

Nagy Darwish received his Ph .D. in Information Systems from Faculty of Computers and Information, Cairo University, Egypt. He is an Associate Professor at Department of Information Systems and Technology, Institute of Statistical Studies and Researches, Cairo University. He is a reviewer in the International Journal of Computer Science and Information Security (IJCSIS), International Journal of Advanced Computer Science and Applications (IJACSA), and International Journal of Advanced Research in Artificial Intelligence (IJARAI). He is a reviewer in many national and international conferences. He is a member of editorial board of Circulation in Computer Science (CCS Archive).

Gerardo De Canio is a Mechanic Engineer and Ph.D in Aerospace Engineering is the head of the Sustainable Innovation Technologies at the ENEA Casaccia R.C., Roma, Italy. His research activities are: the dynamic characterization and experimental verification of technologies and materials for earthquake protection and conservation of immovable and movable Cultural Heritage; the study and design of anti-seismic basements made by marble, ceramics and steel ceramics for high vulnerable statues, museum techs and vulnerable objects. He conducted several in situ mechanical identification together with laboratory shaking table test campaign for macro elements of historical structures. He also made the design and installation of support for the statue "Maestà con Baldacchino ed Angeli Reggicoertina" on the facade of the Orvieto Cathedral, Italy; the design and installation of the marble anti seismic basements for the Bronzes of Riace at the archaeologic museum of Reggio Calabria, Italy; the design and installation of the anti-seismic basement for the bronze statue of "San Michele Arcangelo e drago" at the Museum of the Opera el Duomo di Orvieto (MODO); the design and realization of the transport tool for the statue of the emperor Augustus known as "Augusto da Prima Porta" from the Vatican Museum to Scuderie del Quirinale, Roma, Italy and from Vatican Museum to Grand Palais, Paris, France; the transport tool for the statue of the "Generale da Tivoli" from the Museo Nazionale Romano, Roma, Italy, to the Metropolitan Museum of New York City, NJ, USA.

Sergey V. Dvoinishnikov was born 09.22.1983 in the Berdsk town, Novosibirsk region, Russian Federation. He received a master's degree in physics in Novosibirsk State University (2006). In 2009 he received a PhD degree for the development of optoelectronic systems for measuring three-dimensional geometry of large objects. He is Dr. of Science (2016), Laureate of Russian Government Award in Science and Engineering for young scientists (2017). Dr. S.V. Dvoynishnikov is Senior Scientific Researcher of Kutateladze Institute of Thermophysics Siberian Branch of Russian Academy of Science, Senior Lecturer of Novosibirsk State University. Area of his scientific interests is development of optoelectronic measuring systems for scientific researches and industrial technologies.

Vincenzo Antonio Fioriti has received a Laurea in Controlli Automatici v.o. (Automatic Control Engineering Degree) from La Sapienza University of Rome. He has worked for many industrial firms as data analyst and pattern recognition designer. After that, he worked as researcher at the Dpt. of Energy La Sapienza University, simulating complex systems and at the ENEA CASACCIA facility in the field of Critical Infrastructure Protection (CIP), smart-grid stability and the European Projects competitions. Currently he works at the ENEA SITEC Laboratory, in the field of seismic protection.

Hesham A. Hefny received the B.Sc., M.Sc. and Ph.D. all in Electronics and Communication Engineering from Cairo University in 1987, 1991 and 1998 respectively. He is currently a professor of Computer Science at the Institute of Statistical Studies and Researches (ISSR), Cairo University. He is also the vice dean of graduate studies and researches of ISSR. Prof. Hefny has authored more than 160 papers in international conferences, journals and book chapters. His major research interests include: computational intelligence (neural networks – Fuzzy systems-genetic algorithms – swarm intelligence), data mining, uncertain decision making. He is a member in the following professional societies: IEEE Computer, IEEE Computational Intelligence, and IEEE System, Man and Cybernetics.

Daniel Hernández-Balbuena was born in July, 25, 1971. He received the B.S. degree from Puebla Autonomous University, Puebla, México, in 1996 and the M.S degree from Ensenada Center for Scientific Research and Higher Education, Baja California, México, in 1999. He received the Ph.D. degree in Baja California Autonomous University in 2010. His research interests are in the areas of time and frequency metrology, design and characterization of microwave devices and systems RF measurements, research applications of unmanned aerial vehicles and image digital processing.

Mykhailo Ivanov was born in July,18, 1989. He received the B.S. and M.S. degrees in Kharkov National Aerospace University ''KhAI'', Kharkiv, Ukraine, in 2010, 2012, respectively. He has 11 papers. Since 2012 till the present time, he is represented by his research works in several international conferences in several International Congresses of IEEE USA, Great Britain and Ukraine. Mikhail Ivanov, in October 2013, he joined the Kharkov National Aerospace University "KhAI" where he holded positions of assistant and vice dean in questions of students recruitment. In 2016 was invited to receive the Ph.D. degree in Engineering Institute, Autonomous University of Baja California, Mexico.

Alark Joshi is an Associate Professor in the Department of Computer Science at the University of San Francisco, where he works on data visualization projects for improved neurosurgical planning and treatment. His research focuses on developing and evaluating the ability of novel visualization techniques to communicate information for effective decision making and discovery. He received his postdoctoral training at Yale University, where he was a core member of the BioImage Suite team, whose mission is to develop and disseminate advanced image analysis and visualization software for widespread use. His work has led to novel visualization techniques in fields as diverse as computational fluid dynamics, atmospheric physics, medical imaging and cell biology. Through the illustration-inspired visualization techniques that he developed, atmospheric physicists are able to visualize the time-varying nature of hurricanes more effectively. Some of his current work deals with developing and evaluating the performance of neurosurgeons on novel visualization and interaction techniques. He received his Ph.D in Computer Science from the University of Maryland Baltimore County, his M.S. in Computer Science from the University of Minnesota and his B.S. degree from the University of Pune, India. He owes his interest in medical visualization to his stints at VitalImages Inc. and Siemens Corporate Research.

Vladimir Kartashov was born in Ukraine, on July 03, 1958. He received the M.S. degree in radio engineering from the Kharkov University of Radio Electronics, Kharkov in 1980. He is working in Kharkov University of Radio Electronics since 1984 to present time. He received the Ph.D. degree in radars and TV systems from the Kharkov University of Radio Electronics, Kharkov, in 1990. Since 1986 he has worked as a lecturer (docent). During this time he received the Dr. Sci. degree and Professor (2004). He has published over 96 science papers in radars theory, TV systems, digital signal processing and applications. His fields of research interests are radars, digital signal and image processing.

Lars Lindner was born on July 20th 1981 in Dresden, Germany. He received his M.S. degree in mechatronics engineering from the TU Dresden University in January 2009. He was working as graduate assistant during his studies at the Fraunhofer Institute for Integrated Circuits EAS in Dresden and also made his master thesis there. After finishing his career, he moved to Mexico and started teaching engineering classes at different universities in Mexicali. Since August 2013 he began his PhD at the Engineering Institute of Autonomous University of Baja California in Mexicali and deals since with the development of an optoelectronic prototype for measuring 3D coordinates using dynamic triangulation.

Phan Luu is a psychologist who use EEG to study human brain function, including learning and action regulation. He also engages in research and development of advanced EEG technology and analytic methods.

Arjuna Marzuki received the B.Eng. degree (Hons) in electronics engineering (Com) from the University of Sheffield, Sheffield,U.K., in 1997; the M.Sc. degree from Universiti Sains Malaysia, Penang, Malaysia, in 2004; and the Ph.D. degree in microelectronics engineering from Universiti Malaysia Perlis, Arau, Malaysia, in 2010. From 1999 to 2006, he was an Integrated Circuit Design Engineer with Hewlett-Packard/Agilent/Avago and IC Microsystem in the United States and Malaysia. He is currently a Lecturer with the School of Electrical and Electronic Engineering, Universiti Sains Malaysia. Dr. Marzuki is a registered Professional Engineer with the Society of Professional Engineers UK. He attained Chartered Engineer (Engineering Council, UK) status through the Institution of Engineering and Technology. He was a recipient of IETE J C Bose Memorial Award for the year 2010.

Vladimir G. Meledin was born 11.24.1959 in the Samara city, Russia. He received a master's degree in physics in Novosibirsk State University in 1981, PhD (1989), Dr. of Science (1996), Professor of Novosibirsk State Technical University (1998-2003), Professor (2013). Chief Researcher of Kutateladze Institute of Thermophysics of Siberian Branch of Russian Academy of Science and Novosibirsk State University. Prof. V.G. Meledin is Academician of Russian Academy of Engineering and A.M. Prokhorov Academy of Engineering Sciences, member of ISA. He is a Laureate of Russian Government Award in Science and Engineering (2014). General Director of Institute of Optic-Electronic Information Technologies, Novosibirsk, Russia. Area of his scientific interests is optoelectronic information diagnostics for science and industry.

Jesus Elias Miranda Vega was born in Sinaloa, Mexico, and received the B.E. degree in electrical and electronic engineering from Los Mochis Institute Technology in, Sinaloa, Mexico in 2007 and master's degree in electronic engineering from the Mexicali Institute of Technology, Mexicali, Mexico in 2014, he joined engineering institute at the Autonomous University Baja California (UABC) opto-electronics lab as a Phd student, Mexicali, Mexico in Aug 2016. His current research interest include machine vision, data signal processing, the theory and optoelectronics devices, and their applications.

Marialuisa Mongelli's research activities are mainly directed to the field of the seismic protection of civil and historic structures to develop anti-seismic devices and retrofitting systems within Italian and European research programs. She is involved in finite elemeny analysis, non destructive controls to assess the structural integrity of the elements experimentally verified and to study new methodologies of 3D Motion Capture acquisition data during shaking table tests to validate numerical models.

Fabian N. Murrieta-Rico was born in September the 7th of 1986. He received the B.Sc and M.Sc. degrees from Instituto Tecnológico de Mexicali (ITM) in 2004, and 2013 respectively. In 2017 he received his Ph. D. in Materials Physics from Centro de Investigación Científica y Educación Superior de Ensenada (CICESE). He has worked as automation engineer, systems designer, and as a university professor. His research has been published in different journals, and presented in international conferences since 2009. His research interests are in the field of time and frequency metrology, wireless sensor networks design, automated systems, and highly sensitive chemical detectors. Currently he is involved in development of new frequency measurement systems, and highly sensitive sensors for detection of chemical compounds.

Mohd Tafir Mustaffa graduated with B.Eng. in Electrical and Electronic Engineering from Universiti Sains Malaysia, M. Eng. Sc. and PhD degrees from Victoria University, Melbourne. He is currently a senior lecturer with the School of Electrical and Electronic Engineering, Universiti Sains Malaysia. He is also a senior member of IEEE. Current research area are: RFIC, Analog IC, and RF MEMS.

Norlaili Mohd Noh graduated with B.Eng. Electrical Engineering (Honours) from Universiti Teknologi Malaysia, and both MSc. (Electrical and Electronic Eng.) and Ph.D (Integrated Circuit Design) from Universiti Sains Malaysia. She is currently an Associate Professor with the School of Electrical and Electronic Engineering, Universiti Sains Malaysia. Her specialization is in Analog RFIC Design. She has published more than 85 journal and conference papers and has successfully supervised more than 10 PhD and MSc students. She is also a professional engineer registered with the Board of Engineers Malaysia and a Chartered Engineer registered with UK Engineering Council.

Vitalii Petranovskii received his PhD in Physical chemistry from Moscow Institute of Crystallography in 1988. In 1993-1994 he worked as invited scientist at National Institute of Materials and Chemical Research, Japan. Since 1995, he is working at "Centro de Nanociencias y Nanotecnología, Universidad Nacional Autónoma de México" (2006-2014 – as the Nanocatalysis department chair). His research interests include synthesis and properties of nanoparticles supported over zeolite matrices. He is a member of Mexican Academy of Sciences, International Zeolite Association, and Mendeleev Russian Chemical Society. He has published over 100 papers in peer-reviewed journals and 4 invited book chapters. Also he is co-author of monograph "Clusters and matrix isolated cluster superstructures", SPb, 1995.

Oscar Real Real received the B.Eng. from Baja California Autonomous University, Mexico, in 2014. He is currently a Masters student of Electronic- Instrumentation Topics with the Engineering Faculty, Autonomous University of Baja California, where he is involved in the development of 3D laser Scanner based on Dynamic Triangulation in the Applied Electronic-Instrumentation Department.

Ivan Roselli works on research studies and advanced services in the fields of seismic and vibration engineering. He works in the seismic Hall laboratory at the Enea Casaccia Research Center, where a shaking table facility among the largest in Europe is located. The Enea Casaccia shaking table facility was the first in the world provided with a passive 3D motion capture system dedicated to seismic tests (named 3DVision system) whose installation and development he personally followed. He has been scientific responsible of shaking table qualification tests and experiments. He follows the design of shaking table tests, with particular focus on the acquisition and processing of data.

Basir Saibon received the B. Eng degree in electrical and electronic design systems from Universiti Kebangsaan Malaysia (UKM) in 1992 and MSc. ECE degree from University of Salford at Greater Manchester United Kingdom in 1994. Upon his graduation he worked at Standard Industrial and Research Institute of Malaysia (SIRIM Bhd.), where he involved in motion control detection systems design with IPA-Fraunhoffer GmBh, Germany and MTS System Corp. Minneapolis, Minnesota USA. In 1997 Mr. Saibon moved to Hewlett Packard Inc. where he focused in the development of optical sensor and encoder development. The works include comprehensive semiconductor prototyping to production cycle, from Front End to Back End design, IC fabrication, Device Characterization for Final Package Evaluation (EVB), and NPI to mass production. Mr. Saibon also involved in various semiconductor product development at Hewlett Packard (HP) Penang, HP Singapore, HP San Jose, and Agilent Santa Clara in Silicon Valley California. Mr. Saibon is a certified expert for optical motion encoder characterization where he was the R&D Program Engineering Manager to maintain characterization program for Motion Control and Imaging Division of Agilent Technologies at Bowers Avenue in Santa Clara, USA from 1999-2002. As the Senior Staff Engineer and Program Engineering Manager he owned engineering and technical responsibility to maintain standard product engineering of IC and Characterization for R&D and NPI compliance to Japan Nintei, US FCC, and European CE standard. Basir Saibon was a recipient of Hewlett Packard Star Award from CEO Office in 1999 for major accomplishments of Optical Encoder; 200mm-150mm Die Shrink Conversion for ST Microelectronics and HP proto built wafer fabrication. Mr. Saibon was also awarded Agilent Technologies Rank1 (The highest achievement rank) for 3 consecutive years for his contributions and efforts in related technical areas. In 2002 Basir Saibon worked at Universiti Sains Malaysia and co founded Universiti Sains Malaysia RF IC Design Center (CEDEC) focuses in the development of RF IC, wafer probing and IC measurements. Due to his contribution to CEDEC development, Mr. Saibon is appointed as the Research Fellow of CEDEC in 2014. His previous works at Advanced Technology Research (ATR) Motorola Inc. Florida and Penang Design Center are to manage and support the development of future generation radios IC. This includes Multi-Band Automated Quadrature Modulator for Software Define Radio (SDR), using IBM7WL-fT60GHz-SiGe-BiCMOS 180nm and 90nm to comply with US FCC, TETRA and TETRA2 standards. He also involved in the development of multiband receiver and frequency generator and amplifier for 4G RF IC Radio-Transceiver IC. He is extensively working on VLSI schematic and layout design, performing technology conversion schematic into layout geometry at primitive cell/top level hierarchy/floor planning, layout verification LVS, and DRC. Expert of EDA tools simulators applications such as Spectre, SpectreRF, Berkeley Design Automation (BDA), and Agilent ADS/RFDE. Mr. Saibon is well exposed to IBM 7WL SiGe BiCMOS Triple Well design kit. Basir Saibon was also a Program manager for Motorola-University Research Collaboration where he holds a position as visiting lecturer for research, Board Committee of School of Microelectronic & Photonics, and Board of Academic for Master of Microelectronic Engineering Studies of University Malaysia Perlis (UNIMAP). Since 2012 Mr. Saibon has been with Universiti Kuala Lumpur at Kulim Hi-Tech Park Campus, where he holds the position of Head Department for International, Industrial, and Institutional Partnership (IIIP). This department contributes many collaborations with Industries especially the MNCs and local industrial players in semiconductor and electronics manufacturing. Basir Saibon was also the Vice Chair for IEEE Penang Joint Chapter, IEEE Malaysia ED/MTT/SSC in 2014. He is the co-developer and technical architect for Bach. Eng. Tech in Applied Electronics focuses on the latest electronics technology development. In 2014 Basir Saibon led the IIIP team to develop UNIKL first Industrial Center of Excellence (ICoE), a collaboration effort between UNIKL, Keysight Technologies, and SIRIM SMT and was awarded initial budget and matching grant worth of RM5.5Million to establish RF Microelectronics Laboratory.

Steven Shofner is a senior software engineer with Philips Neuro.

Oleg Sytnik was born in Dneprodzerginsk, Ukraine, on May 17, 1958. He received the M.S. degree in radio engineering from the Kharkov University of Radio Electronics, Kharkov in 1980. He worked in the Design Bureau of Machine-Building Plant of Dnipropetrovsk (1980-1982), and then at the Radio Engineering Department of the Kharkov University of Radio Electronics (1982-1986). He received the Ph.D. degree in radars and navigation from the Kharkov University of Radio Electronics, Kharkov, in 1986. Since 1986 he has worked as a senior researcher at the A. Ya. Usikov Institute for Radio Physics and Electronics under the National Academy of Sciences of Ukraine in the Department of Radiophysical Introscopy and in the A.I. Kalmykov Center for Remote Sensing of the Earth under the Space Agency of Ukraine. During this time he received the Dr. Sci. degree in Radio Physics and Mathematics and Professor (2005). He has published over 163 science papers in radars theory, digital signal processing and applications. His fields of research interests are radars, digital signal and image processing, pattern recognition and stochastic non-stationary processes.

Don M. Tucker is the CEO and Chief Scientist at Electrical Geodesics, Inc. He is Professor of Psychology and Associate Director of the NeuroInformatics Center at the University of Oregon. He has published over the last 40 years on normal and abnormal psychology, brain function, and the electrophysiological methods for investigating brain activity. His 2007 book, Mind From Body, examines how experience can be understood in relation to brain mechanisms. His 2012 book with Phan Luu, Cognition and Neural Development, examines the continuity of neural arousal control from embryonic differentiation through life span self-regulation. His basic research examines self-regulatory mechanisms of the human brain. These mechanisms include motivational and emotional control of cognition as well as neurophysioloigcal control of arousal, sleep, and seizures. His applied research focuses on high- performance computing for analyzing human brain activity with dense arrays of scalp sensors, including electroencephalographic (EEG), electrical impedance tomography (EIT) and near-infrared spectroscopy (NIRS). Based on Tucker's invention of the Geodesic Sensor Net, the dense array (64 to 256 channel) EEG systems made by Electrical Geodesics, Inc are now used in 1000 laboratories world wide, with 2000 publications with this technology now in the scientific literature.

Vera (Vira) V. Tyrsa was born on July 26, 1971. She received the B. S. and M. S. degrees in Kharkov National University of Automobiles and Highways, Kharkov, Ukraine, in 1991, 1993, respectively (Honoris Causa). She received the Ph.D. degree in Kharkov National Polytechnic University on specialty "Electric machines, systems and networks, elements and devices of computer technics" in 1996. She has written 1 book, 7 book chapters, and more than 50 papers. She holds one patent of Ukraine and one patent of Mexico. From 1994 till the present time, she is represented by her research works in international congresses in USA, England, Italy, Japan, Ukraine, and Mexico. In April 1996, she joined the Kharkov National University of Automobiles and Highways, where she holds the position of associated professor of Electrical Engineering Department (1998–2006). In 2006–2011, she was invited by Polytechnic University of Baja California, Mexico for professor and researcher position. Currently, she is a professor of electronic topics with the Engineering Faculty, Autonomous University of Baja California. Her current research interests include automated metrology, machine vision systems, fast electrical measurements, control systems, robot navigation and 3D laser scanners.

S. Vasavi is working as a Professor in Computer Science & Engineering Department with 20 years of experience. She pursued her MS from BITS, Pilani in the year 1999 and PhD from Acharya Nagarjuna University in the year 2010. She currently holds R&D projects from UGC and ISRO-ADRIN. She published 44 papers in various Scopus indexed, Google Scholar Indexed conferences and journals. She filed two patents. She is the recipient of UGC International travel grant in the year 2015, for her visit to ICOIP 2015,USA and TEQIP grant for her visit to ICICT 2016, Thailand. She is the Keynote speaker for the IEEE international conference ICOIP conducted at Singapore in July 2017. She visited reputed universities at U.S.A, Singapore and Thailand. She also Visited Argonne National Laboratory, She is an IEEE member, life member of Computer society of India (CSI), Member Machine Intelligence Research Labs, Washington, USA. Her Research Areas are Bigdata analytics: Image object Classification. Reviewer for Scopus Indexed journals and conferences. Received Best Teacher Award, Vishitta Mahila Award, Conferred Outstanding Women in Engineering Award.

Xiao-Wei Ye is currently an Associate Professor in the Department of Civil Engineering at Zhejiang University, China. His research interests include structural health monitoring, fatigue of steel structures, and structural reliability. Dr. Ye received his Ph.D. degree from the Department of Civil and Environmental Engineering at The Hong Kong Polytechnic University in 2010. He has served as PI or Co-PI for more than 20 research projects. He has published over 50 peer-reviewed journal articles, 2 books, 1 book chapter and 3 international conference proceedings. He also has 6 Chinese patents and 4 Chinese computer software copyrights.

Rosario Isidro Yocupicio-Gaxiola was born in Los Mochis, Sinaloa, Mexico. He studied at the Instituto Tecnológico de Los Mochis and completed a PhD at Centro de Nanociencias y Nanotecnología of Universidad Nacional Autónoma de México (CNyN.UNAM) in 2017. Actually, he is a posdoctoral researcher at CNyN-UNAM. His research interests lie in the synthesis, analysis and applications of porous solids, particularly the study of zeolites as microporous and mesoporous systems. His research is focused in the use of microporous materials as catalysts, supports of catalysts, sorbents and opticals devices.

Bassem Zohdy received his Masters degree from Arab Academy for Science and Technology (AAST), he is a Ph.D. Candidate and a researcher at The Institute of Statistical Studies and Research (ISSR), Cairo University.

Index

2.5D Vision 186-187
3D Motion Capture 269, 271, 273, 277, 282-283, 289
3D Vision 134-136, 138-139, 172, 186

A

Actor Prolog 135-137, 139-143, 146-148, 151-155, 157-162, 165-166, 168, 171, 173-174, 177-178, 186-187
Agent Logic Programming 136, 156-157
Analyte 1-5, 8, 10-11, 16
Anomalous Behavior Detection 134

B

Backpropagation 112, 133
Bezier curve 357-361
Blob 141-142, 144, 146, 148-150, 152, 171-172, 186, 275
bridge inspection and evaluation 256-257, 266

C

Cepstrum 373-380, 385-390
Cladding 3, 6-7, 16
Cloudy triangulation 49-51, 53-57, 61, 63-65, 69, 71-73
CMOS 17, 19, 21, 23-28, 33, 36, 41-44, 99, 256, 262, 277
CMOS Image Sensor 17, 19, 23-25, 27
Coating 4, 6-8, 16, 272
Coefficient of Determination (R2) 187
Color space 168, 190-191, 207, 209, 220
computer vision technology 256-257, 259-260, 262, 267
concrete cracking 256
Continuous field-of-view 347
Convolutional Neural Networks 106-107, 109, 111, 114, 133, 237

D

Deep Learning 98, 107, 110, 114, 117, 121, 128-129, 131, 237
Defocus 89, 93-94, 100
Depth 87, 89-94, 96-100, 111, 130, 173, 187, 197, 215, 235, 243, 246, 348, 376, 379, 381, 383, 389
Dynamic Behavior 278, 280-281, 283-284, 286, 290, 293
Dynamic Triangulation 100, 324-325, 338-340

E

Electroencephalogram (EEG) 133
ENEA Casaccia 277, 289

F

False Alarm 137, 380-381, 390
Fiber Optic 7, 303
Filtering Techniques 271, 275
Focus 36, 56, 85-86, 88-89, 93-94, 96, 100, 110, 158, 269, 271, 273, 297
foreign object intrusion 256

G

Gaussian Mixture Models 260

H

high-precision 49-50

I

Inhomogeneities 51, 65, 379, 390
Integrated Circuit 20
Intelligent Visual Surveillance 134-140, 144, 153, 156, 165-166, 172, 177-178, 187

K

Keras 128
Key frame extraction 189, 194-195

L

Laser 3, 6-9, 49-51, 53-59, 69, 71-74, 81-86, 89, 98-
100, 108, 135, 243-244, 246-247, 270, 285, 305,
312, 338-340, 342-344, 348, 354
Laser Scanning System 339
linear discriminant analysis 322
linear predictive coding 317-318

M

Machine Learning 107-108, 117, 235, 237, 239-240,
251, 302
Machine Vision Applications 234, 248
mel-frequency cepstral coefficients 318, 320
Meta data 206, 208, 210, 228
Molecular Adsorption 6-7
Motion Magnification 269, 272, 290-291, 293
Motion planning 340, 353, 356, 362-365, 367
Moving Average 60, 311-312, 327-329, 334
Moving Object Classification 188-189, 198, 200-201,
204, 206-208, 216, 218, 228

N

Non-Contact 79-81, 97-98, 101, 272

O

Object-Oriented Logic Programming 134-136, 139,
153, 178, 187
Optical Properties 2-4, 10

P

Passive Markers 271, 273, 278, 280
phase inhomogeneities 51, 65
Photogrammetry 108, 118, 133, 243, 246
Photostability 9-10, 16
Profilometry 89, 97, 101
Projective Transform Matrix 146, 187
Prolog to Java 153

Q

Quantum Yield 9, 16

R

Rayleigh Resolution Criteria 374, 390
ReLU 109, 133
RGB-D 187
Robotic group 353, 362, 367

S

Scanner 100, 345
Scene detection 189, 195-196
Segmentation 62, 101, 111, 114-115, 118-120, 122,
126-131, 194-196, 201, 236-238, 240, 260
Sensor 2-3, 6-11, 16-19, 21-25, 27-29, 44, 85-86, 88-
89, 107-108, 113-122, 127-130, 133, 200, 234,
244, 248-250, 272, 277, 285, 291, 302-303, 316,
324, 339, 343-344, 353
Sensors 1-2, 4-7, 9-11, 17-18, 21, 25, 83, 85, 98, 108,
112-122, 127-131, 133, 135, 235-236, 247-248,
257, 260, 262, 267, 270, 277-278, 281-283, 285,
291, 293-294, 297, 302-303, 316, 324, 339, 353
Shading 90, 92-93, 100, 197
Shape 4, 81, 85, 89-93, 100-101, 107, 128, 199-200,
204, 206, 216, 238, 246, 256, 272-273, 279, 302,
316, 350, 357, 373, 375, 378-379, 381
Skeleton 137, 173-174, 187
Skewness 143, 149, 187
spectrum 6-8, 20, 64, 204, 238, 278, 313, 373-376,
378, 383, 386-390
Statistical Criterion 374, 391
steel production 73
Stereophotogrammetry 89, 95-96, 101
Structural health monitoring 80-81, 85, 256-257, 267,
296, 301-303, 315
subsurface ground layers 373

T

targets 99, 246, 270, 324, 373-374, 378-380, 391
Technical Vision System 338-341
TensorFlow 128
texture 81, 93, 97-98, 119, 127, 189, 194, 198, 200,
204, 206-207, 216, 238, 272, 294, 373
TFLearn 128

thickness measurements 49, 57-58, 60, 69, 72, 74
Thin Films 2, 6
Threshold Activation Function 109, 133
Time-of-Flight (ToF) 135-136, 172, 187, 339
training sample 112, 380-381
Triangulation 49-51, 53-57, 61, 63-66, 68-73, 80, 83, 85-89, 95, 97-100, 130, 243, 246-247, 274, 324-325, 338-340, 342-343, 345

U

unmanned aerial vehicle 256-257

V

video pulse 373, 376, 378-379, 381, 383
Vision system 79-81, 86-88, 98, 101, 235, 237, 240, 256-257, 338-341

Z

Zeolite 4, 7-11, 16

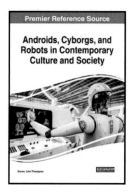

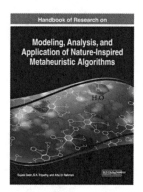

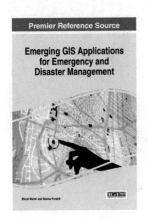
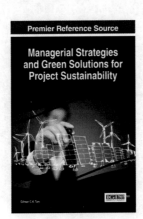

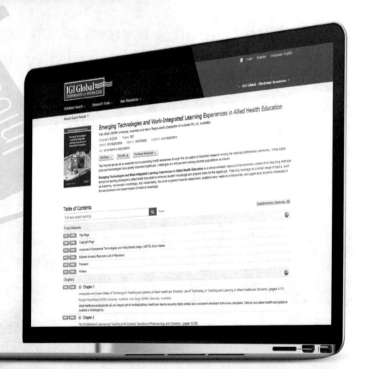